# Outer Space—
# A New Dimension of the Arms Race

sipri

**Stockholm International Peace Research Institute**

SIPRI is an independent institute for research into problems of peace and conflict, especially those of disarmament and arms regulation. It was established in 1966 to commemorate Sweden's 150 years of unbroken peace.

The institute is financed by the Swedish Parliament. The staff, the Governing Board and the Scientific Council are international.




sipri

**Stockholm International Peace Research Institute**
Bergshamra, S-171 73 Solna, Sweden
Cable: Peaceresearch, Stockholm
Telephone: 08-55 97 00

# Outer Space—A New Dimension of the Arms Race



Edited by
Bhupendra Jasani

sipri
Stockholm International Peace Research Institute

ALERE FLAMMAM

Taylor & Francis Ltd
London

First published 1982 by Taylor & Francis Ltd
4 John Street, London WC1N 2ET



Distributed in the United States of America by
Oelgeschlager, Gunn & Hain, Inc.
1278 Massachusetts Avenue, Harvard Square,
Cambridge, MA 02138
and in Scandinavia by
Almqvist & Wiksell International,
26 Gamla Brogatan,
S-101 20 Stockholm, Sweden

**British Library Cataloguing in Publication Data**

Jasani, B.
Outer space—a new dimension of the arms race.
1. Space sciences—Congresses
2. Technology transfer—Congresses
I. Title II. Stockholm International Peace Research Institute
623.4′194 TI74.3

ISBN 0 85066 231 1

Typeset by Red Lion Setters, London WC1N 2LA
Printed and bound in the United Kingdom by
Taylor & Francis (Printers) Ltd, Rankine Road,
Basingstoke, Hampshire RG24 0PR

# Preface

During the last quarter of a century space technology has made tremendous advances, especially in the military field, and is now spreading from the great powers to many more nations. This book is primarily directed to the reader concerned about the various aspects of the arms race, particularly its extension into outer space. The book will also serve as useful background material for the United Nations Second Special Session on Disarmament in New York in June–July 1982, and for the Second United Nations Conference on the Exploration and Peaceful Uses of Outer Space in Vienna in August 1982.

As a contribution to future discussions to take place on issues related to outer space, SIPRI organized a symposium on "Outer Space—A New Dimension of the Arms Race" at which 20 eminent scientists, lawyers and diplomats from 12 different countries discussed the consequences of the militarization of outer space as well as further arms control and disarmament measures. The meeting took place in Stockholm between 17 and 20 November 1981. This book contains the papers presented at the symposium and an introductory section of seven chapters in which I have drawn on the discussions at the conference, and on other material, to present a general background to the issues covered by the symposium papers.

*Acknowledgements*
The invaluable editorial assistance of Barbara Adams and Gillian Stanbridge is gratefully acknowledged.

SIPRI
1982

*Bhupendra Jasani*
Editor

# Contents

PART II

*Space technology, crisis monitoring and arms control*

# Abstracts of papers

## Paper 1. Reconnaissance satellites

B.G. BLAIR

This paper describes the technological evolution in satellite reconnaissance and examines the main technical aspects and capabilities of spacecraft and sensors. As public sources of information provide the background, the author has focused mainly on US early warning and photo-reconnaissance spacecraft. The intention of the discussion is to give the reader sufficient technical knowledge to enable him to appreciate the implications that the various uses of reconnaissance satellites can have on national security and international stability.

## Paper 2. Identification of military components within the Soviet space programme

G.E. PERRY

Systematic observations of radio signals and analysis of orbital data for many years have revealed components of the Cosmos programme which can be interpreted as performing military missions in the fields of photographic reconnaissance, electronic intelligence gathering, early warning detection, and ocean surveillance by radar techniques. Decoding of telemetry from navigation satellites has shown the existence of separate civilian and military systems within the Cosmos programme and a similar dichotomy is found to exist within the Molniya-1 communications satellites. Results of observations of the anti-satellite test of 2 February 1981 are discussed.

## Paper 3. Navigation satellite systems: their characteristics, potential and military applications

K.D. McDONALD

The evolution and maturing of space system technology has seen an increasing interest and commitment directed towards exploiting the capabilities and potential of satellite-based systems for navigation and related applications. The proven utility of satellite systems has been extended from communications, surveillance, and surface navigation into the fields of aeronautics and space; to other functional areas such as mapping and geodesy, precise time and time transfer; and also to a wide range of military applications. This paper provides a description of the principal navigation satellite systems, either in development or deployed, which have military implications. The systems discussed are the operational US Navy TRANSIT Navigation

Satellite System, and the developmental Department of Defense NAVSTAR Global Positioning System (GPS). A review of early navigation satellite developments is provided along with a description of the TRANSIT and NAVSTAR GPS system concepts, performance characteristics, principles of operation, and implementation features. The current status and plans for both TRANSIT and GPS are discussed with particular attention to their military uses. Some of the more significant advantages, limitations, and military benefits of navigation satellite systems are described. Future trends, issue areas, and transition plans are discussed for the use of navigation satellite systems for military and other applications.

## Paper 4. System aspects in military satellite communications

E. BIGLIERI

Satellite communications systems exhibit a number of features that are highly desirable for military applications. In turn, military demands place a number of characteristic requirements on the system: these include a low probability of detection and location of mobile terminals by means of the signals which they emit, and protection against electronic measures such as jamming signals intended to disrupt communications. The military satellite communications system aspects that derive from these requirements are discussed in this paper. In particular, use of spread-spectrum techniques, adaptive antenna nulling, and extra high frequency spectrum allocation are discussed in some detail.

## Paper 5. Image analysis and sensor technology for satellite monitoring

T. SAKATA and H. SHIMODA

This paper describes the image analysis of satellite data and sensor technology. The specific information about monitoring can be extracted from satellite-acquired data using visual, analogue and computer systems. Analogue and digital processing for remotely sensed data is done mainly by a computer system which consists of various kinds of input, output, and software systems. The image processing system can improve the quality and accuracy of ground objects. Also, sensor technology strongly affects the image quality. However, the total performance of the monitoring system cannot be determined by the sensor technology alone. Rather, it is determined by the matching of each subsystem, including data transmission subsystems and the optimum strategic way of satellite operations.

## Paper 6. Information extraction from images

T. ORHAUG and G. FORSSELL

The physical basis for collecting information on the physical environment by various remote sensing methods is briefly discussed and the role of imaging sensors is shown. Image interpretation is described in terms of a 'concise description' of the scene represented by the image and the importance of *a priori* information or knowledge is pointed out. The relative importance of temporal, spectral and spatial data for image interpretation is discussed. Imagery data are often used in digital form. Imaging systems must therefore be characterized in terms of their resolving power and their geometrical and radiometrical quantization. Some of the effects caused by quantizations are briefly discussed, and the necessity of formulating spatial resolution as a statistical problem is stressed. A broad description of a study using computer-simulated imagery is given. A photographic image with good ground resolution (approximately 10 cm) was digitized and filtered using a digital computer. The resulting imagery data were used for generating photographic products having ground resolutions ranging from 0.2 m to 80 m. These products were used both for comparison with satellite (Landsat) images and for conducting interpretation tests.

## Paper 7. The prospects for beam weapons

A.M. DIN

Weapons capable of hitting distant targets within a split second are being developed in a number of military laboratories. These beam or directed energy weapons encompass lasers and charged as well as neutral particle accelerators. When (or if) operative, they may be stationed in outer space, in the atmosphere or on the Earth's surface and be used for attacking a variety of objects. The status of the underlying technology is described and the tactical and strategic implications assessed. Possible control measures are discussed.

## Paper 8. Orbital BMD and the space patrol

P.J. NAHIN

The evolutionary trend of ballistic missile defence (BMD) from a ground-based mode to higher and higher altitudes is discussed. Then, making the final leap to a BMD in the Earth's orbit (a weapon development driven by tremendous technical forces), a case is made for the possible desirability of such a system. The key concept is that it not be done unilaterally but as jointly constructed US/USSR orbital battle stations staffed by crews selected world-wide. This paper advocates the creation of a space patrol, charged with the single task of preventing intercontinental ballistic missile warfare. It is argued that only through such a radically new arms control idea can the imaginations of all of the populations of the Earth be captured (along with their enthusiastic support).

## Paper 9. Some remarks on Soviet and US strategies concerning manned activities in outer space

H. KAUTZLEBEN

There is serious concern about the new round of the arms race in which outer space activities play an important part. It is submitted that the US Space Shuttle programme envisages the use of manned space flights for missions with clearly military aims; such aims did not exist in the previous development of manned space travel. The Soviet programme, based upon the use of orbital stations in connection with service satellites, is outlined. Particular attention is paid to the inter-Cosmos programme in which nine international crews have participated in the exploration of outer space for peaceful purposes. The only guarantee that manned space flights will be used exclusively for peaceful purposes is to conclude further international agreements against the militarization of outer space. In this respect the Soviet proposal submitted to the 36th Session of the UN General Assembly is an important step.

## Paper 10. Some recent data on the utilization of space for military purposes and on second generation nuclear weapons

M. FELDEN

There is an evolutionary logic in the development of nuclear weapons, whereby one of the effects of a nuclear explosion is enhanced in relation to the others. Such a logic has led to the development of weapons with reinforced radiation (the neutron bomb), as well as to research on an electromagnetic pulse (EMP) bomb and on high-energy beam weapons (using different technologies). The author also submits that the latter should constitute a second generation of nuclear weapons and become actual before the turn of the century. Initially they would be defensive in application, but, because high-energy beam weapons could so undermine the offensive potential of standard nuclear weapons, it could be perceived necessary to enhance the war-fighting character of the beam weapons. This would lead to a whole new conception of nuclear war. Some of the difficulties in relation to the development of beam weapons are outlined, but are not seen to be insurmountable. In addition, the mechanism and effects of the EMP bomb are described. Three such bombs might be sufficient to totally cripple the military and civilian systems of Europe, the USA, or the USSR. The implications of EMP are one of the major elements determining the necessity to revise current defence policies.

## Paper 11. Use of satellites in crisis monitoring

K. SANTHANAM

This paper compares and contrasts aerial and space reconnaissance in the context of crisis monitoring. Limitations in space reconnaissance are indicated. Inferred instances of the use of satellites in military conflicts during the period covering 1960 through 1980 are mentioned. Questions regarding their effectiveness in preventing the outbreak of hostilities and conflict resolution are addressed.

## Paper 12. Verification using satellites, feasibility of an international or multinational agency

M. ABDEL-HADY and A. SADEK

An international or multinational satellite-monitoring agency (ISMA) can make a valuable contribution to the verification of compliance with certain arms control and disarmament agreements, as well as for monitoring of crisis areas, thus contributing to confidence building among nations. From a technical point of view, the idea of an ISMA is both possible and feasible, and can be established in logical steps compatible with available technical and financial resources. Such an agency can make a significant contribution to serving the interests of the international community, particularly developing non-space-power countries, on a non-discriminatory basis. From a legal point of view, there is no provision in international law that prohibits the establishment of an international governmental organization for satellite monitoring, which would not interfere in internal affairs of states and would abide by operational rules worked out by the international community.

## Paper 13. What additional arms control measures related to outer space could be proposed?

D. GOEDHUIS

Attention is given to the considerable divergencies in the interpretation of some of the rules so far adopted that are aimed at a limitation of the military uses of outer space. The need is stressed for a clarification of the meaning and content of certain ambiguous clauses laid down in four of the international instruments containing arms control measures in outer space. It is submitted that there is a strong need to establish —in the framework of the United Nations—a special forum composed of representatives of the bodies concerned with halting the militarization of outer space. As outer space has so far remained free from "kill-mechanisms", it can be assumed that the problems which have to be overcome in order to arrive at further arms limitations in this medium would be comparatively less arduous than those arising in other fields where unbridled military competition has already occurred. Consequently it appears undesirable to include the issue of arms control in space in the discussions on arms control in general. Furthermore it is suggested that proposals regarding a complete demilitarization in space would be ineffective and even counter-productive. At present the most urgent task would be the establishment of a ban on anti-satellite weapons. As a general consensus on a precise definition of the controversial term 'peaceful purposes' is unlikely to be achieved, an effort should be made to identify activities which could be considered as not being conducted for such purposes. Paramount importance is stressed on arousing public consciousness about the disastrous consequences which would result from an intensification of the arms race in space.

## Paper 14. Anti-satellite weapons: the prospects for arms control

D.L. HAFNER

The United States and the Soviet Union made important progress towards common positions on anti-satellite activities (ASAT) arms control in their negotiations during 1978–79. Those talks are now in abeyance, but in the interim neither side has advanced its ASAT weapon programme significantly. Thus the prospects for ASAT arms control remain good. While the body of current international law protecting satellites is substantial, it cannot halt ASAT competition for it does not prohibit all ASAT testing and deployment. Satellites are attractive targets because they are relatively few in number, are vital components of military forces, and are relatively vulnerable to a variety of weapons. Both the USA and the USSR have justified their ASAT programmes by citing the need to attack satellites engaged in hostile acts and to retaliate against the other's ASAT attacks. It is not feasible to prohibit satellites from performing all military-related functions or hostile acts. If ASAT competition is to be averted, the USA and the USSR must be convinced that competition would be potentially destabilizing and unjustifiably costly, and that ASATs are neither an adequate way nor the only way to cope with threats from satellites. In drafting an ASAT treaty, a series of issues must be resolved, including the scope of an ASAT ban, provisions regarding hostile acts by satellites, opening the accord to third parties, and verification of compliance and dismantling. A Soviet draft treaty banning all weapons from orbit, placed before the 36th UN General Assembly, has its flaws. But as an indicator of continuing US and Soviet interests in the ASAT threat, it is an encouraging sign; the prospects for ASAT arms control remain good.

## Paper 15. Banning all kinds of weapons in outer space

O.V. BOGDANOV

Generally speaking, the field of international law has seen rapid progress in directions aimed to cover some of the new aspects of possible extensions of the arms race. During the last 15 years a number of treaties on outer space have been introduced to the international community. This paper looks at existing international treaties and outlines a new Soviet proposal for a treaty to ban the deployment of all types of weapons in outer space.

# PART I

# Military space technology and its implications

by

Bhupendra Jasani

# Chapter 1. Introduction

Man's curiosity about the 'heavens' dates back, perhaps, to the beginning of history. Initially, he fulfilled a desire to see into the depths of space by using a variety of telescopes, but the ultimate desire has been to launch himself into space. The realization of this dream did not depend only on the availability of powerful rockets but also on the understanding of the laws governing the motions of bodies round the planets.

Nearly three and a half centuries ago, the work of two scientists radically changed the thoughts of mankind in many respects. These men were Johann Kepler and Isaac Newton. Kepler, a German astronomer and mathematician, worked out the mathematical laws that govern the motion of bodies in orbit round the Earth and other planets and stars. His work was published between 1609 and 1619.

While Kepler's laws described the planetary motions they did not explain them. It took the genius of Newton to unravel the mysteries of planetary motion. Between 1665 and 1667 he worked out the law of gravitation, the laws of motion and the basic notions of differential calculus. However, this work was only published 20 years later and it was not until some two and a half decades ago that the technology became available for applying the laws evolved by Kepler and Newton to artificial Earth satellites.

With the launch of the Soviet Sputnik in 1957, the space age dawned. Since then spacecraft have turned out to be more important as military tools than as a means of satisfying man's desire to explore our solar system. In fact, space has become a crucial environment for the military—so much so that if the satellites of a space power were to be destroyed its military capability would be drastically reduced. This is because artificial Earth satellites are the essential eyes, ears and nerves of the fighting forces of today. For example, satellites orbiting close to the Earth identify potential military targets and determine their precise location. Spacecraft not only relay military messages over long (or short) distances but are also able to guide modern missiles, aircraft and naval vessels carrying lethal nuclear and conventional weapons to their targets with almost pin-point accuracies.

Satellites are so much a part of modern weapon systems that they have been added to the already long list of potential military targets in any future war between the two powers. They are such important targets that the crippling or destruction of one or more satellites by the enemy would be tantamount to a declaration of war; war on Earth could thus start in space. Both sides have some conventional anti-satellite (ASAT) weapons and both are investigating various other methods of destroying each other's satellites, including weapons which up to now were strictly in the province of science fiction writers. Among these are high-energy laser and particle-beam weapons. Neither side has deployed such weapons in space. However, because there appears to be a sound scientific basis for such systems and since considerable technological advances have already been made in some necessary areas, these exotic weapons could actually be orbiting the Earth in the foreseeable future.

There is no doubt that the advances in space technology have added considerable impetus to the arms race but, in addition, these advances are contributing significantly to the current change from policies of 'MAD' (Mutual Assured Destruction) to those which actually consider fighting a nuclear war. These implications are discussed in Chapter 5.

First, however, for a better appreciation of the threat inherent in the military use of space, a brief account of space technology and its applications is given. In order to allow for a better understanding of this technology, Chapter 2 discusses some of the basic concepts of space flight in near-Earth orbit. A brief review of various modes of propulsion and development of launch vehicles in various countries is given in Chapter 3. Chapter 4 provides an overview of the types of military activity in outer space, including manned flights and ASAT systems, while a more peaceful activity, the monitoring of crises from outer space, is discussed in Chapter 6. All military satellites launched between 1977 and 1981 are listed in appendix 1A (those launched between 1958 and 1976 can be found in reference 1).

Apart from having military utility, spacecraft as a treaty verification tool have already become embodied in arms control agreements such as the Strategic Arms Limitation Talks (SALT I and II), and this practice will almost certainly continue.

Growing militarization of outer space has prompted suggestions to demilitarize this environment either completely or partially. In 1979, Italy presented a memorandum to the Committee on Disarmament which states: "Indeed, it seems advisable, in the interest of international security, to impose a total ban on military activities, other than peaceful, in outer space ..." [2]. As a first step, on a limited basis, Italy proposed an Additional Protocol to the 1967 Outer Space Treaty extending the prohibition contained in Article IV of the Treaty.[1] The proposed Protocol was intended,

---

[1] For further comments on the Italian proposal see Part II, Paper 13.

among other things, to prevent the development and use of Earth- or space-based systems to damage, destroy or interfere with the functioning of any spacecraft.

In addition to this, in August 1981 the Soviet Union proposed to the United Nations a new treaty banning the placement of any kind of weapon into orbit around the Earth [3].

Moreover, between June 1978 and 1979 the USA and the USSR discussed the control of their anti-satellite activities. Both the verification aspect as well as the arms control issues relating to outer space are dealt with in Chapter 6.

## References


1. SIPRI, *Outer Space—Battlefield of the Future?* (Taylor & Francis, London, 1978).
2. Committee on Disarmament document CD/9, 26 March 1979.
3. UN document A/36/192, 20 August 1981.

# Chapter 2. Some basic orbital concepts

...If we now imagine bodies to be projected in the direction of lines parallel to the horizon from greater heights... those bodies, according to their different velocity, and the different force of gravity in different height, will describe arcs either concentric with the earth, or variously eccentric, and go on revolving through the heavens in those orbits just as the planets do in their orbits.

Isaac Newton

## *I. Orbit*

If an object above the surface of the Earth is given sufficiently high horizontal velocity, it will not fall back to Earth but instead continue in a circular or an elliptical path round the Earth. The object is said to be in an orbit. In the case of a spacecraft, it is raised to a height and given horizontal velocity using a launcher which usually has more than one stage. In the initial phase of the flight, the launcher accelerates through the atmosphere in a vertical direction and gradually bends over into a horizontal path (figure 2.1). The vertical phase lasts for a very short time (about 10 seconds) and up to an altitude of about 130 m. The first stage of the launcher takes the spacecraft

**Figure 2.1. Flight sequence for a spacecraft launcher**

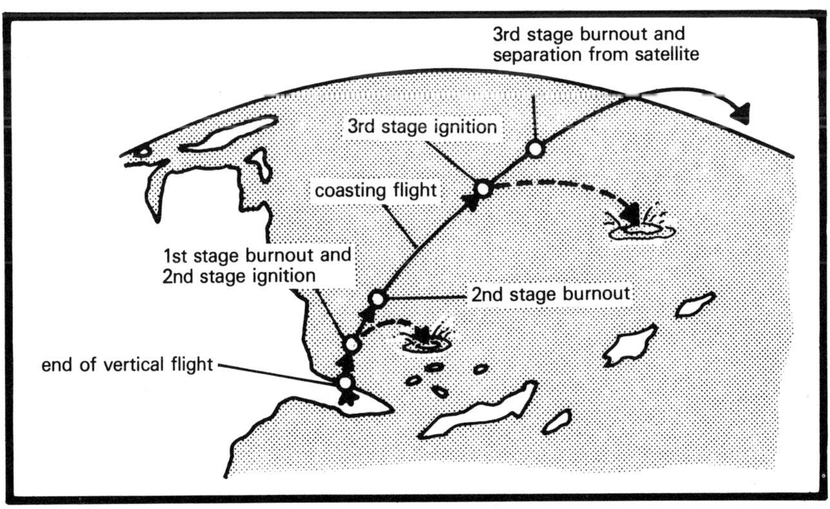

**Figure 2.2. Different orbits for a range of injection velocities**
$v_i$ is the injection velocity for a circular orbit.

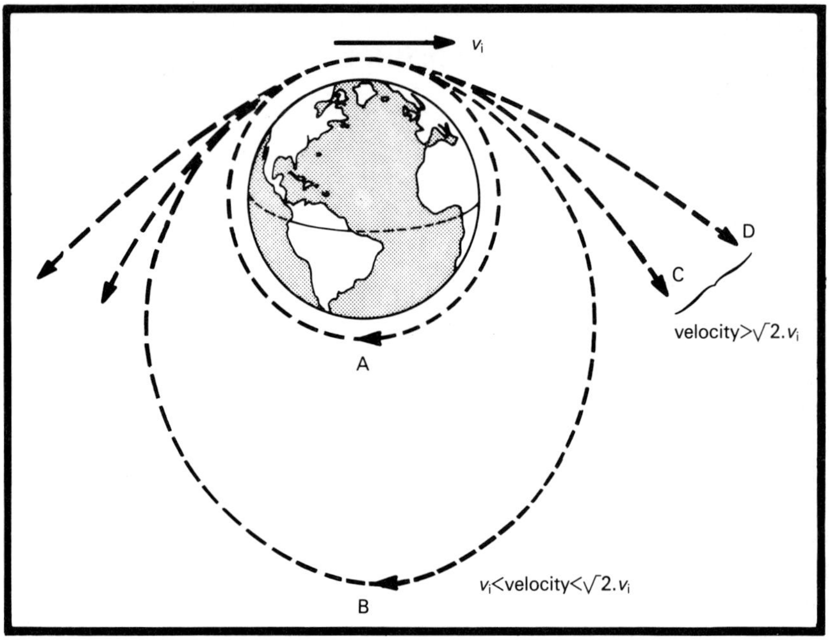


to the upper atmosphere at an altitude of about 60 km. The spacecraft is then gradually brought to its orbital height by the second and the third stage of its launcher. Once this is achieved, the spacecraft is turned into a horizontal position and given a horizontal velocity into its orbit. This velocity is known as the injection velocity.

Assuming that the shape of the Earth is perfectly spherical and that its gravitational field is uniform, and ignoring the effects of the atmospheric drag on the spacecraft, then with an injection velocity ($v_i$) such that the centrifugal force on the spacecraft balances the gravitational force, the spacecraft will move round the Earth in a circular orbit (orbit A in figure 2.2); if the injection velocity lies between $v_i$ and $\sqrt{2} \cdot v_i$, the orbit will be an ellipse (orbit B in figure 2.2); or, the shape of the orbit may not be circular because the direction of injection velocity is either above or below the horizontal. Thus a satellite usually describes an elliptical orbit. If the injection velocity is equal to or greater than $\sqrt{2} \cdot v_i$, the orbit of the spacecraft will be a parabola (curves C and D in figure 2.2) and eventually it will escape from the Earth's gravitational field and move away into space. If, however, the value of the injection velocity is less than $v_i$, the satellite will fall towards the Earth along a curved path.

## II. *Orbital characteristics*

A satellite orbit is usually elliptical with the centre of the Earth at one of the foci of the ellipse (figure 2.3a). There are six basic parameters, called the orbital elements, which completely describe the size, shape and orientation of an orbit of a satellite in space. These are: $a$, the semi-major axis of the satellite's orbit; $e$, eccentricity of the orbit; $T$, time of perigee passage; $\Omega$, the right ascension of the ascending node; $i$, the orbital inclination; and $\omega$, argument of perigee. These are described below.

### Semi-major axis, eccentricity and time of perigee passage

The size and shape of the ellipse are determined by its semi-major axis, $a$, and its eccentricity, $e$. If the distance between the foci of the ellipse is $2c$ (figure 2.3a), then the eccentricity is defined as the ratio $c/a$. These are two of the six orbital elements.

During the lifetime of a satellite, most of the orbital elements are continuously changing, so that the values of all the elements must be given for a particular time. Often this time is chosen to be the passage time of the satellite through the perigee point. This time, $T$, is the third orbital element and, together with $a$ and $e$, defines the ellipse in a plane. The perigee is the shortest distance between the satellite and the Earth, and the apogee the longest distance (see figure 2.3a).

### The right ascension of the ascending node

The position of the orbital plane in space is usually given in a terrestrial sidereal rectangular co-ordinate system, the axes of which are shown in figure 2.3b. The origin, O, of the co-ordinate system is the centre of the Earth; the z-axis is oriented towards the North Pole and the Earth's equatorial plane is in the xy-plane. The x-axis is oriented towards the Vernal equinox or the first point of Aries.[1] The point of intersection of the satellite ground track with the equator is known as the ascending node. In the figure, it is marked N. The angle $\Omega$ between the x-axis and the line, ON, defines the fourth orbital element and it is called the right ascension of the ascending node. The longitude of a satellite's ground track at a particular time can be deduced from the knowledge of $\Omega$ for the satellite and the Greenwich meridian. During the lifetime of a satellite $\Omega$ does not remain fixed. This is because the orbital plane

---

[1] The orbital plane of the Earth is called the plane of the ecliptic. The angle between the Earth's equatorial plane and the ecliptic plane is 23° 7′. The line of intersection of these two planes defines an important direction in space, known as the first point of Aries. At any particular time this direction is exactly known and all celestial directions can conveniently be referred to it.

**Figure 2.3. Geometry of a satellite in orbit**

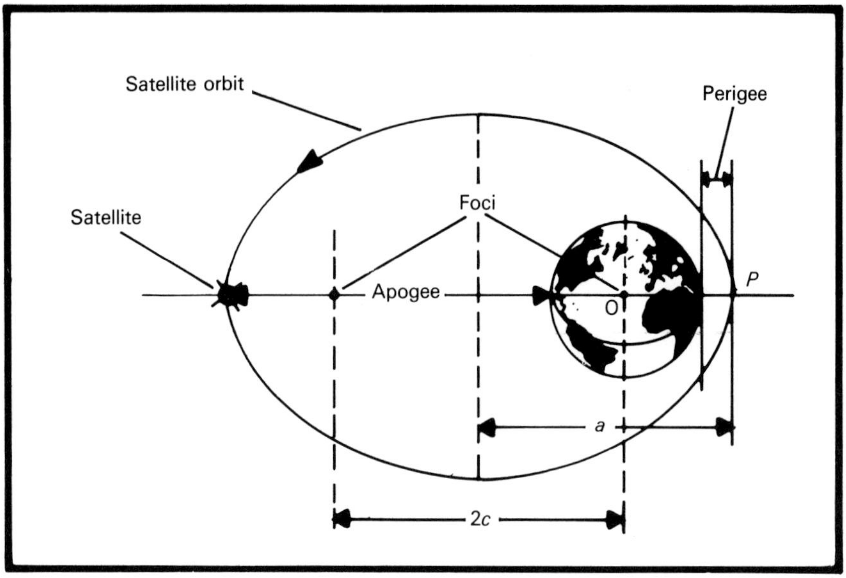


(a)

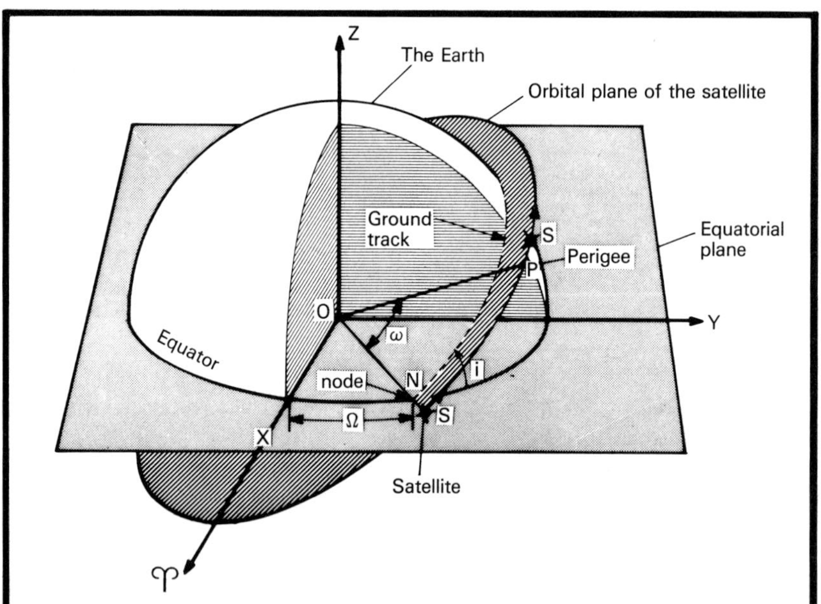


(b)

turns round the Earth's axis while keeping the orbital inclination constant (the angle between the orbital plane of the satellite and the equatorial plane of the Earth). The change in $\Omega$ is known as precession; this occurs because the Earth is not spherical so that its centre of gravity is not at the same point as its centre of mass. Moreover, the mass is not uniformly distributed. The rate of change of $\Omega$ is approximately given by the following equation:

$$\frac{d\Omega}{dt} = \frac{-10g^{-3.5}\cos i}{(1-e^2)^2} \qquad \text{degrees/day}$$

For low altitude orbits with small orbital inclinations, $\Omega$ can change as much as 9° per day. However, for an orbital inclination, $i$, of 90°, cos 90° is zero, so that, from the above equation, the rate of change of $\Omega$ is zero. Thus $\Omega$ does not change for orbits with an orbital inclination of 90°.

Another characteristic of an orbit with 90° inclination (i.e., a polar orbit) is that the satellite will pass over a region of interest on the Earth at the same time of day throughout the active lifetime of the satellite, and changes in activity in the region can therefore be compared. Such an orbit is called a sun-synchronous orbit.

## Orbital inclination

The fifth and perhaps the most important of the orbital elements is the orbital inclination, $i$. This is the angle between the orbital plane of the satellite and the equatorial plane of the Earth (see figure 2.3b). It is the value of $i$ which determines the range of latitudes over which a satellite travels during each revolution. For observations of an area situated at high latitudes, for example, a satellite with an orbital inclination of 90° would be used as it passes over the polar regions. With the knowledge of $\Omega$ and $i$, it is now possible to specify the position of the satellite orbital plane relative to the coordinate system considered. The orbital element $i$ is the only element which remains practically constant during the lifetime of a satellite.

## Argument of perigee

The orientation of the ellipse within the orbital plane is given by the sixth orbital element, $\omega$, known as the argument of perigee. This is the angle between the line ON (see figure 2.3b) in the equatorial plane and the line OP, where P is the perigee. The line OP always lies along the major axis of the ellipse (see figure 2.3a). The oblateness of the Earth also affects $\omega$, which for all except one value of orbital inclination is not constant. This means that the major axis of the orbit rotates in the orbital plane in the direction of motion of the satellite. The rate of change of $\omega$ is given by the following equation:

$$\frac{d\omega}{dt} = \frac{4.98\, a^{3.5}\,(5\cos^2 i - 1)}{(1 - e^2)}$$

For $\omega$ to be stationary, the rate of change $\frac{d\omega}{dt}$ has to be zero so that the above equation becomes

$$\begin{aligned} 5\cos^2 i - 1 &= 0 \\ \text{or} \qquad i &= \cos^{-1}\sqrt{1/5} \\ &= 63.4° \end{aligned}$$

This means that if a satellite is launched at an orbital inclination of 63.4°, its perigee will remain stable along any chosen latitude. For other values of orbital inclination $\omega$ could vary up to 5° per day.

The variations, or perturbations as they are called, of $\Omega$ and $\omega$ are due to the oblate shape of the Earth. However, there are other sources and types of orbital perturbation. These are briefly summarized below.

**Other orbital perturbations**

Among several perturbation effects, the most significant is that due to the Earth's rotation. However, this effect is an apparent one. The Earth rotates eastwards on its axis at the rate of about 15° per hour. For a terrestrial observer, therefore, the satellite orbit appears to shift continuously westward.

Another important effect on the orbit of a satellite is that due to the Earth's atmosphere. This effect, known as atmospheric drag, causes the satellite to decelerate. As a result, the centrifugal force on the satellite decreases and the Earth's gravity pulls the satellite further into the atmosphere. The net effect is the eventual decay of the satellite unless a propulsive energy is given to it.

While the gravitational fields of the Sun and the Moon can be ignored in comparison with the Earth's gravitational field for satellites in low orbits, they become significant for spacecraft in highly eccentric elliptical orbit. One of the effects is to depress the perigee height of a satellite, thus shortening its lifetime.

The other orbital perturbation effect that should be mentioned here is that due to the Earth's magnetic field. For large satellites the drag caused by the Earth's magnetic field could be significant. At altitudes greater than 1 200 km, the magnetic drag can exceed aerodynamic drag.

# Chapter 3. Space transportation vehicles

> Mankind will not remain for ever on the earth, but, in a quest for light and space, will, at first timidly, penetrate the atmosphere, and later conquer the whole of the solar system.
>
> Konstantin Tsiolkovsky (1903)

## *I. Background*

The roar of a rocket standing vertically on its launch pad is simply the ejection of hot gases through its exhaust nozzle at great speed. The force of the gas at first slowly lifts the rocket off the ground and eventually imparts enough speed to the spacecraft mounted on top of the launcher to place it in orbit round the Earth or even to escape the Earth's force of gravity.

The concept of the rocket is centuries old. Probably as early as 3 000 B.C.—and certainly by 1 000 B.C.—the Chinese had developed rockets for, among other things, use in warfare. It is the latter use which has given the greatest impetus to modern rocket technology. While the idea of rockets sophisticated enough to lift men to the Moon has captured people's imagination, the development of this technology to hurl devastating weapons long distances and to launch military Earth satellites into space has gone on virtually unnoticed.

A practical demonstration of long-range rockets was given during the last couple of years of World War II, when the V-2 missiles were developed in Germany. This development has formed the basis of modern intercontinental ballistic missile (ICBM) technology. The ICBMs used to transport nuclear weapons subsequently formed the basis for launchers that can take artificial Earth satellites and man into space.

In the following sections, a brief description of various modes of propulsion is given. Development of launch vehicles in various countries is also briefly reviewed.

## *II. Some basic concepts*

A launcher consists of (*a*) a single booster rocket or a rocket with more stages, depending on the weight of the satellite; and (*b*) the satellite, housed in a protective structure placed on top of the launcher. The function of the first stage of a multi-stage rocket is to provide enough propelling 'force of thrust' to lift the full mass of the launcher and the payload up to a required height. Because of the application of continuous thrust by the rocket engine and continuously reducing mass as the fuel burns, the speed of the vehicle increases steadily and reaches a maximum when the propellant is finally consumed. The multi-stage launcher has been developed because a single rocket cannot reach the height required to orbit the satellite when the payload is too heavy. As soon as the propellant in the first stage is consumed, it is separated from the launcher and the second stage is ignited and so on. Usually a maximum of three stages is used.

During the initial phase of the flight, the trajectory of the launcher is determined by various forces acting on it. The launcher is accelerated by the propulsive force of exhaust gases of the burning propellant. While travelling through the atmosphere, the motion of the launcher is affected by large aerodynamic forces of drag and lift. The other force acting on the launcher is that for steering the vehicle by means of a thrust-vector-control system which controls the direction of the launcher as well as stabilizing it to prevent it from tumbling under the high forces of engine thrust. The system consists of an engine swivel or gimbal arrangement to deflect the forces of engine thrust from the centre of mass. Throughout the trajectory of the launcher and the satellite in its orbit, the vehicles are pulled towards the centre of the Earth by its force of gravity.

It can be seen that of all the forces acting on the launcher and its payload, only the force of the rocket engine thrust can be controlled by man. Therefore, it is useful to consider briefly some of the basic principles of how this force is produced.

## *III. Types of rocket engine*

The concept of rocket propulsion is based on Newton's Third Law of Motion, which states that for every action there is an equal and opposite reaction. Thus, in the case of rocket propulsion, a forward thrust is realized by the rearward ejection of high velocity combustion gases.

There are two most commonly used devices for generating the forward thrust in a rocket, namely, the liquid-propellant engine and the solid-propellant engine. Other types of engine using nuclear reactors, electric propulsion and laser propulsion have been investigated. The latter is the least developed one.

## Liquid-propellant rockets

A liquid-propellant rocket may be either a monopropellant, a bipropellant or a tripropellant. In a monopropellant rocket, a propellant (e.g., hydrazine) reacts with a catalyst which promotes the decomposition of the propellant. The decomposition results in the production of heat. In a bipropellant rocket, two chemicals (e.g., a fuel and an oxidizing agent) burn together, whereas in a tripropellant rocket engine, three different chemicals (e.g., hydrogen, oxygen and beryllium) are mixed in a combustion chamber and are burned together.

The principle of a liquid bipropellant engine is shown in figure 3.1. The main features are: (*a*) the injector through which the propellants enter the combustion chamber; (*b*) the combustion chamber where the propellants are mixed and are ignited; and (*c*) the exhaust nozzle. Propellants, such as a combination of oxygen and kerosene, are ignited by means of a spark plug, whereas a mixture of nitrogen tetroxide and hydrazine ignites spontaneously on mixing.

As the propellants burn, very hot gases are produced causing a high temperature and thereby increasing the pressure of these gases in the combustion chamber. At high pressures, the gases pass through the narrow neck of the exhaust nozzle, which accelerates them to sonic speeds. On diverging in the exhaust nozzle (figure 3.1), the gases expand and accelerate to supersonic speeds.

**Figure 3.1. Liquid-bipropellant rocket engine**

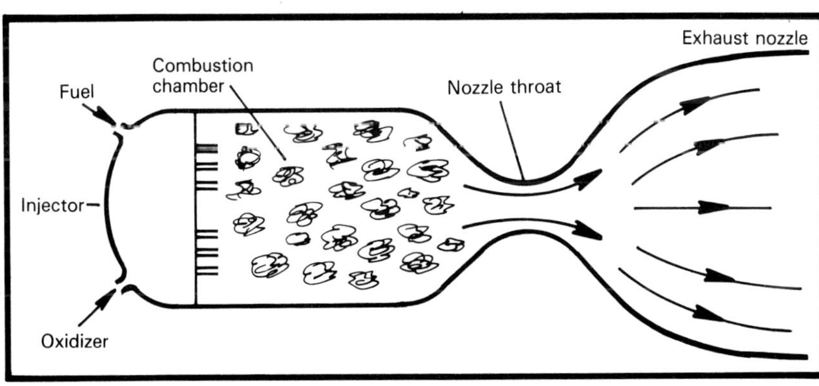


Broadly there are two types of liquid propellant, namely mono- and bipropellants (storable at normal temperatures, about 16°C) and cryogenic propellants. Characteristics of some of the common liquid propellants are shown in tables 3.1, 3.2 and 3.3. The parameter specific impulse in the table

**Table 3.1. Liquid monopropellants**

| Propellant | Maximum specific impulse (seconds) |
|---|---|
| Hydrogen peroxide | 165 |
| Hydrazine | 199 |
| Nitromethane | 245 |
| Ethylene oxide | 199 |

**Table 3.2. Storable liquid bipropellants**

| Fuel | Oxidizer | Maximum specific impulse (seconds) |
|---|---|---|
| Hydrazine/unsymmetrical dimethylhydrazine (50/50) | Nitrogen tetroxide ($N_2O_4$) | 289 |
| Unsymmetricai dimethylhydrazine | $N_2O_4$ | 286 |
| Unsymmetrical dimethylhydrazine | Red fuming nitric acid (15% $NO_2$) | 266 |
| Unsymmetrical dimethylhydrazine | Maximum density red fuming nitric acid with $N_2O_4$ | 278 |
| Unsymmetrical dimethylhydrazine | $N_2O_4$ + NO (15%) | 288 |
| Unsymmetrical dimethylhydrazine | Hydrogen peroxide | 284 |
| Unsymmetrical dimethylhydrazine | Chlorine trifluoride | 280 |
| Unsymmetrical dimethylhydrazine | Chlorine pentafluoride | 297 |
| Hydrazine | $N_2O_4$ | 292 |
| Hydrazine | Hydrogen peroxide | 287 |
| Hydrazine | Chlorine trifluoride | 295 |
| Hydrazine | Chlorine pentafluoride | 313 |
| Kerosene-type fuel | Red fuming nitric acid (15% $NO_2$) | 257 |
| Pentaborane | Hydrazine | 318 |

is a measure of the efficiency of the rocket engine. It is usually defined as the ratio of the thrust produced to the rate at which the propellant is used up (i.e., pounds force/pounds per second).

## Solid-propellant rockets

Owing to the ease with which solid fuels can be handled and used, it is not surprising that such fuels are widely used in military rockets. For example, many air-to-air missiles, ground-to-air missiles, ICBMs (such as Minuteman) and space boosters (such as Titan III) use solid fuels in their rocket engines. In a solid-propellant rocket, a solid-fuel-oxidizer combination is used.

The propellant called "grain" is a solid mixture of an oxidizer and a fuel, and once ignited it is capable of sustained combustion. As in the case of

**Table 3.3. Cryogenic liquid bipropellants**[a]

| Fuel | Oxidizer | Maximum specific impulse (seconds) |
|---|---|---|
| Kerosene-type fuel | Liquid oxygen | 300 |
| Kerosene-type fuel | Liquid fluorine | 322 |
| Unsymmetrical dimethyl-hydrazine | Liquid oxygen | 310 |
| Hydrazine | Liquid oxygen | 313 |
| Hydrazine | Liquid fluorine | 365 |
| Ammonia | Liquid oxygen | 294 |
| Ethyl alcohol | Liquid oxygen | 290 |
| Methane | Liquid oxygen | 311 |
| Liquid hydrogen | Liquid oxygen | 290 |
| Liquid hydrogen | Liquid fluorine | 412 |

[a] At least one propellant is cryogenic

a liquid-propellant rocket, gases at high temperature are produced in the combustion chamber and expand through the nozzle throat into the exhaust nozzle (see figure 3.2). The combustion of the fuel is started by an igniter.

**Figure 3.2. Solid-propellant rocket engine**

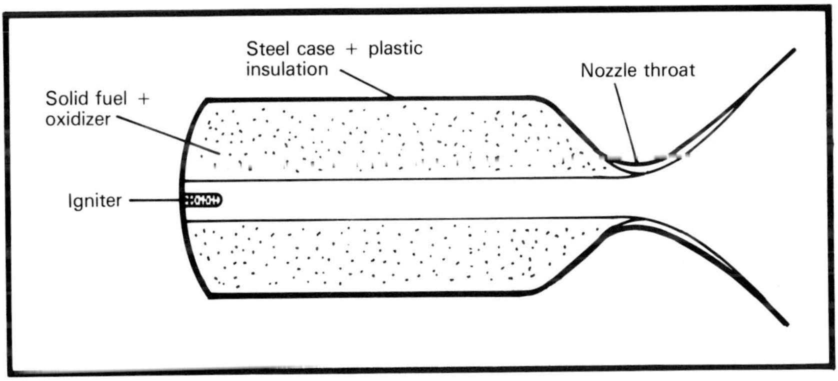


Most modern solid fuels are of two types, double-base or composite. The former is a mixture of two compounds, either one of which on its own would be a rocket propellant. Usually the mixture contains nitroglycerin and nitrocellulose. The fuel (carbon and hydrogen) and the oxidizer (oxygen) are contained in each of these compounds. Both these compounds are monopropellants.

A composite fuel is formed by a mixture of two or more unlike compounds into a composite material. None of the constituents would make a good propellant on its own. Modern composite propellants use a polymer based on a synthetic rubber, such as polybutadiene or polysulphide. This fuel acts as a binder also for the crumbly oxidizer, such as finely ground potassium nitrate or ammonium perchlorate crystals. Often 10 to 20 per cent by weight of aluminium or beryllium powder is added to a polymer-based grain to increase the efficiency with which the energy is released.

Properties of some of the solid propellants are given in table 3.4.

**Table 3.4. Some solid propellants**

| Propellant | Maximum specific impulse (seconds) |
|---|---|
| DB | 255 |
| DB/AP/Al[a] | 258 |
| DB/AP-HMX/Al[a] | 272 |
| PVC/AP[b] | 239 |
| PVC/AP/Al[c] | 253 |
| PBAN/AP/Al[c] | 265 |
| PU/AP/Al[c] | 263 |
| CTPB/AP/Al[c] | 265 |
| HTPB/AP/Al[c] | 264 |
| PBAA/AP/Al[c] | 265 |

*Key*

| | |
|---|---|
| Al | Aluminium |
| AP | Ammonium perchlorate |
| CTPB | Carboxy-terminated polybutadiene |
| DB | Double base (e.g., nitrocellulose–nitroglycerine) |
| HMX | Cyclotetramethylene tetranitramine |
| HTPB | Hydroxy-terminated polybutadiene |
| PBAA | Polybutadiene–acrylic acid polymer |
| PBAN | Polybutadiene–acrylic acid–acrylonitrite terpolymer |
| PU | Polyurethane |
| PVC | Polyvinyl chloride |

[a] AP/Al with 40% DB as binder
[b] AP/Al with 20% binder
[c] AP/Al with 15% binder

## Nuclear rockets

The main difference between the rockets described above and a nuclear rocket is that in the above rockets the heat energy used to produce thrust is derived from burning the fuel with an oxidizer in a combustion chamber, whereas the heat energy in a nuclear rocket is derived from nuclear reactions in a nuclear reactor (see figure 3.3). The nuclear rocket uses a single propellant which does not undergo any chemical reaction. However, the conversion of the thermal energy into kinetic energy is made using the same technique as before, that is, a narrow-necked exhaust nozzle.

The specific impulse ($I_{sp}$) in the case of a nuclear rocket is given by

$$I_{sp} \propto \sqrt{\frac{t}{m}}$$

where $t$ is the temperature of the propellant at the nozzle inlet and $m$ is its molecular weight [1]. It can be seen from the above relationship that the specific impulse increases with increasing propellant temperature and decreasing

**Figure 3.3. Nuclear rocket engine**

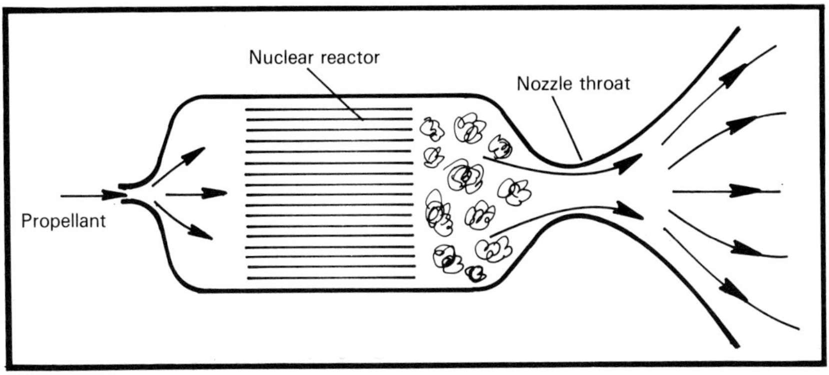


molecular weight. It is the latter factor which gives the nuclear rocket its advantage since the best propellant, hydrogen, is the one with the lowest molecular weight. In a nuclear rocket, liquid hydrogen is pumped into the reactor where it is heated and ejected as a gas by expanding it through a nozzle, thrust thus being developed. For a given propellant temperature ($t$), the specific impulse for a nuclear reactor is about three times that for a chemical propellant. In practice a specific impulse would be about 1 000 pounds of thrust per pound of hydrogen flow per second, a value more than twice that of the best chemical propellants [1].

## Electric propulsion

While considerable progress has been made in the field of electric propulsion, much research and development still remains to be done, particularly on systems based on conversion of nuclear energy into electrical energy. Nuclear reactors probably offer the best compact energy source. However, small electric propulsion systems used for station-keeping of a satellite have already been developed and deployed. These, the electric thrusters, convert the electric power and propellant into thrust.

In an electric thruster, the electric power is used to accelerate the propellant to a high velocity through the nozzle. There are three types of electric thrusters, namely, electrothermal, electromagnetic and electrostatic thrusters.

### *Electrothermal thrusters*

In this system, the propellant gas is heated to high temperatures either by an electric arc or by a heating element. The gas is then allowed to expand through a nozzle to produce thrust. The latter system, called a resistojet, has already been used in a satellite station-keeping application. Such a system can produce larger exhaust gas velocities than those in chemical rockets.

*Electromagnetic thrusters*
In this case the propellant gas is ionized to form a plasma which is then accelerated by electric and magnetic fields through the nozzle. Such a plasma thruster is still in the developmental stage. However, a more promising system is the magnetoplasmadynamic (MPD) system in which the propellant is allowed to pass between two electrodes at a low pressure. An electric current flows between the electrodes. Electromagnetic force is the main mechanism by which thrust is produced, a magnetic field being produced either due to the current between the two electrodes or by a separate field coil. An advantage of the MPD thruster lies in the simplicity of its system. However, much work still needs to be done before such a system becomes useful for space applications.

*Electrostatic thrusters*
The propellant in this system is also ionized either by heating it, using, for example, hot tungsten, or by bombarding the propellant with electrons. Initial acceleration of the ions is achieved because like charges repel each other and the ions are further accelerated using an oppositely charged accelerator grid. This combination of repulsion and attraction ejects the ions with high exhaust velocities through the nozzle. A thruster in which electrons are used to ionize the propellant was tested in space for a short time in 1964 by the USA in the SERT 1 (Space Electric Rocket Test) programme [2].

## Laser propulsion

The principle of laser propulsion is illustrated in figure 3.4. A laser beam is focused into a plasma chamber containing hydrogen propellant. The laser heats the core of the propellant fluid to a plasma state and then the plasma is allowed to expand through a nozzle to produce thrust. The laser propulsion system considered differs fundamentally from the other systems discussed so far in that the energy source for the propellant is outside the rocket, that is, the laser is remotely placed. It is envisaged that a laser is either placed on the ground or in orbit and the energy is beamed to the rocket [3]. If placed in orbit, it would be used to propel a spacecraft from a lower orbit to a higher one or for a deep-space mission [4]. For a space-based system, the weight of the power source for the laser would be carried not by the launcher but by an orbital transmission station. An Earth-based laser propulsion would have to be a pulsed laser beam of very high energy to minimize the problems of atmospheric transmission. An average power of more than 1 000 MW would be required to launch a 1 000 kg payload [5].

In the USA both the Defense Advanced Research Projects Agency (DARPA) and NASA are investigating the use of high-energy lasers for rocket propulsion. Under the DARPA project, Earth-based carbon-dioxide electric-discharge lasers are being considered. It is expected that such a device would produce a specific impulse of 1 000 seconds, over three times that available with normal chemical propellants [6]. However, there are a

**Figure 3.4. Laser propulsion rocket**

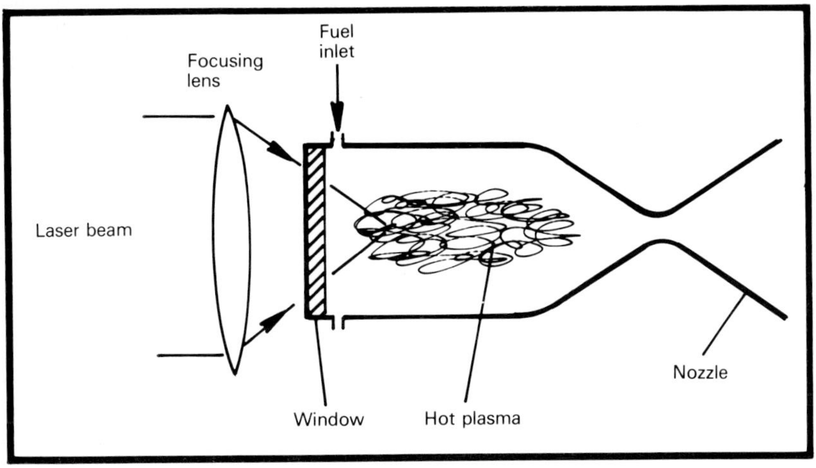


number of problems with such a device, the major one being transmission of a laser beam through the atmosphere. This could be minimized by, for example, using a pulsed laser, placing it on top of a mountain and using adaptive optics.

Under the NASA project, lasers for orbit-to-orbit propulsion are being investigated. In this case a 30-kW continuous-wave carbon-dioxide laser would orbit at an altitude of 500 km. Testing such a device on the ground by NASA has begun. It is expected that such a system may achieve a specific impulse of 2 000 seconds compared to the maximum of 430 seconds achievable by chemical propellants [7]. However, the first test flight is not expected until the year 2004. In the Soviet Union a jet engine using a continuous-wave laser is being considered [8].

With this very brief survey of some of the main methods of propelling objects into outer space, it is useful to consider the spread of launcher technology throughout the world. In the following sections developments of manned as well as unmanned launchers for space travel in various countries are briefly reviewed.

## *IV. Development of national launch vehicles*

### USA

By the end of 1981, some 10 non-reusable launchers had been developed by the USA, of which at least three—Thor, Atlas and Titan—were derived from ballistic missiles. Some of the US launch vehicles were developed as joint NASA and Department of Defense (DoD) projects. As can be seen from

## Table 3.5. Number of US launches per year, 1958–81

| Launches | 1958 | 1959 | 1960 | 1961 | 1962 | 1963 | 1964 | 1965 | 1966 | 1967 | 1968 | 1969 | 1970 | 1971 | 1972 | 1973 | 1974 | 1975 | 1976 | 1977 | 1978 | 1979 | 1980 | 1981 |
|---|---|---|---|---|---|---|---|---|---|---|---|---|---|---|---|---|---|---|---|---|---|---|---|---|
| Vanguard | 1 | 2 | | | | | | | | | | | | | | | | | | | | | | |
| Jupiter C | 3 | | | | | | | | | | | | | | | | | | | | | | | |
| Thor-Able | 1 | 1 | 2 | | | | | | | | | | | | | | | | | | | | | |
| Atlas B | 1 | | | | | | | | | | | | | | | | | | | | | | | |
| Thor-Agena A | | 6 | 5 | | | | | | | | | | | | | | | | | | | | | |
| Juno II | | 2 | 1 | 1 | | | | | | | | | | | | | | | | | | | | |
| Thor-Able Star | | | 3 | 3 | 1 | 2 | 2 | 3 | | | | | | | | | | | | | | | | |
| Atlas-Agena A | | | 1 | 1 | | | | | | | | | | | | | | | | | | | | |
| Delta | | | 2 | 3 | 9 | 7 | 3 | 5 | 2 | 2 | | 2 | | | 7 | 6 | | 10 | 9 | 10 | 11 | 3 | 3 | 3 |
| Atlas-Able | | 1 | 2 | | | | | | | | | | | | | | | | | | | | | |
| Thor-Agena B | | | 2 | 12 | 19 | | 2 | 1 | | | | | | | | | | | | | | | | |
| Scout | | | | 2 | 3 | 4 | 7 | 5 | 8 | 8 | 6 | 1 | 3 | 5 | 5 | | 6 | 2 | 2 | 1 | 1 | 1 | | |
| Redstone | | | | 3 | | | | | | | | | | | | | | | | | | | | |
| Atlas-Agena B | | | | 5 | 12 | 3 | 2 | 2 | 1 | | | | | | | | | | | | | | | |
| Atlas D | | | | 2 | 3 | 1 | | 3 | 4 | 1 | | | | | | | | | | | | | | |
| Thor-Agena D | | | | | 7 | 8 | 2 | 2 | | 1 | | | 2 | | | | | | | | | | | |
| TAT-Agena D | | | | | | 9 | 19 | 15 | 11 | 5 | 3 | | | | | | | | | | | | | |
| Atlas-Agena D | | | | | | 5 | 14 | 12 | 24 | 9 | 1 | 1 | 1 | | | 1 | | | | 1 | 2 | | | |
| Atlas-Centaur | | | | | | 1 | 1 | 2 | 4 | 4 | 3 | 3 | 1 | 4 | 4 | 3 | 1 | 2 | 3 | 3 | 7 | 2 | 3 | 3 |
| Saturn I | | | | | | | 3 | 3 | 1 | | 1 | | | | | | | | | | | | | |
| Titan II | | | | | | | 1 | 5 | 5 | | | | | | | | | | | | | | | |
| TAD | | | | | | | 1 | 3 | 6 | 10 | | | | | | | | | | | | | | |
| Titan III A | | | | | | | 1 | 2 | | | | | | | | | | | | | | | | |
| Thor-Altair | | | | | | | | 4 | 2 | | | | | | | | | | | | | | | |
| Titan III C | | | | | | | | 3 | 3 | 3 | 2 | 2 | 2 | 2 | 1 | 2 | 1 | 2 | 2 | 2 | 3 | 3 | | 2 |
| TAT-Agena B | | | | | | | | | 1 | | | | | | | | | | | | | | | |
| Titan III B-Agena D | | | | | | | | | 3 | 7 | 8 | 6 | 6 | 5 | 2 | 4 | 3 | | 3 | | 2 | | 1 | 2 |
| Thorad-Agena D | | | | | | | | | 1 | | 1 | 7 | 4 | 3 | | | | | | | | | | |
| Thor-Burner II | | | | | | | | | 1 | 4 | 2 | 1 | 2 | 2 | 2 | 1 | | | 2 | 1 | 1 | 1 | | |
| LTTAT-Agena D | | | | | | | | | | 6 | 6 | 3 | | 1 | | | | | | | | | | |
| Saturn V | | | | | | | | | | 1 | 2 | 4 | 1 | 2 | 1 | 1 | | | | | | | | |
| TAID | | | | | | | | | | | 4 | 3 | | | | | | | | | | | | |
| Atlas F | | | | | | | | | | | 2 | 1 | | 1 | | | 1 | 1 | | 2 | 5 | 1 | 4 | 1 |
| Atlas-Agena | | | | | | | | | | | 1 | | | | | | | 1 | | 1 | | | | |
| LTTAT-Burner II | | | | | | | | | | | 1 | | | | | | | | | | | | | |
| LTD | | | | | | | | | | | 4 | | | | | | | | | | | | | |
| Atlas-Burner II | | | | | | | | | | | 1 | | | | | | | | | | | | | |
| Saturn IB | | | | | | | | | | | 1 | | | | | 3 | | 1 | | | | | | |
| LTTA-Delta | | | | | | | | | | | | 4 | | 1 | | | | | | | | | | |
| LTTA-Delta N | | | | | | | | | | | | 2 | | | | | | | | | | | | |
| TAT-Delta | | | | | | | | | | | | | 5 | 1 | | | | | | | | | | |
| TAT-Delta M | | | | | | | | | | | | | 1 | | | | | | | | | | | |
| TAT-Delta-Agena D | | | | | | | | | | | | | 1 | | | | | | | | | | | |
| Long Tank TAT Delta | | | | | | | | | | | | | 1 | | | | | | | | | | | |
| Thor-Agena | | | | | | | | | | | | | | 1 | | | | | | | | | | |
| Long Tank Thor-Delta | | | | | | | | | | | | | | 1 | | | | | | | | | | |
| Long Tank Thorad-Agena D | | | | | | | | | | | | | | 2 | | | | | | | | | | |
| Thor-Delta | | | | | | | | | | | | | | 1 | | | | | | | | | | 2 |
| Titan III D | | | | | | | | | | | | | | 1 | 4 | 3 | 2 | 2 | 2 | 1 | 2 | 1 | 2 | 1 |
| Delta N | | | | | | | | | | | | | | 1 | | | | | | | | | | |
| Atlas-Centaur-Agena D | | | | | | | | | | | | | | 1 | | | | | | | | | | |
| Titan III B | | | | | | | | | | | | | | | 1 | | 1 | 3 | 1 | 2 | | 1 | | |
| Thorad-Agena | | | | | | | | | | | | | | | 3 | | | | | | | | | |
| Scout-Altair | | | | | | | | | | | | | | | | 1 | | | | | | | | |
| Thorad-Delta | | | | | | | | | | | | | | | | | 7 | | | | | | | |
| Titan III E-Centaur | | | | | | | | | | | | | | | | | 2 | 2 | | 2 | | | | |
| Thor-Burner IIA | | | | | | | | | | | | | | | | | 2 | | | | | | | |
| Thor-Burner | | | | | | | | | | | | | | | | | | 1 | | | | | | |
| Titan-Centaur | | | | | | | | | | | | | | | | | | | 1 | | | | | |
| Atlas | | | | | | | | | | | | | | | | | | | 1 | | | 1 | | |
| Scout D | | | | | | | | | | | | | | | | | | | | | | 1 | | |
| Scout G | | | | | | | | | | | | | | | | | | | | | | 1 | | |

table 3.5, the Scout, Thor, Atlas and Titan vehicles have formed the backbone of the US space activities.

Although the first successful artificial Earth satellite was launched in

1958, the concept of satellites was already being formulated by the mid-1940s. In early 1945, the US Navy had begun a satellite study and in May 1946, Project RAND presented a study containing designs of satellites to the US Army Air Force. In late 1945, the US Navy Bureau of Aeronautics established a Committee for Evaluating the Feasibility of Space Rocketry (CEFSR) [9]. In early 1946, the Navy CEFSR and the Army Air Force discussed the possibility of establishing a joint Earth satellite project. But this co-operation never materialized, and an official go-ahead was not given to the US satellite programme for about a decade.

However, the general conclusion from the various studies was that a satellite launcher with three stages, using a hydrogen-oxygen propellant, should be developed. Such a system would orbit a payload weighing about 37 000 kg at an altitude of about 600 km. In 1955, the US announced a programme to launch a satellite as a part of the International Geophysical Year (IGY). While this was a joint programme, with the National Science Foundation, the National Academy of Sciences and the DoD as the participants, the programme was mainly financed by the latter. By early 1956, the design of the Vanguard vehicle was completed. This three-stage launcher was largely based on a vehicle developed earlier.

While the Vanguard launcher was not used extensively (see table 3.5), it was later incorporated into many subsequent launchers. For example, the US Air Force Thor and Atlas series of Able launchers used the Vanguard for the second and third stages, the NASA Scout vehicle still uses the Vanguard for the third stage and the Delta launch vehicle uses Thor for the first stage and an improved Vanguard for the second and third stages. Liquid fuel was used as the propellant in the Vanguard launchers. In fact, it was not until the introduction of the Scout vehicles that solid propellants were used.

*Scout*

The development of the Scout, the first US launcher to use solid propellant, began in early 1958. Its four-stage configuration is still retained although all its stages have been uprated over the years. Since its development, NASA, agencies of the DoD and a number of other countries have used the Scout mainly for launching satellites. Development of the DoD vehicle began in 1959 and now it is mainly used for launching satellites. Characteristics of this and other US vehicles are given in table 3.6.

*Thor*

The DoD, during 1955, concluded that it was desirable to develop an intermediate-range ballistic missile (IRBM). In 1955 the US Air Force was granted resources for the Thor programme and by the end of 1958, the missile was deployed. Apart from its role as an IRBM, the Thor, in conjunction with other launchers, such as the Vanguard, was used extensively for orbiting various types of satellite. With the Vanguard used for the second stage, the launcher was named Thor-Able and was used to launch Pioneer 1, the first

**Table 3.6. Characteristics of some of the US launch vehicles**

| Vehicles | Height (excl. payload) (m) | Maximum diameter (m) | Launch weight (kg) | Payload (kg) | Orbit (km) | Stage number | Propellant | Thrust (kg) | Date of first launch of original version |
|---|---|---|---|---|---|---|---|---|---|
| *Basic vehicles* | | | | | | | | | |
| Scout | 22.9 | 1.1 | 21 410 | 180 | 555 | 1 | Solid | 48 534 | 1960 |
| SLV-1A | | | | | | 2 | Solid | 28 032 | |
| | | | | | | 3 | Solid | 12 700 | |
| | | | | | | 4 | Solid | 2 586 | |
| TAT[1] | 18.4 | 2.4 | 61 689 | 817 | 555 | 1 | LOX/RP | 78 018 | 1963 |
| LTAT[1] | 22.6 | 2.4 | 83 227 | | | 1 | LOX/RP | 78 018 | 1967 |
| TAD[1] | 24.6 | 2.4 | 99 790 | 907 | 555 | 1 | LOX/RP | 78 018 | 1964 |
| | | | | | | 2 | IRFNA/UDMH | 3 538 | |
| | | | | | | 3 | Solid | 2 603/4 536 | |
| Atlas F | | 3.1 | 119 068 | 612 | 185 | 1/2 | LOX/RP-1 | 149 686 | 1968 |
| | | | | 3 265 | 185 × 20 187 | 1 | LOX/RP-1 | 25 855 | |
| Titan III B | 23.4 | 3.1 | 175 268 | 3 629 | 185 | 1 | $N_2O_4/N_2H_4$-UDMH | 235 868 | 1966 |
| | | | | | | 2 | $N_2O_4/N_2H_4$-UDMH | 45 813 | |
| Titan III C | 32.9 | 3.1 | 631 400 | 13 154 | 185 | 0 | Solid | 1 088 622 | 1965 |
| | | | | or | or | | | | |
| | | | | 1 429 | 36 000 | | | | |
| | 22.3 | 3.1 | | | | 1 | $N_2O_4/N_2H_4$-UDMH | 235 868 | |
| | 8.8 | 3.1 | | | | 2 | $N_2O_4/N_2H_4$-UDMH | 45 813 | |
| | 4.6 | 3.1 | | | | 3 | $N_2O_4/N_2H_4$-UDMH | 7 258 | |
| Titan III D | 32.9 | 3.1 | 589 670 | 13 608 | 185 | 0 | Solid | 1 088 622 | 1971 |
| | 22.3 | | | | | 1 | $N_2O_4/N_2H_4$-UDMH | 235 868 | |
| | 7.9 | 3.1 | | | | 2 | $N_2O_4/N_2H_4$-UDMH | 45 813 | |

| | | | | | | | | | |
|---|---|---|---|---|---|---|---|---|---|
| Titan-34D (IUS) | 27.7 | 3.1 | 670 636 | 46 465 | | 0 | Solid | 1 088 622 | Planned for 1981 |
| | 23.7 | 3.2 | | | | 1 | $N_2O_4/N_2H_4$-UDMH | 235 868 | |
| | 7.9 | 3.1 | | | | 2 | $N_2O_4/N_2H_4$-UDMH | 45 813 | |
| | 4.6 | 9.1 | | 1 905 | 36 000 | Solid | | | |
| *Upper stages* | | | | | | | | | |
| Agena D[2] | 7.1 | 1.5 | 6 821 | 3 810<br>1 225 | 185<br>36 000 | Varies | IRFNA/UDMH | 7 258 | 1962 |
| Centaur | 9.1 | 3.1 | 15 876 | 5 126<br>907 | 185<br>36 000 | Varies | LOX/$LH_2$ | 13 608 | 1963 |
| Burner II | 1.8 | 1.6 | 653 | 1 360 | 550 | Varies | Solid | 4 536 | 1966 |
| STS-SSUS-D/PAM | 2.0 | 1.2 | 2 794 | 111 | 550 | Varies | Solid | 5 897 | 1981 |
| IUS | 5.0 | 2.9 | 14 656 | 2 268-<br>2 722 | 36 000<br>or planetary | 1-2 | Solid | 20 000 | |

[1] This has three solid fuel strap-on boosters on the first stage w th a thrust of 73 390 kg.
[2] Payloads shown are with Atlas SLV-3A booster.

*Abbreviations*

TAT = Thrust-Augmented-Thor
LTTAT = Long-Tank-Thrust-Augmented-Thor
LOX = Liquid Oxygen
RP = Rocket Propellant (hydrocarbon)
TAD = Thrust-Augmented-Delta
IRFNA = Inhibited Red Fuming Nitric Acid
UDMH = Unsymmetrical Dimethylhydrazine
IUS = Inertial Upper Stage
$LH_2$ = Liquid Hydrogen
SSUS = Spin-Stabilized Upper Stage
STS = Space Transportation System
PAM = Payload Assist Module

US deep-space probe. After some modifications, Thor-Able Star was used to launch the first US Transit navigation satellite.

The Thor booster was used in a number of other combinations, for example, Thor-Delta, Thor-Agena D and Thor-Burner II. Until about 1972, the vehicle was also used in Thrust-Augmented-Thor (TAT) and Long-Tank-Thor (LTT) combinations. The TAT was equipped with three solid-propellant strap-on motors and the LTT combined all the TAT features with additional liquid-propellant capacity. The latter permitted a larger burning time for the main engine, making it possible to carry 30 per cent more payload, but did not increase thrust. NASA had also used Thor for the first stage of Delta and for the Thrust-Augmented-Delta (TAD).

*Atlas*

The original design consideration for an ICBM which was later named Atlas began in 1945, but the funding for the ICBM was discontinued two years later. However, the programme was revived after a decade and the development of the Atlas began in 1955. Atlas was the first of the modified ICBMs to appear on a NASA (essentially a civilian arm of the US space programme) launching pad. This booster has been used very extensively to launch DoD satellites into space—Atlas placed many of these military satellites in orbit with Agena for the final stage. And the man-rated version of Atlas was the booster for the US manned Mercury flights.

*Titan*

By about 1953, considerable advances had been made in nuclear physics resulting in the feasibility of relatively light-weight thermonuclear weapons. This, among other factors, gave an impetus to the design of a follow-on ICBM to the Atlas and in 1955 development of the two-stage Titan began. The first of the Titan I series was successfully tested in early 1959 and by 1962 these missiles were deployed. However, Titan I was superseded by Titan II and Minuteman missiles.

The development of the most powerful US ICBM, the Titan II, began in mid-1960. By late 1961 the Gemini manned programme was initiated with the decision to launch it using the Titan II missile. Around the same time, the concept of the first DoD launch vehicle for military satellites, Titan III, was formulated and its development began in late 1962. A number of versions were developed including Titan III M which was to be used for launching the Manned Orbiting Laboratory.

The Titan III vehicles have become the standard US heavy-duty vehicle and launchers of most military satellites. Table 3.6 gives the main characteristics of the widely used Titan vehicles.

The above review of US vehicles does not exhaust the types of launcher developed. However, all the spacecraft launchers developed by 1972 had one thing in common—none of the launchers had been designed to be used again. But the situation has now changed. The era of the reusable launcher

Figure 3.5. Some of the US space transportation systems

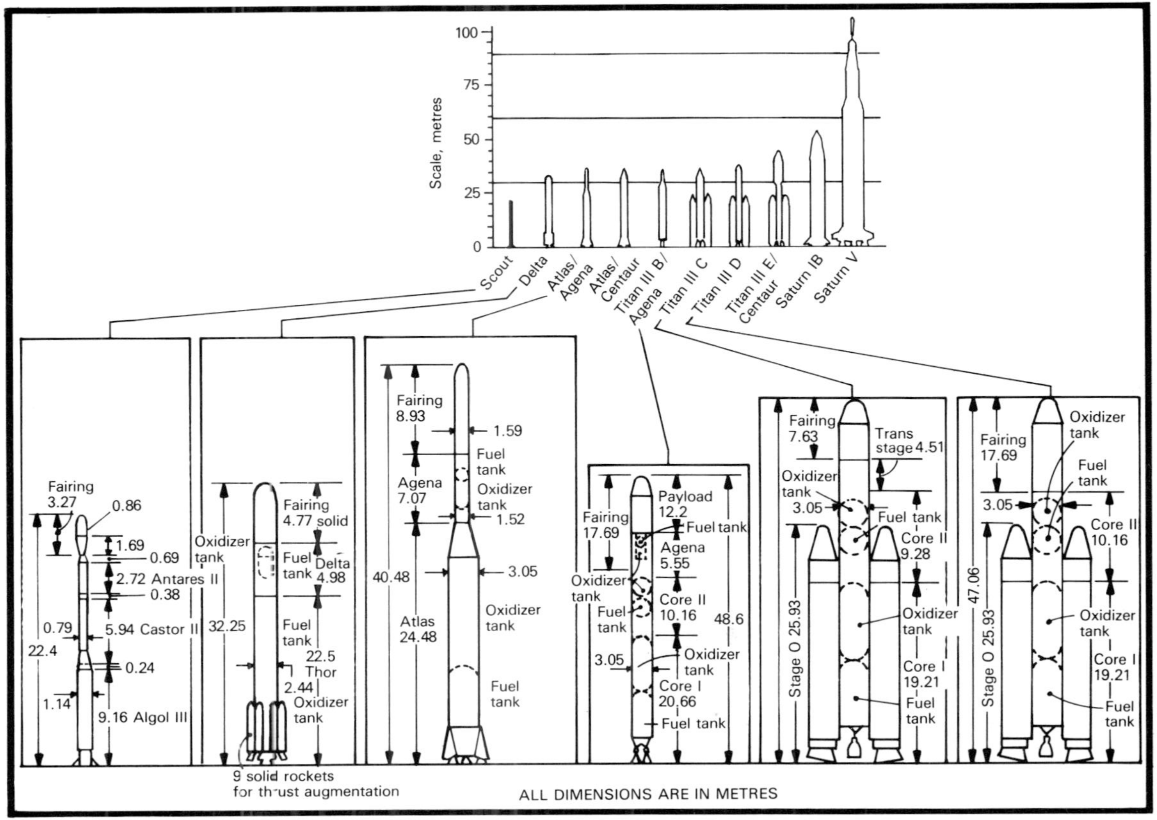

dawned when on 12 April 1981 the Space Shuttle was launched. A brief description of this system and its development is given below.

*Space Shuttle Transportation System (SSTS)*

The development of a partially reusable space transportation system was given the go-ahead at the beginning of 1972. This system was developed for use by both the DoD and NASA. It consists of an orbiter similar in configuration to a delta-wing aircraft but powered by rockets using liquid propellant. The fuel is contained in a large tank attached to the orbiter at lift-off. Two booster rockets, which are powered by solid propellant, are mounted on each side of the fuel tank.

The SSTS is launched vertically with all engines, both in the orbiter and the boosters, firing at the lift-off. At about 40 km above the surface of the Earth, the booster stages separate from the fuel tank and are recovered from the ocean from a predetermined area. These are used again. The fuel tank is separated from the orbiter just before the latter attains its orbit. The orbiter's attitude is controlled by a series of smaller rocket engines which are also used to manoeuvre the vehicle.

The SSTS is envisaged mainly for placing satellites into orbit, retrieving satellites from orbit and repair and service of satellites in orbit. It could also be used to place a satellite with a propulsive rocket in a low orbit for subsequent transfer into higher orbits, including synchronous orbit, or to an 'escape' trajectory for deep-space travel. The orbiter could accommodate up to seven astronauts.

The characteristics of the SSTS and other launchers are summarized in table 3.6. The table also indicates the parameters of various upper stages. In figure 3.5 configuration and relative sizes of various US launchers are shown.

## USSR

Interest in rockets in the Soviet Union dates back to the 17th century, at least; and in 1829, during the war between Turkey and Russia, rockets were used for the first time by the military [10]. Formulation of ideas on spaceflight dates back to last century when, among others, K.I. Konstantinov (1817–71) began publishing his ideas on the subject and laid the scientific foundation for Soviet military missiles. Some of the models evolved by Konstantinov were quite advanced for that time and had a range of four to five kilometres [11]. The founder of space engineering, however, was K.E. Tsiolkovsky, who in 1883 described an interplanetary craft with a rocket engine. Between 1903 and 1914 he formulated a theory on rocket flight and laid down basic principles of space rocket design. In 1929, Tsiolkovsky formulated his concept of a multi-stage rocket [12].

The years between 1929 and 1933 were, on the whole, more productive in terms of new ideas. After this period a group of experts on rockets, V.P.

Glushko, S.P. Korolev and A.A. Kostikov, to name a few, laid the foundation for the Soviet ICBM. In fact, Korolev's group designed an ICBM, the successful launch of which was announced in 1957.

In both the USSR and the USA, development in long-range missiles essentially produced most of the space launch vehicles. Whereas all Soviet spacecraft were launched using an ICBM, the first few US spacecraft were launched using essentially non-military launch vehicles.

In the following sections a brief review of the Soviet launchers is given. Because of the absence of an adequate Soviet nomenclature for these launchers, certain DoD and NATO names, as well as a nomenclature devised arbitrarily, are used. For a description of this nomenclature see reference 13.

*The "A" series*

This series is based on the first Soviet ICBM which was initially tested in August 1957. Two months later, the ICBM designated as the SS-6 Sapwood was used to launch the first man-made Earth satellite, Sputnik 1. The vehicle, designated as "A", is classed either as a one-, one-and-a-half- or two-stage launcher. The A launcher is propelled using liquid propellant consisting of liquid oxygen (LOX) as the oxidizer and kerosene as the fuel. While this launcher has not been used extensively, other launchers based on A have formed the backbone of the Soviet vehicles (see table 3.7).

**Table 3.7. Number of Soviet launches per year, 1957–81**

| Year / Launches | 1957 | 1958 | 1959 | 1960 | 1961 | 1962 | 1963 | 1964 | 1965 | 1966 | 1967 | 1968 | 1969 | 1970 | 1971 | 1972 | 1973 | 1974 | 1975 | 1976 | 1977 | 1978 | 1979 | 1980 | 1981 |
|---|---|---|---|---|---|---|---|---|---|---|---|---|---|---|---|---|---|---|---|---|---|---|---|---|---|
| A | 2 | 1 | | | | | | | | | | | | | | | | | | | | | | | |
| A-1 | | | 2 | 3 | 4 | 7 | 6 | 1 | | | | | 1 | 5 | 6 | 5 | 3 | 5 | 6 | 5 | 7 | 5 | 8 | 6 | 6 |
| A-2 | | | 1 | | | | | 2 | 2 | 2 | 3 | 5 | 28 | 30 | 31 | 31 | 39 | 35 | 40 | 38 | 37 | 46 | 45 | 38 | 39 |
| A-2-e | | | | 2 | 2 | 6 | 3 | 4 | 11 | 8 | 8 | 4 | 4 | 6 | 3 | 11 | 11 | 7 | 12 | 11 | 10 | 8 | 7 | 12 | 14 |
| B-1 | | | | | | 7 | 4 | 7 | 7 | 7 | 13 | 16 | 14 | 19 | 12 | 12 | 10 | 6 | 5 | 4 | 2 | | | | |
| A-1/2 | | | | | | | 3 | 14 | 19 | 23 | 25 | 32 | 10 | | | | | | | | | | | | |
| A-m | | | | | | | 1 | 1 | | | | | | | | | | | | | | | | | |
| C-1 | | | | | | | | 1 | 6 | | 4 | 6 | 6 | 9 | 17 | 13 | 15 | 17 | 18 | 29 | 28 | 21 | 18 | 16 | 19 |
| D | | | | | | | | | 2 | 1 | | 1 | | | | | | | | | | | | | |
| A-x | | | | | | | | | 1 | 1 | | | | | | | | | | | | | | | |
| F-1-r | | | | | | | | | | 2 | 9 | 2 | 1 | 2 | 2 | | | | | | | | | | |
| D-1 | | | | | | | | | | | 2 | | | 2 | 2 | | 2 | 2 | | 3 | 2 | 1 | 2 | | 1 |
| F-1-m | | | | | | | | | | | 2 | 5 | 2 | 3 | 5 | 1 | 1 | 3 | 4 | 8 | 7 | 1 | 2 | 4 | 3 |
| D-1-e | | | | | | | | | | | | 3 | 4 | 3 | 5 | 1 | 5 | 4 | 4 | 2 | 2 | | | 1 | |
| A-1-e | | | | | | | | | | | | | | 1 | | | | | | | | | | | |
| X-1 | | | | | | | | | | | | | | 1 | | | | | | | | | | | |
| F-2 | | | | | | | | | | | | | | | | | | | | | | 4 | 1 | 8 | 4 |
| D-1-E | | | | | | | | | | | | | | | | | | | | | | 4 | 5 | 4 | 6 |
| F-1 | | | | | | | | | | | | | | | | | | | | | | | | | 6 |

*A-1.* This vehicle consists of the A launcher as the first stage and a new upper stage which uses liquid propellants. This combination was first used in 1959 when Luna 1 was launched using it. However, later it was used to put the Vostok satellites into orbit. These were the precursors of manned flights.

*A-2.* The use of a more powerful stage on the A launcher became known in 1961 when Venus probes (Voskhod) were launched. The A-2 was also used to launch the Soyuz 2 spacecraft.

*A-2-e.* An added stage to the A-2 launcher, the 'escape' stage or the e-stage, is used so that after putting the rocket stage into an orbit close to the Earth, the escape stage is fired to accelerate the probe out into deep space. Thus the A-2-e is a four-stage launcher.

*The "B" series*

Launchers in this series were developed for putting a smaller payload of a few hundred kilograms into orbit. While the large ICBM-type of A launcher has been launching heavy payloads from the Tyuratam launch site, Kapustin Yar on the Volga has been used to launch smaller payloads of 130–450 kg using the B launchers. It is possible that the SS-4 Sandal, an intermediate range missile, was used for the first stage in the B series of launchers.

*B-1.* This launcher was powered in the first stage by burning kerosene and nitric acid. The development of the launcher took place between 1952 and 1957 and it was in use from 1957 to 1977. In the second stage, LOX and unsymmetrical dimethylhydrazine were burnt [14]. The complete launcher was first used in 1952, and between 1966 and 1977 it had been used to launch satellites from the newly built Plesetsk launch site. Since 1977 no further launches appear to have taken place (see table 3.7).

*The "C" series*

It has been suggested that the first stage of this vehicle is probably a medium-range ballistic missile, SS-5 Skean [13].

*C-1.* The first and the upper second stages of this launcher use liquid propellants. The launcher was first used in 1964 to orbit multiple payloads. While initially satellites were launched using C-1 from Tyuratam, since 1967 the Plesetsk launch site has been used more frequently.

*The "D" series*

This launcher may have been developed as a satellite launcher *per se* rather than as a missile (which was then later adapted for launching satellites). It was first used in 1965 (see table 3.7).

*D-1.* The first stage has six strap-on boosters surrounding the central core. Above the central core is mounted the so-called 'Proton stage'. The basic launcher (D) is often referred to as the Proton vehicle since its first payload was the Proton-1 satellite.

*D-1-E.* For interplanetary travel, a fourth stage is added to the D-1. This escape stage is ignited before the spacecraft has completed one revolution round the Earth. This is to put itself and its payload on an interplanetary trajectory. In another version, the E-stage has been replaced by a more powerful fourth H-stage. Thus the new version is designated D-1-H.

*D-1-E-e.* This, a five-stage version of the D vehicle, was first launched in 1974. The escape stages of both the D and A launchers (E- and e-stages) are used. The E-stage places the combined e-stage/payload into a transfer orbit and the e-stage places the payload into its final orbit.

*The "F" series*
The F series of launchers are based on the SS-9 Scarp missile, which has been publicly exhibited by the Soviet Union. Four versions of the Scarp have been identified but only two appear to be used as spacecraft launchers. These two versions are described below.

*F-1-m.* This launcher is based on the first two stages of the SS-9 Scarp missile. It is believed that storable liquid propellants are used to power the launcher. Details on the upper stage have not been publicized. However, it is known that when the vehicle is used to launch an interceptor satellite the second stage and the payload together with the m-stage are placed in orbit. The payload remains attached to the m-stage until after the interception is made, when the payload is put into a higher elliptical orbit [14].

When an ocean surveillance satellite is launched using this launcher, both the first and the second stages appear to be orbited. The m-stage is orbited together with the payload and a platform. The payload is attached to a rocket which is used to manoeuvre the satellite into a higher orbit.

*F-1-r.* In the case of a FOBS flight, a platform and an integrated re-entry rocket/payload are placed above the first two stages of the F vehicle. Once the second stage is in orbit, the platform/re-entry rocket/payload is separated. The platform orientates the assembly and before the assembly has completed one revolution round the Earth, the re-entry rocket/payload is released and the rocket fired to put the payload on a re-entry trajectory. This re-entry rocket is the r-stage.

*The "G" series*
While it has been known, since the mid-1960s, that the Soviet Union has been developing a large US Saturn V type vehicle, not much information exists on the launcher. The following types of launcher have been tentatively suggested [14].

*G-1.* This is supposed to be a two-stage vehicle without the strap-on boosters used for A and D launchers. The second stage of G-1 might be the up-rated central core of the D launcher [14].

*G-1-T.* Usually this launcher is designated as G-1 or G-1-E. However,

**Table 3.8. Characteristics of some of the Soviet launch vehicles**

| Vehicles | Height (excl. payload) (m) | Maximum diameter (m) | Payload (kg) | Orbit (km) | Stage | Propellant | Thrust (kg) | Date of first launch of original version |
|---|---|---|---|---|---|---|---|---|
| *Basic vehicles* | | | | | | | | |
| A | 28 | 2.95 | | | Core | Kerosene-LOX | 96 000 | 1957 |
| | 19 | 3 | | | 4 boosters | Kerosene-LOX | 408 000 | |
| A-1 | 28 | 2.95 | 5 000 | 200 | Core | Kerosene-LOX | 96 000 | 1959 |
| | | | 2 825 | 600 | | | | |
| | | | 400 | Lunar | | | | |
| | 19 | 3 | | | 4 boosters | Kerosene-LOX | 408 000 | |
| | 3.1 | 2.6 | | | Upper | | | 1960 |
| A-2 | 28 | 2.95 | 6 000 | 250 | Core | Kerosene-LOX | 96 000 | 1963 |
| | | | 4 350 | 1 000 | | | | |
| | | | 900 | Mars | | | | |
| | 19 | 3 | | | 4 boosters | Kerosene-LOX | 408 000 | |
| | 8 | 2.6 | | | Upper | | 30 000 | |
| A-2-e | 28 | 2.95 | 7 500 | 200 | Core | Kerosene-LOX | 96 000 | 1961 |
| | | | 1 250 | 19 000 | | | | |
| | 19 | 3 | | | 4 boosters | Kerosene-LOX | 408 000 | |
| | 8 | 2.6 | | | Upper | | 30 000 | |
| | 2 | 2 | | | Escape | | | |
| B-1 | 20.3 | 1.65 | 420 | 200 | 1 | Nitric acid + refined kerosene | 72 000 | 1962 |
| | | | 310 | 1 000 | | | | |
| | 8.5 | 1.65 | | | 2 | UDMH | 11 000 | |
| C-1 | 19.8 | 2.5 | 1 050 | 200 | 1 | | 118 000? | 1964 |
| | 8.4 | 2.5 | 525 | 1 455 | 2 | | | |
| D-1 | 31 | 4.2 | 22 500 | 200 | Core | | 400 000 | 1965 |
| | | | 13 000 | 1 000 | | | | |
| | | | 6 125 | Lunar | | | | |
| | 30 | 3.1 | | | 6 boosters | | 1 500 000 | |
| | | | | | Core | | | |
| D-1-E | 31 | 4.2 | 2 880 | 35 850 | | | | |
| | | | 6 600 | Mars | | | | |
| | 30 | 3.1 | | | 6 boosters | | 1 500 000 | |
| | | | | | Escape | | | |

| | | | | | | |
|---|---|---|---|---|---|---|
| F-1-m or | 20.7 | 3 | 4 700 | 200 | 1 | 1966 |
| F-1-r | | | 2 585 | 1 200 | | |
| | | 3 | | | 2 | |
| | | | | | 3 | |
| G-1-e | | | 31 750 | Lunar | 1 | |
| | | | | | 2 | |
| | | | | | 3 | |
| *Upper stages* | | | | | | |
| A | 28 | 2.95 | 2 000 | | 2 | |
| | | | | | (SS-6 Sapwood) | |
| A-1 | 3.1 | 2.6 | 5 000 | | SS-6 + Luna 3rd stage | |
| A-m | 5.0 | 2.0 | 3 000 | | SS-6 + manoeuvre | |
| A-1-m | 5.0 | 2.0 | 4 750 | | SS-6 + Luna 3rd | |
| | | | | | stage + manoeuvre | |
| A-2 | 8.0 | 2.58 | 7 500 | | SS-6 + Venera | |
| | | | | | 3rd stage | |
| A-2-e | 8.0 | 2.58 | | | SS-6 + Venera 3rd | |
| | 2.5 | 2.0 | 7 500 | | stage + probe rocket | |
| A-2-m | 8.0 | 2.58 | 7 500 | | SS-6 + Venera 3rd | |
| | 5 | 2.0 | | | stage + manoeuvre | |
| B-1 | 8.5 | 1.65 | 450 | | SS-4 Sandal + | |
| | | | | | Cosmos 2nd stage | |
| C-1 | 8.4 | 2.5 | 1 100 | | SS-5 Skean + | |
| | | | | | restartable 2nd stage | |
| D-1 | 12.0 | 4.0 | 22 500 | | Proton + 3rd stage | |
| D-1-E | 12.0 | 4.0 | | | Proton + 3rd stage + | |
| | 4.0 | 3.9 | 22 500 | | probe rocket | |
| D-1-m | 12.0 | 4.0 | | | Proton + 3rd stage + | |
| | 7.5 | 3.9 | 22 500 | | manoeuvring stage | |
| F-1-m | 6.0 | 2.5 | | | SS-9 Scarp + 3rd stage | |
| | 5.0 | 2.5 | 4 700 | | + manoeuvring stage | |
| F-1-r | 6.0 | 2.5 | | | SS-9 Scarp + 3rd stage | |
| | 8.0 | 2.0 | 4 700 | | + retro stage | |

*Source:* Reference [15a]

Figure 3.6. Some of the Soviet space transportation vehicles; approximate sizes of various vehicles are indicated by the scale

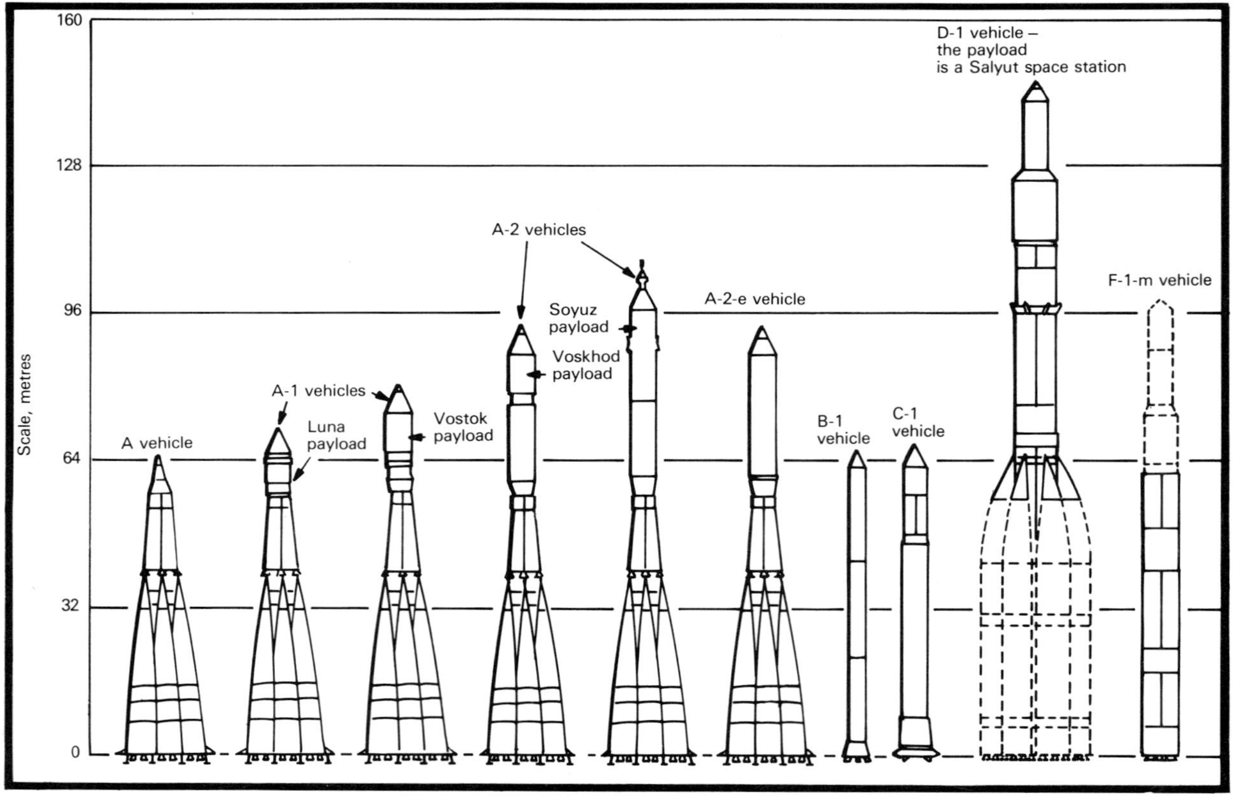


*Source:* Reference [15b]

neither the e-stage of the A-2 launcher nor the e-stage of the D-1 is likely to be used. Therefore the new trans-lunar or escape-stage is designated as the T-stage [14].

The characteristics of some of the Soviet launch vehicles are given in table 3.8, and their configuration and relative sizes are shown in figure 3.6.

*Soviet space shuttle?*
In the Soviet Union the possibility of reusable launch vehicles has been studied since the late 1960s but serious design development appears to have begun in the early 1970s. A vehicle, called Kosmolyot, has been designed with two stages, a recoverable booster and the orbiter itself [16]. More recently it has been reported that the Soviet Union has carried out some tests with its re-usable launcher [17]. It appears that a winged orbiter is designed to dock with the orbiting space station [18].

## People's Republic of China

The People's Republic of China was the fifth nation to launch a satellite using its own launcher. It is interesting to note that, so far, China's is the heaviest of any state's first satellite: China-1 had a payload of about 170 kg as compared with the payload of the first Soviet satellite, Sputnik 1 (84 kg), and the payload of the first US satellite, Explorer 1 (14 kg).

Research on rockets probably began in the late 1950s and by the end of 1966, IRBMs and medium-range ballistic missiles (MRBMs), probably based on the Soviet SS-3 and SS-4 (Sandal) missiles, were tested. It is not clear whether this first launcher was a solid-fuelled multi-stage rocket and to what extent its development was indigenous. China-1 and China-2 satellites were launched using this first type of rocket.

A second type of heavy rocket was used to launch China-3, -4, -5 and -7 satellites. This launcher is believed to be a modified version of the Chinese long-range missile, the surface-to-surface X-3 (CSS-X-3) [19]. The CSS-X-3 ICBM used for launching satellites is designated by the Chinese as the FB-1 and may be similar to the US Titan family of missiles.

During their first triple payload launched in September 1981, the Chinese tested for the first time two micro-propulsion electric rockets for satellite station-keeping and attitude control [20].

## Europe

Although the European Launcher Development Organization (ELDO) and the European Space Research Organization (ESRO) were established in 1962, it was not until 1975 that a coherent single programme was developed. The newly established European Space Agency (ESA) became responsible

**Figure 3.7. Some of the ESA, French, Indian and Japanese launch vehicles**
All dimensions are in metres.

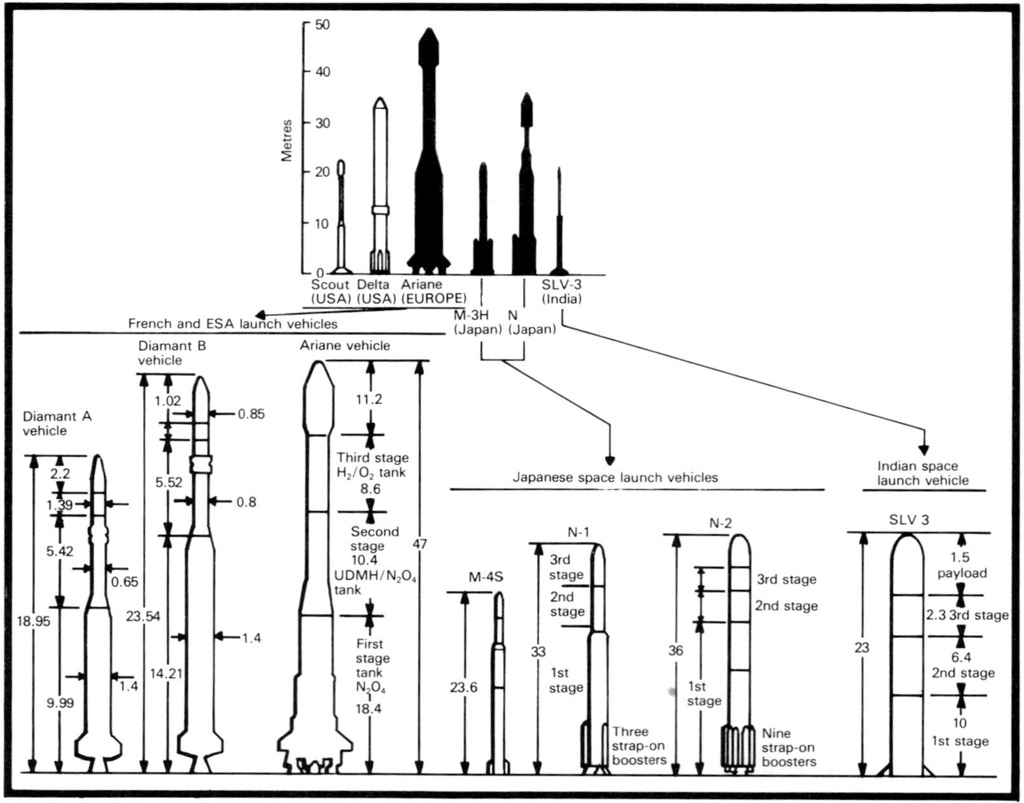

for a complete satellite and launcher programme. With the experience of Diamant, a very successful French-developed launcher in the 1960s, France assumed a major role in the development of Ariane, an ESA launcher.

Ariane has three stages (see figure 3.7). The first stage is propelled using liquid-fuel nitrogen peroxide and unsymmetrical dimethylhydrazine. The latter is the fuel, while the former is the oxidizer. The third stage is propelled by oxygen and hydrogen fuel, the most powerful oxidizer/fuel combination known.

## India

The Indian rocket programme began in about 1967 with the development of sounding rockets. India's first indigenously built sounding rocket was Rohini-75, which was first launched in 1967. A number of other improved sounding rockets have been developed but work on a small orbital launcher only began in 1973 with the development of a solid-propellant SLV-3 rocket (see figure 3.7).

The SLV-3 launcher consists of four solid-propellant stages and it is designed to place a 40-kg satellite in low orbit round the Earth. The solid propellant is a mixture of about 70 per cent ammonium perchlorate oxidizer and about 30 per cent aluminium powder together with binder [21].

India has launched two satellites, Rohini satellite-1 (RS-1) and RS-D1, in July 1980 and May 1981, respectively. During the launch of the latter satellite, the first stage of SLV-3 developed a problem so that RS-D1 did not achieve the required orbit and the satellite re-entered the atmosphere 10 days after launch. The configuration of SLV-3 is shown in figure 3.7.

## Japan

A number of launch vehicles have been developed by Japan.

*Lambda 4S (L-4S)*
This rocket, based on the sounding rocket Lambda 3, was a four-stage launcher using solid propellant. The fourth stage was added to the three-stage Lambda 3 sounding rocket. The first successful flight of L-4S was carried out in 1970 when a 24-kg payload was orbited.

*M-4S.* This, the second type of orbital launch vehicle, was based on the Lambda series of launchers. Essentially, M-4S comprised a new larger first stage using solid propellant with three Lambda stages placed on it. The first successful launch of the M-4S occurred in 1971.

*M-3C.* An improved version of the M-4 launcher, M-3C, had three stages which used solid fuel. It used liquid-injection thrust vector control on

**Table 3.9. Characteristics of some of the Chinese, ESA, French, Indian and Japanese launch vehicles**

| Vehicles | Height (excl. payload) (m) | Maximum diameter (m) | Launch weight (kg) | Payload (kg) | Orbit (km) | Stage number | Propellant | Thrust (kg) | Date of first launch of original version |
|---|---|---|---|---|---|---|---|---|---|
| *People's Republic of China* | | | | | | | | | |
| FB-1 | | | 199 581 | 2 000 | 200 | 1 | $N_2O_4$-UDMH | 279 867 | 1970 |
| | | | | | | 2 | $N_2O_4$-UDMH | 74 512 | |
| | | | | | | 3 | LOS/$LH_2$ | | |
| *ESA* | | | | | | | | | |
| Ariane | 18.4 | 3.8 | 159 665 | 2 750 | 550 | 1 | $N_2O_4$-UDMH | 249 023 | 1979 |
| | | | | 2 699 | 870 | | | | |
| | | | | 970 | 36 000 | | | | |
| | 10.4 | 2.6 | 36 301 | | | 2 | $N_2O_4$-UDMH | 68 946 | |
| | 8.6 | 2.6 | 9 389 | | | 3 | LOX/$LH_2$ | 6 124 | |
| *France* | | | | | | | | | |
| Diamant B | 14.1 | 1.4 | 24 403 | 109 | 550 | 1 | $N_2O_4$-UDMH | 35 607 | 1965 |
| | 4.7 | 0.8 | | | | 2 | Solid | 15 330 | |
| | 2.1 | 0.8 | | | | 3 | Solid | 4 627 | |
| *India* | | | | | | | | | |
| SLV-3 | 10.0 | 1.0 | 17 000 | 40 | 308 × 986 | 1 | Solid | 55 300 | 1980 |
| | 6.4 | 0.8 | | | | 2 | Solid | 23 500 | |
| | 2.3 | 0.8 | | | | 3 | Solid | 8 100 | |
| | 1.5 | 0.7 | | | | 4 | Solid | 2 200 | |
| *Japan* | | | | | | | | | |
| L-4S | 16.4 | 0.74 | 9 500 | 12 | 500 × 300 | 1 | Solid | 36 288 | 1966 |
| | | 0.31 | | | | 2 | Solid | 13 426 | |
| | | 0.5 | | | | 3 | Solid | 6 577 | |
| | | 0.48 | | | | 4 | Solid | 998 | |

| | | | | | | | | | |
|---|---|---|---|---|---|---|---|---|---|
| M-4S | 12.9 | 1.41 | 26 300 | 75 | 450 × 1 500 | 1 | Solid | 85 000 | 1971 |
| | 4.8 | 1.41 | 9 400 | | | 2 | Solid | 29 000 | |
| | 4.02 | 0.86 | 2 700 | | | 3 | Solid | 13 200 | |
| | 1.86 | 0.79 | 508 | | | 4 | Solid | 2 700 | |
| M-3C | 12.9 | 1.41 | 26 300 | 160 | 250 | 1 | Solid | 88 300 | 1974 |
| | 5.3 | 1.41 | 9 800 | | | 2 | Solid | 28 400 | |
| | 2.33 | 1.14 | 1 310 | | | 3 | Solid | 5 800 | |
| M-3H/M-3S | 14.9 | 1.41 | 54 000 | 270/263 | | 1 | Solid | 97 296/ 103 873 | |
| | 4.6 | 1.41 | | | | 2 | Solid | 28 372 | |
| | 2.0 | 1.14 | | | | 3 | Solid | 5 806 | |
| N-1 | 21.5 | 2.44 | 90 401 | 400 | 1 000 | 1 | LOX/RJ-1 | 78 018 | 1975 |
| | 7.25 | 0.79 | | | | (3 × strap-on boosters) | Solid | 70 760 | |
| | 5.43 | 1.62 | | | | 2 | $N_2O_4$/A-50 | | |
| | 1.74 | 1.65 | | | | 3 | Solid | | |
| N-2 | 22.7 | 2.44 | 134 990 | 1 100 | 1 000 | 1 | LOX/RJ-1 | 78 018 | 1981 |
| | 7.25 | 0.79 | | | | (9 × strap-boosters) | Solid | 212 282 | |
| | 5.8 | 2.44 | | | | 2 | $N_2O_4$/A-50 | 4 355 | |
| | 1.74 | 2.44 | | | | 3 | Solid | 6 804 | |

the first- and the second-stage engines. In 1974 the launcher successfully put a satellite into orbit.

*M-3H and M-3S.* These were essentially similar to the M-3C launcher. The main difference was the more powerful engines in the second and the third stages.

*"N" series*
The basic N launcher uses solid propellants in its first, second and fourth stages and liquid in the third stage. These launchers are based on the US Delta launcher.

*N-1.* The N-1 has three solid-fuel strap-on boosters clustered around the base of the first stage. The first stage uses LOX/RJ-1 as the propellant while the propellant on the second stage is a $N_2O_4$/A-50 combination. The third stage uses a solid propellant [22]. The N-1 was used for the first time in 1975.

*N-2.* This launcher is capable of launching into a geostationary orbit nearly three times the payload of the N-1 launcher. There are nine solid-propellant strap-on boosters on the first stage of the N-2. The first stage uses LOX/RJ-1 propellant, while the second stage is propelled by an $N_2O_4$/A-50 combination. The launcher successfully placed a satellite in geosynchronous orbit in August 1981.

*H-1*
This is a new generation of satellite launcher which will be able to place in geosynchronous orbit a payload four times larger than that which could be orbited by an N-1 vehicle. Testing of this launcher will begin in 1986 [23]. A liquid propellant will be used for the first stage, which will have nine solid-propellant strap-on boosters. For the second stage, a liquid-oxygen and liquid-hydrogen combination will be used as a propellant. The third stage will be propelled using a solid propellant.

Initially H-1 will have a two-stage configuration. The configurations of some of the Japanese launchers are shown in figure 3.7. Characteristics of the Chinese, ESA, French, Indian and Japanese launchers are given in table 3.9.

## *V. Summary*

There is no doubt that there has been a clear link between strategic missiles and space launchers. The space launchers developed in France, the People's Republic of China, the USA and the USSR were all based on their respective military missiles. Furthermore, all of these nations have launched military satellites which support their military operations to a varying degree. In fact, military satellites are becoming an integral part of nuclear and other weapon systems.

In the cases of Japan and India—the other two nations which have launched satellites using their own space launchers—missiles were not developed first. In Japan the launchers are based on the indigenously developed technology for sounding rockets used for scientific experiments. There are a number of other countries, for example, Brazil, which have advanced sounding rocket programmes. These nations will, no doubt, eventually develop their own launchers. Moreover, there is no reason why a nation cannot develop an ICBM capability once it has developed space launcher technology.

It can be seen that a number of countries—China, the European nations of the ESA organization, France, India, Japan, the USA and the USSR—possess space launching technology and the trend is for this technology to spread.

## References


1. Lietzke, A.F., 'Nuclear rockets', in *Exploring in Aerospace Rocketry*, National Aeronautics and Space Administration (US Government Printing Office, Washington, D.C., 1971), pp. 301–12.
2. Kaugman, H., 'Electric propulsion', in *Exploring in Aerospace Rocketry*, National Aeronautics and Space Administration (US Government Printing Office, Washington, D.C., 1971), p. 322.
3. Kautrowitz, A., 'Propulsion to orbit by ground-based lasers', *Astronautics & Aeronautics*, Vol. 10, May 1972, pp. 74–76.
4. Nored, D.L., 'Application of high power lasers to space power and propulsion', *Second NASA Conference on Laser Energy Conversion*, NASA Report SP-394, January 1975, pp. 95–107.
5. Hertzberg, A., 'High-power gas lasers: applications and future developments', *Journal of Energy*, Vol. 1, No. 6, November–December 1977, pp. 331–346.
6. 'Space vehicles could be propelled by remote lasers', *Physics Today*, Vol. 30, No. 8, August 1977, pp. 17–70.
7. 'Laser propulsion testing begins', *Laser Focus*, Vol. 17, No. 10, October 1981, pp. 36–37.
8. Bunkin, F.V. and Prokhorov, A.M., *Soviet Physics Uspekhi*, Vol. 19, 1977, p. 561.
9. Hall, R.C., 'Early US satellite proposals', in *The History of Rocket Technology*, ed. E.M. Emme (Wayne State University Press, Detroit, Michigan, 1964), pp. 67–93.
10. Winick, L.E., 'Birth of the Russian space programme', *Spaceflight*, Vol. 20, No. 5, May 1978, pp. 162–73.
11. Glushko, V.P., *Development of Rocketry and Space Technology in the USSR* (Novosti Press Publishing House, Moscow, 1973).
12. Tokaty, G.A., 'Soviet rocket technology', in *The History of Rocket Technology*,

ed. E.M. Emme (Wayne State University Press, Detroit, Michigan, 1974), pp. 271–84.

13. Sheldon, C.S., II, 'The Soviet space program—a growing enterprise', *TRW Space Log*, Vol. 8, No. 4, Winter 1968/69, pp. 8–19.
14. Gibbons, R.F., 'Soviet launch vehicle designation', *Spaceflight*, Vol. 19, No. 2, February 1977, pp. 54–80.
15. *Soviet Space Programs*, 1971–75, Vol. I, Staff Report, Committee on Aeronautical and Space Sciences, US Senate, 30 August 1976 (US Government Printing Office, Washington, D.C., 1976).
    (*a*) —, pp. 43–47.
    (*b*) —, pp. 611–13.
16. Gatland, K., 'Soviet Space Shuttle', *Spaceflight*, Vol. 19, No. 6, 6 June 1977, p. 211.
17. Williams, T., 'Soviet re-entry tests: a winged vehicle?', *Spaceflight*, Vol. 22, No. 5, 5 May 1980, pp. 213–14.
18. Petrov, B.N., Lopatin, V.I., Mitroskin, E.I., Vasilév, V.A. and Pavlenko, A.I., 'Stochastic terminal control of shuttle', Paper prepared for Automatic Control in Space, International Astronautical Federation Congress Symposium, 17–21 May 1976, Rottach-Egern, FR Germany.
19. Pirard, T., 'Chinese "Secrets" orbiting the Earth', *Spaceflight*, Vol. 19, No. 10, October 1977, pp. 355–61.
20. 'China constructs micro propulsion rockets', *Xinhua*, 14 January 1982.
21. Velupillai, D., 'ISRO: India's ambitious space agency', *Flight International*, Vol. 117, No. 3612, 28 June 1980, pp. 1466–70.
22. Kidger, N., 'Japan's new launcher', *Spaceflight*, Vol. 23, No. 8, October 1981, pp. 244–45.
23. Lenorovitz, J.M., 'Japan tests satellite launcher', *Aviation Week & Space Technology*, Vol. 115, No. 11, 14 September 1981, pp. 18–19.

# Chapter 4. Military activities in outer space

> ...control of space will be decided in the next decade. If the Soviets control space, they can control Earth, as in past centuries the nation that controlled the seas dominated the continents.
>
> John F. Kennedy (1960)

## *I. Background*

Man's military ambition has extended his armed activities from land to the high seas, into the air and more recently beyond into outer space. Man has sought to enhance his military aims on Earth by deploying a number of types of satellite, including navigation, communications, weather, geodetic, reconnaissance and early warning satellites. Between 1957 and the end of 1981 some 1 917 such satellites had been launched, which constitutes about 75 per cent of all satellites orbited. The yearly launches of civilian and military satellites are plotted in figure 4.1. For purposes of comparison the annual launches of US and Soviet military satellites are also plotted. It can be seen that during the first decade the military satellites were launched at an increasing rate; this has now become constant at about 100 satellites a year. About 85 satellites are launched every year by the Soviet Union and about 15 by the USA. The difference occurs merely because the US satellites tend to be much longer-lived.

The military importance of space is further emphasized by the fact that the US DoD budget request for 1982 was about $5.9 billion for its space programme where the NASA figure for the same year was about $5.6 billion. Normally, civilian programmes tend to be costly since some projects, such as the space shuttle, involve manned space flights, which require complex and more safety-oriented space systems. Moreover, the above DoD budget does not include costs needed for some of the most secret programmes, which could amount to some $2.5 billion* [1]. If the same amount were added to the 1982 DoD budget, it would come to at least $8.4 billion. This is one and a half times the budget of the civilian programme. The yearly budgets for the

---

* Throughout the book 1 billion is equivalent to one thousand million

**Figure 4.1. Military and civilian satellites launched between 1957 and 1981**

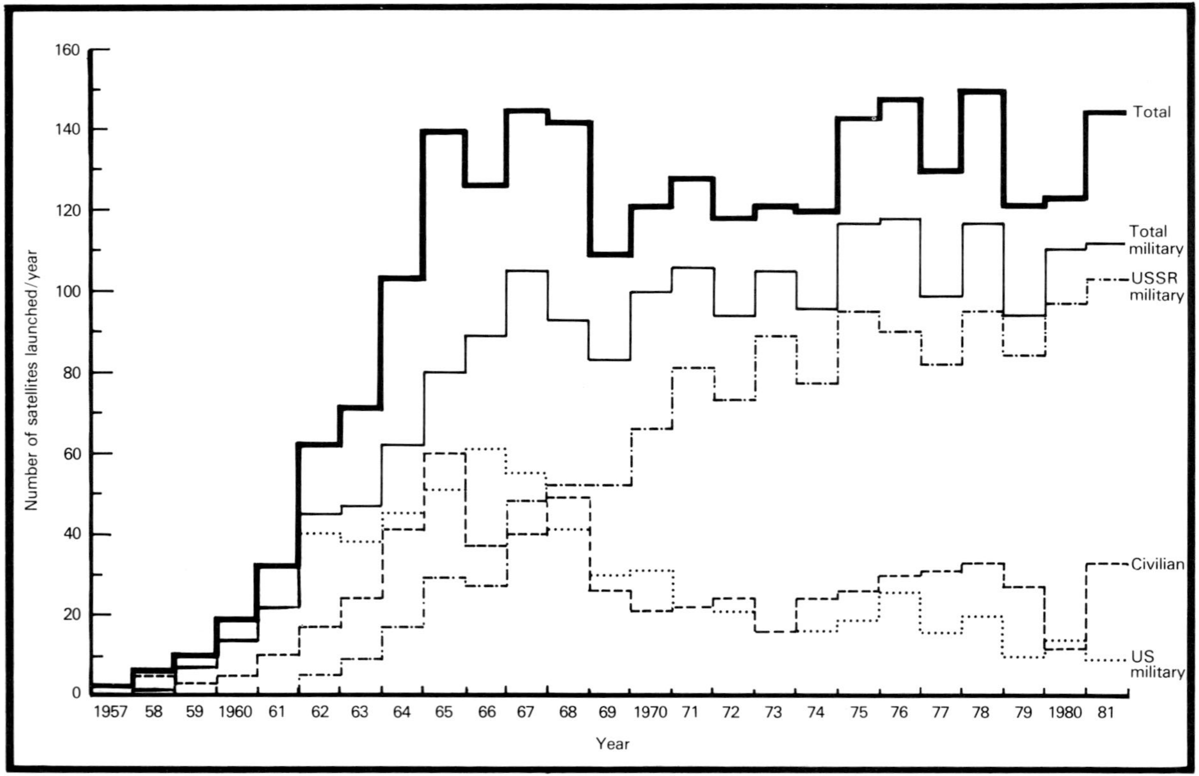

US DoD and NASA over the past 25 years are shown in figure 4.2. The situation in the Soviet Union is probably not very different.

Another indication of the military significance of satellites is that they have been placed on the US target list (and probably the Soviet one) and much effort is being devoted to ASAT systems for disabling and even destroying enemy satellite systems. This was highlighted during a US congressional hearing in 1979 when Senator Levin questioned General Stafford: "The target list was the basis for cost-benefit tradeoffs which proved that the miniature [deleted] vehicle system was most cost effective, according to your statement. . . . Provide details of the other systems considered as possible ASAT systems for the purpose of attacking this target list." General Stafford's answer was considerably censored but some glimpses of the kind of ASAT systems involved and their targets is obtained from his answer which was: "This study indicated the air launched miniature vehicle was the most cost effective weapon against [deleted] low altitude [deleted] satellites. . . . We also have a low risk, although more limited in capability, conventional high explosive [deleted] anti-satellite system in conceptual design" [2].

Of the several types of military satellite launched each year, three kinds are perhaps the most important to the US and Soviet armed forces: reconnaissance, navigation and communications satellites. In the following sections a brief review of these is given. Both the Soviet Union and the USA are developing weapons for destroying satellites. Such a programme has essentially three components, namely, space surveillance, satellite survivability and anti-satellite weapons. Some of the developments in these fields are also reviewed briefly below.

## *II. Reconnaissance satellites*

These can be divided into four types—photographic, electronic, ocean surveillance and early warning satellites. The orbits of each of these types of satellite are optimized for the specific mission of the satellite. For the factors influencing the choice of orbit see reference 1 of chapter 1. A brief review of these missions is given below. The altitudes and the orbital inclinations of US and Soviet reconnaissance satellites are indicated in figure 4.3.

### Photographic reconnaissance satellites

Of the military satellites launched by the People's Republic of China, the Soviet Union and the United States, some 40 per cent have been used for photographic reconnaissance purposes from a low altitude orbit of about 200 km (see figure 4.3). There are two basic types of reconnaissance mission. On area surveillance missions a large area of a particular country is scanned

**Figure 4.2. Yearly US space budget for NASA and the Department of Defense**

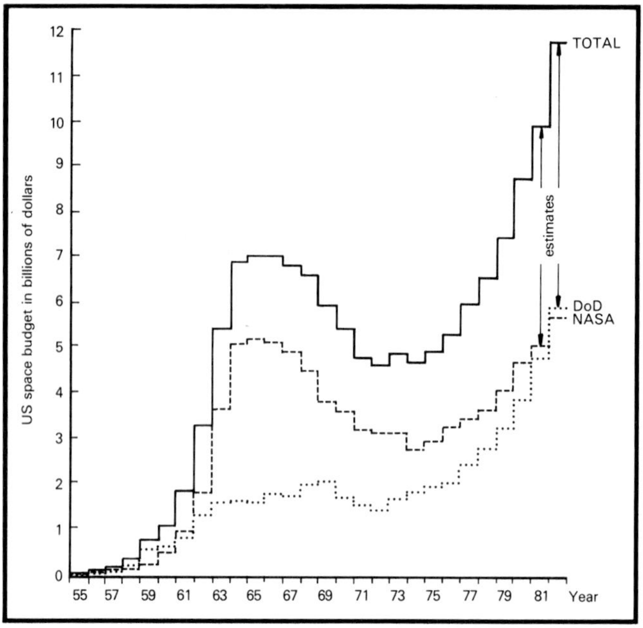


for objects of potential military interest using a wide angle, low resolution camera. For the second type of mission a camera with a high resolution and a relatively narrow field of vision is used to re-photograph areas of particular interest located during the area surveillance mission.

Until recently these two missions were always carried out by different satellites. While this practice continues in the case of the Soviet satellites, with the development of new generation US Big Bird satellites, area surveillance spacecraft have been phased out by the USA. The Big Bird, launched using the Titan-3D launcher, performs both the area surveillance and the close-look types of mission. Photographs with high resolution images taken during the close-look missions are returned to the Earth for processing and analysis, whereas the area surveillance images or films are usually developed and converted into electronic signals aboard the satellite and then transmitted to the Earth. As advances are made in the latter technique, fewer films will be returned to the Earth. Since, in addition, the size of the electronic processing equipment will be reduced, more fuel will be able to be carried on board the spacecraft to maintain its altitude. Thus a trend towards longer-lived satellites can be expected. Such a trend can be seen in figure 4.4. It can be seen that the lifetimes of most of the US photographic reconnaissance satellites have been increasing. For example, the lifetime of the Big Bird satellite has increased from 52 days for the first satellite in 1971 to 261 days for the most recent one launched in 1980.

While the USA still continues to launch its third-generation close-look

**Figure 4.3. Various types of satellite orbits; characteristics of various orbits are given in the box**

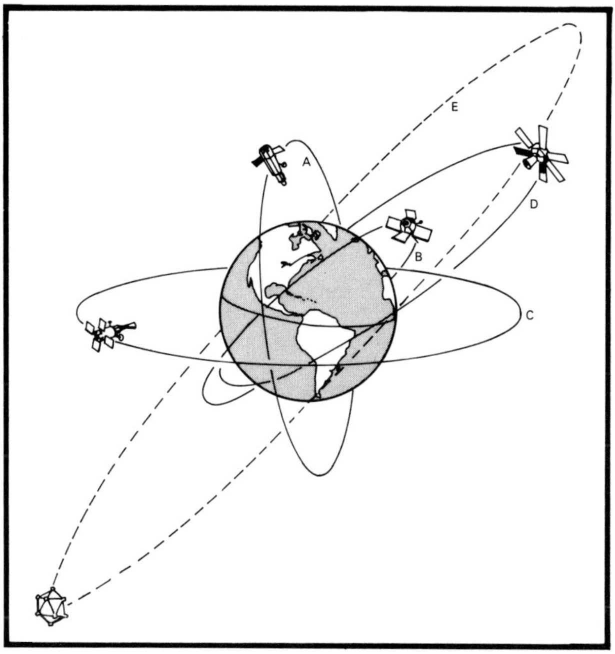


| Satellite | Altitude (km) | Orbital inclination | Type of orbit |
|---|---|---|---|
| Reconnaissance | | | |
| *Photographic* | | | |
| Big Bird | 180–290 | 97° | A |
| KH-11 | 240–530 | 97° | A |
| Cosmos | 180–350 | 62°, 72°, 82°, 67° | B |
| *Electronic* | | | |
| US | 480 | 97° | A |
| Cosmos | { 500 | 74° | B |
| | 650 | 82° | B |
| Ocean surveillance | | | |
| US | 1 100 | 63° | B |
| Cosmos | 250 | 65° | B |
| Early Warning | | | |
| US | 36 000 | 0° | C |
| Cosmos | 688 × 39 000 | 63° | D |
| Nuclear explosion detection | | | |
| US | 110 000 | 35° | E |
| Meteorological | | | |
| US | 36 000 | 0° | C |
| Meteor | { 900 | 81° | B |
| | 610 | 98° | A |
| Communications | | | |
| US-SDS DSCS | 250 × 39 000 | 64° | D |
| US-FLTSATCOM | 36 000 | 0° | C |
| Molniya | 440 × 40 000 | 63° | D |
| Cosmos | 1 400 | 74° | B |
| Navigation | | | |
| US | 20 000 | 64° | B |
| Cosmos | 1 000 | 83° | B |

**Figure 4.4. The orbital lifetimes of US reconnaissance satellites launched between 1976 and 1981; lifetimes of Big Bird and T-3B/A-D satellites are average lifetimes**

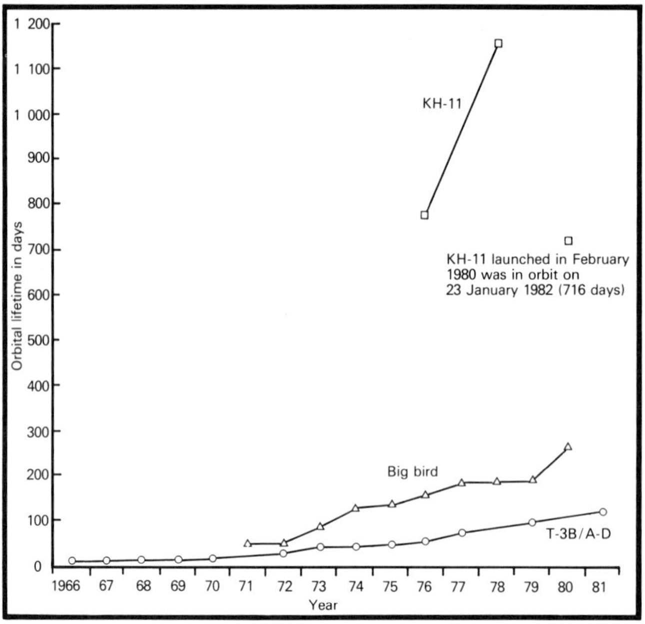


satellites using Titan-3B/Agena-D, the fifth and latest generation of spacecraft, the KH-11, has been developed and orbited. The lifetime of the latter spacecraft has increased from 770 days for the first satellite (launched in 1976) to 1 166 days for a satellite that was launched in 1978. The lifetime of the third-generation satellites has increased from 20 days to 112 days (see figure 4.4). The KH-11 satellites developed by the US Central Intelligence Agency (CIA) (project 1 010) transmit images in real time in digital form to a ground station. The data gathered are transmitted and processed in the same way as occurs with the Landsat satellites [3]. For further discussions on image processing see papers 5 and 6 of part II.

A number of types of sensor are used on reconnaissance satellites. These range from photographic and return-beam vidicon (RBV) television cameras to multispectral scanners (MSS) and microwave radars. Sensors are discussed in more detail in papers 1, 5 and 6 of part II. It is sufficient to note here that the resolution required for the detection, recognition, identification and description of various targets varies, as is indicated in table 4.1. Thus a resolution of, for example, 1.5 m is quite satisfactory for detection of all types of target indicated in the table and is sufficient for recognition of such objects as aircraft, ground-to-ground missiles, troop units, radio communications installations, etc. The details of what can be seen by reconnaissance satellites are still a closely guarded secret, but some estimates have been made indicating that a ground resolution of about 15–30 cm can be obtained by using photographic cameras on board the military reconnaissance satellites (see reference 1 of chapter 1). It has been reported that satellite photography was used by the CIA to observe details during the domestic demonstrations of the late 1960s [4]. The figures for resolution of 15–30 cm are, therefore, not unreasonable.

**Table 4.1. Resolution (in metres) required for interpretation tasks**

| Target | Detection[a] | General identification[b] | Precise identification[c] | Description[d] | Analysis |
|---|---|---|---|---|---|
| Bridge | 6 | 4.6 | 1.5 | 0.9 | 0.3 |
| Communications | | | | | |
| Radar | 3 | 0.9 | 0.3 | 0.15 | 0.04 |
| Radio | 3 | 1.5 | 0.3 | 0.15 | 0.15 |
| Supply dump | 1.5 | 0.6 | 0.3 | 0.03 | 0.03 |
| Troop units | 6 | 2 | 1.2 | 0.3 | 0.08 |
| Airfield facilities | 6 | 4.6 | 3 | 0.3 | 0.15 |
| Rockets and artillery | 0.9 | 0.6 | 0.15 | 0.05 | 0.01 |
| Aircraft | 4.6 | 1.5 | 0.9 | 0.15 | 0.03 |
| Command and control headquarters | 3 | 1.5 | 0.9 | 0.15 | 0.03 |
| Missile sites (SSM/SAM) | 3 | 1.5 | 0.6 | 0.3 | 0.08 |
| Surface ships | 7.6 | 4.6 | 0.6 | 0.3 | 0.08 |
| Nuclear weapon components | 2.4 | 1.5 | 0.3 | 0.03 | 0.01 |
| Vehicles | 1.5 | 0.6 | 0.3 | 0.05 | 0.03 |
| Land minefields | 9 | 6 | 0.9 | 0.03 | – |
| Ports and harbours | 30.5 | 15 | 6 | 3 | 0.3 |
| Coasts and landing beaches | 30.5 | 4.6 | 3 | 1.5 | 0.08 |
| Railway yards and shops | 30.5 | 15 | 6 | 1.5 | 0.6 |
| Roads | 9 | 6 | 1.8 | 0.6 | 0.15 |
| Urban areas | 61 | 30.5 | 3 | 3 | 0.3 |
| Terrain | – | 91 | 4.6 | 1.5 | 0.15 |
| Surfaced submarines | 30.5 | 6 | 1.5 | 0.9 | 0.03 |

*Source:* "Reconnaissance Handy Book", p.125, published by McDonnell Douglas Corporation, USA.

[a] Requires location of a class of units, object or activity of military interest.
[b] Requires determination of general target type.
[c] Requires discrimination within target types of known types.
[d] Requires size/dimension, configuration/layout, components construction, count of equipment, etc.

The extent to which the US KH-11 and the Big Bird satellites have observed the Earth since 1977 is shown in figure 4.5. There is considerable overlap between the coverage of individual KH-11 spacecraft and between that of KH-11 and Big Bird satellites. A reason for such overlaps can be deduced from an analysis of some of the orbital characteristics of these satellites. As was seen in chapter 2, a satellite orbit can be fixed in space by two of the orbital elements, the angle of inclination of the orbital plane and the right

**Figure 4.5. Coverage obtained by the US and the Soviet long-lived**

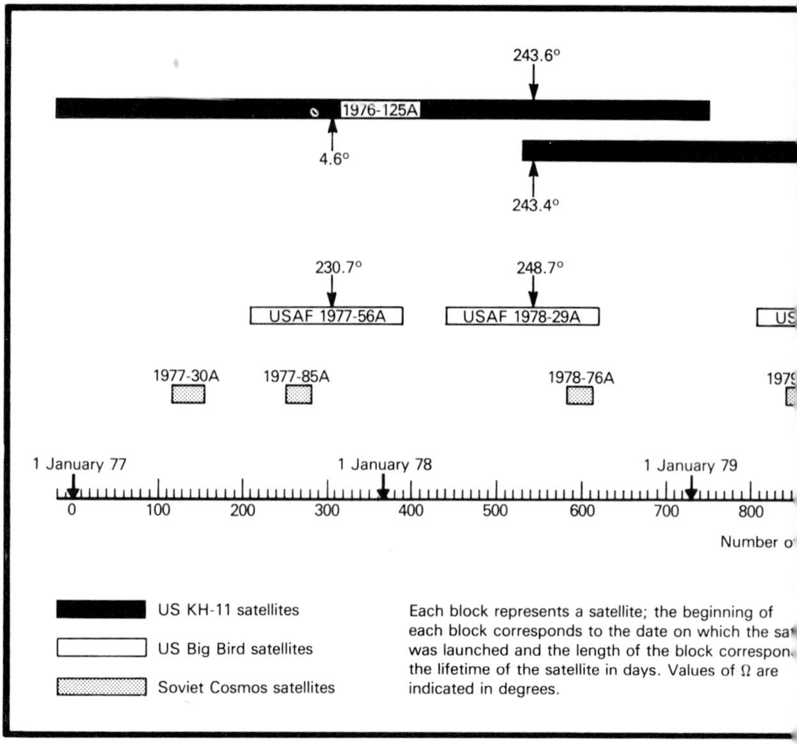


ascension of the ascending node (Ω, see figure 2.3). It was also noted that for an orbital inclination of 90°, the values of Ω do not change through the lifetime of the satellite. For the satellites considered in figure 4.5, the orbital inclinations are very similar, that is, about 97°, so that Ω does not change significantly. The position of the orbital plane relative to the point of Aries can, therefore, be determined by Ω, the values of which are indicated in figure 4.5 for particular times. If one compares Ω for two such satellites, one can therefore get an idea of the relation between the two orbital planes at any one time (e.g., if the difference is zero then the two satellites are in the same orbit, and if the difference is 90° then the two orbital planes are at right angles to each other).

The values of Ω for the KH-11 satellites 1976-125A and 1978-60A are seen from figure 4.5 to be very similar, suggesting that satellite 1978-60A was probably a replacement for 1976-125A. However, the overlap between the KH-11s 1978-60A and 1980-10A occurs for a very long period of time and the difference in Ω for these satellites is 46°. A similar difference (49°) in the values of Ω can be observed between the KH-11s 1980-10A and 1981-85A. The latter satellite was launched some 10 days after 1978-60A had decayed. The difference in the values of Ω for these two satellites is not very large (about 9°) suggesting that 1981-85A may have been a replacement for 1978-60A. Since the launch of

photographic reconnaissance satellites launched during 1977–1981

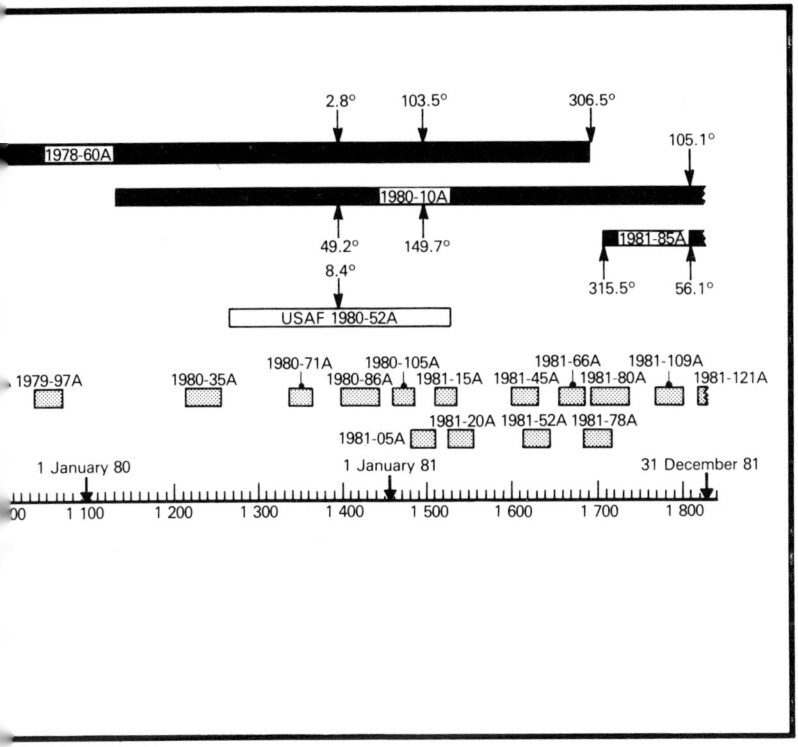


satellite 1980-10A in early 1980, with the exception of some 10 days in 1981, two KH-11 satellites have been in orbit all the time. This may be the beginning of a two-satellite pattern designed to obtain a greater frequency of observation of the Earth's surface. This is apparent from figure 4.6, in which the ground tracks obtained over a period of 24 hours for satellites 1980-10A and 1981-85A are plotted. From the figure it can be seen that the gap between two consecutive tracks of satellite 1980-10A is filled in by a ground track of satellite 1981-85A thus increasing the frequency of observation. A wider coverage of the Earth at any given time is also possible because the orbits are spaced (in this case by 49° and in the case of the KH-11s 1978-60A and 1980-10A by some 46°) such that a pair of satellites can observe different parts of the Earth's surface simultaneously.

From the values of $\Omega$, no clear pattern emerges in the relationship between KH-11 and Big Bird satellites. This may be because KH-11 satellites are operated by the CIA while the US Air Force is in charge of the Big Bird satellites. It is, however, possible that the latter satellites are used to obtain photographs of areas which the sensors on the KH-11s indicate to be of specific interest. While the sensors on board the KH-11s are sophisticated, highest quality images can probably still be obtained only by using photographic equipment.

An advantage of long-lived satellites is that fewer launchings are required. Most of the other types of US satellite are also long-lived, hence the considerable difference in the launch rates between the US and Soviet satellites (see figure 4.1). The Soviet Union has also begun to launch long-lived satellites but the maximum lifetime achieved so far is 42 days. An overlap between some of these Soviet photographic reconnaissance satellites is also beginning to take place (see figure 4.5). However, the above type of analysis is not very useful in the case of these reconnaissance satellites since they are launched at an orbital inclination of about 67° where the rate of change of $\Omega$ is not negligible. Also the extent of overlap is relatively small. The way these satellites are used is discussed more fully in paper 2 of part II.

After the USA and the USSR, the People's Republic of China was the third country to launch a reconnaissance satellite. The first photographic reconnaissance satellite may have been China 3, launched on 26 July 1975. The ability to recover on Earth payloads ejected from a spacecraft was first demonstrated in 1975 when a capsule from China 4 was recovered. China 6 was placed in a similar orbit to that of China 4. The satellite was reported to have been recovered [5], but there is some doubt as to whether it was in fact the satellite or part of its payload. The last reconnaissance satellite to be launched was China 8 in January 1978.

Two more countries are now planning to launch reconnaissance satellites —France and Japan. France is the third nation, after the USA and the USSR, to have launched its own satellite with its own launcher. It was in 1973 that France announced its interest in developing military reconnaissance satellites [6]. In about 1978, with the development of its nuclear weapons and MIRVed (multiple independently targetable re-entry vehicle) missiles, France began planning its reconnaissance satellite system for such uses as the identification of potential targets [7]. No doubt this programme will acquire considerable impetus because of the successful launches of Ariane in 1981. The military reconnaissance satellite is expected to be based on the civilian Earth observation satellite, SPOT (Systéme Probatoire d'Observation de la Terre) which is partly military-financed [8, 9]. The military version of SPOT is called SAMRO. The SAMRO programme is expected to cost about $340 million [10, 11].

Japan plans to launch an Earth resources satellite by about 1985. It has been suggested that Japan is interested in developing a military reconnaissance satellite as well [12, 13]. A possible collaboration with the USA is being considered.

### Electronic reconnaissance satellites

While quite a lot is known about photographic reconnaissance satellites, comparatively little data is available on electronic surveillance satellites—the 'ears' in space. Such spacecraft carry equipment designed to detect and monitor radio signals generated by the enemy's military activities. Signals

Figure 4.6. Ground tracks of two US KH-11 satellites (1980-10A and 1981-85A) for a period of 24 hours; an indication of increased frequency of observation using two satellites

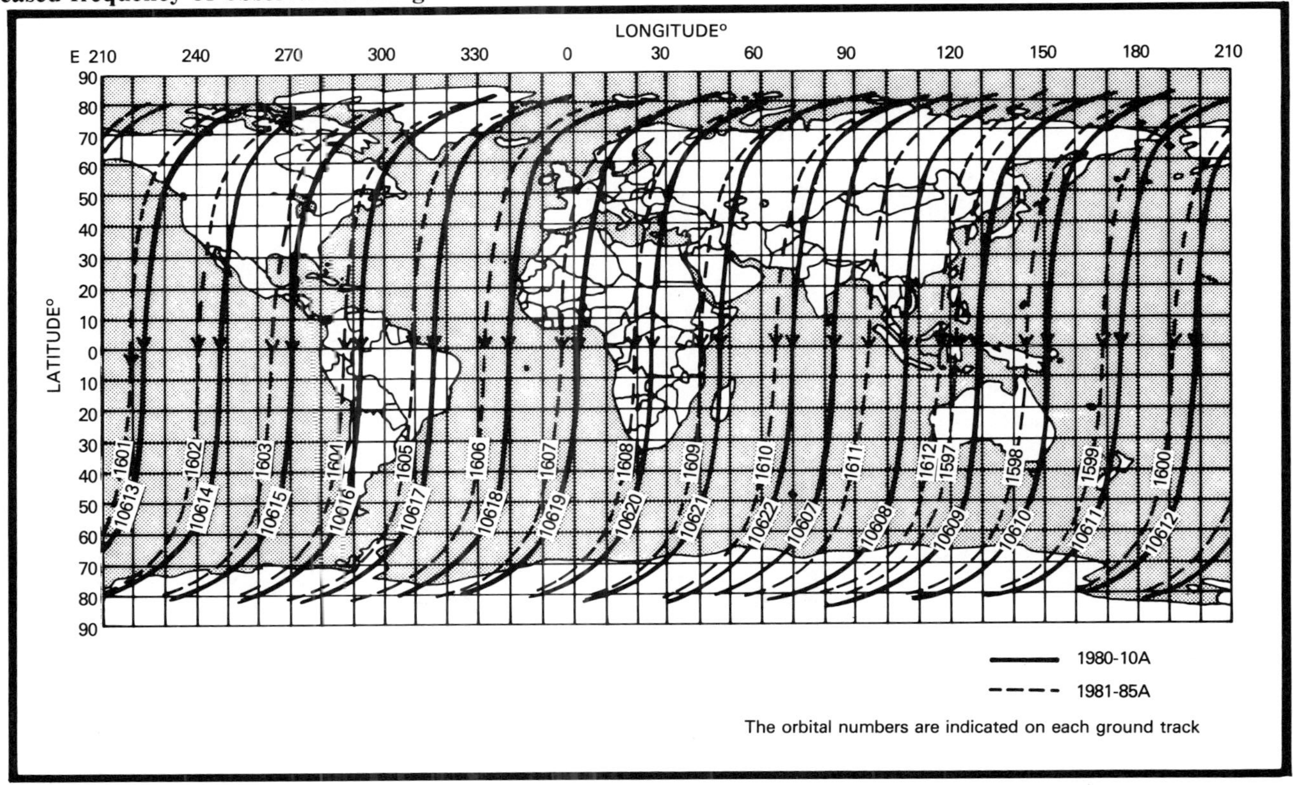

Figure 4.7a. Seasat SAR image showing the city of Cologne. An enlargement of the blocked section is shown in figure 4.7b (below)

Figure 4.7b. Enlargement of the blocked section in figure 4.7a; bridges, roads, the airport, and small boats along the Rhine River can clearly be seen (right)

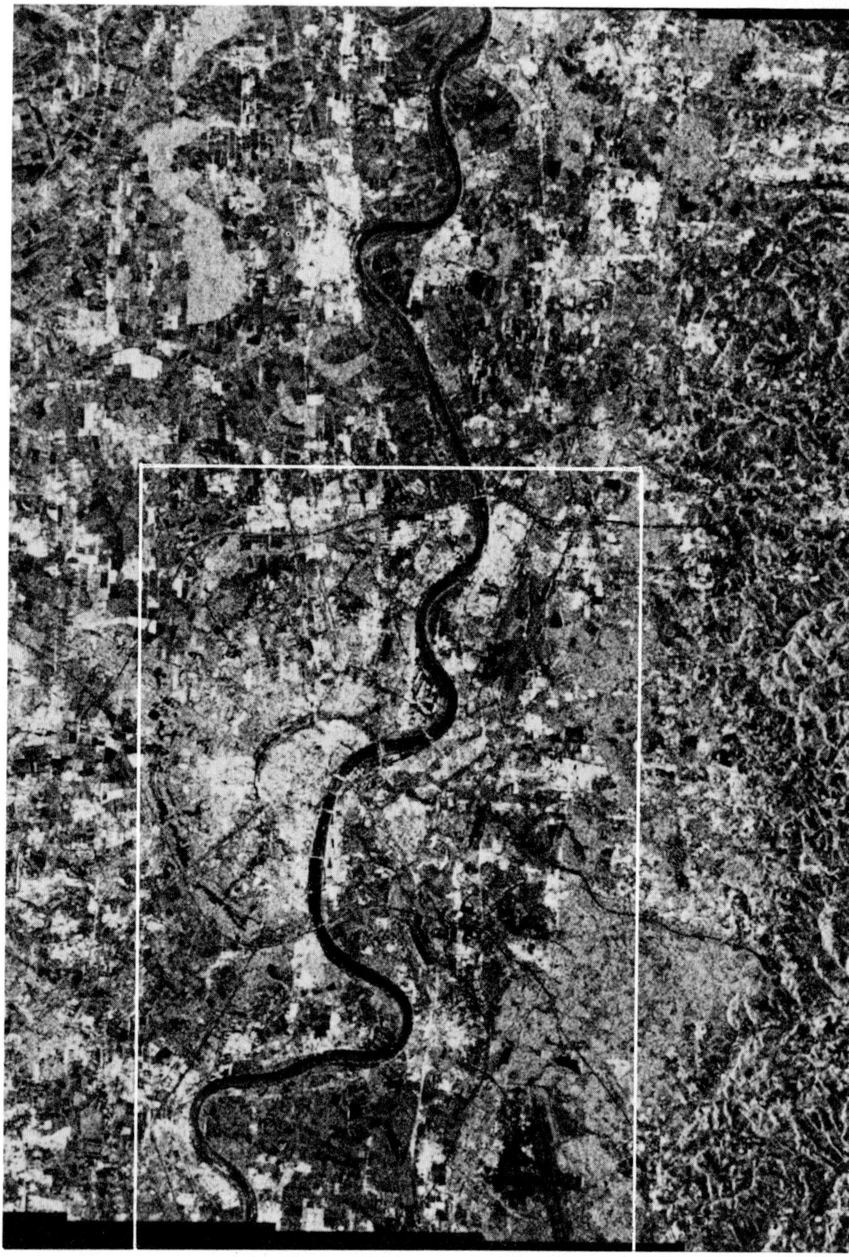

originate from military communications between bases, from early warning radars, air-defence and missile-defence radars or from those used for missile control (see paper 1 of part II). These satellites also gather data on missile testing, new radars and many other types of communications traffic. Not only do they locate systems producing electronic signals but also measure the characteristics of the signals so as to be able to plan penetration of defences. The precise location of sources of signals intercepted could be performed by navigation satellites by a method resembling hyperbolic navigation [14]. In this system the process is reversed and instead of using four transmitters on four satellites to establish the position of a known object, four receivers on four satellites are used to locate the position of an unknown transmitter on the Earth's surface. Such a method has been used by the US Navy to identify and determine the positions of naval vessels. This method is described below.

Both the USA and the Soviet Union are known to have launched electronic reconnaissance satellites. (The Soviet satellites are discussed in detail in paper 2 of part II.) Whereas the US electronic reconnaissance satellites have mostly been carried into orbit by Big Bird satellites, the Soviet satellites are launched using independent launchers.

The types of orbit of electronic reconnaissance satellites are indicated in figure 4.3.

## Ocean surveillance and oceanographic satellites

During the 1970s, two types of satellite were developed to monitor the oceans. One, the ocean surveillance satellite, is used to detect and track military surface ships while the other, the oceanographic satellite, is used to determine various ocean properties.

### *Ocean surveillance satellites*

Considerable progress has been made in the development of space-based sensors to monitor the oceans including, for example, radars able to detect military surface vessels [15]: the US Seasat 1 satellite, launched in June 1978 and orbited at an altitude of about 500 km, carried, among other sensors, a synthetic aperture radar (SAR). This is a side-looking radar with a relatively short antenna which is made to behave like a very long one with a narrow beam. A long antenna can be synthesized by taking advantage of the motion of the satellite in its orbit (see paper 5 of part II). Such a radar could provide all-weather photographs. An example of the quality of such photographs is shown in figures 4.7a and 4.7b.

The first US satellite designed to track and locate surface ships was launched in April 1976 under the project White Cloud. The satellite carried three small sub-satellites which were placed into circular orbits similar to that of the main satellite. These satellites were identified as ELINT Ocean

Reconnaissance Satellites or EORSATs. Under the so-called Clipper Bow programme the US Navy is also planning to launch satellites with high resolution radar. Once these Radar Ocean Reconnaissance Satellites, or RORSATs, are fully operational, the EORSATs will, from the detection of electronic signals emitted from surface ships, help the RORSATs identify surface vessels.

**Figure 4.8. Seasat SAR photograph of French side of Strait of Dover**
Wave formations and several ships can be seen. Photograph taken at 06.45 hours on 19 August 1978

The first operational Soviet ocean surveillance satellite was launched in May 1974. Such Soviet satellites carry radar sensors and perform their missions in pairs. The radar is powered by a small nuclear reactor. After the accident in early 1978 of Cosmos 954, which scattered radioactive debris from the burnt-up reactor as it entered the Earth's atmosphere, there was a short lull before launches of ocean surveillance satellites were resumed in 1980. For more details on these satellites, see paper 2 of part II.

*Oceanographic satellites*

Detection of submerged submarines from space is still difficult, but considerable effort is being put into the development of space sensors to measure as many ocean properties as possible so as to facilitate submarine detection by non-space-based sensors. Knowledge of properties—such as, the difference between the ocean surface and the geoid, the height of waves, the strength and direction of ocean currents and surface winds, temperatures at and below the sea surface, salinity and coastal features—is important both for the use of some weapons and for the detection of submarines. Even for the detection of small surface vessels, the knowledge of the state of the sea surface is useful. For example, a radar altimeter on board the Seasat 1 was used to measure the wave heights on the sea surface with an accuracy of better than 10 cm [17]. Even with the SAR, waves could be seen very clearly (see figure 4.8).

Among the factors influencing the accuracy of a submarine-launched ballistic missile (SLBM) are the water vapour content in the atmosphere close to the region of the launch and the wind velocity at the sea surface. These have been measured using the scanning multi-channel microwave radiometer and radar altimeter on board the Seasat 1 satellite. Water vapour of less than 2 g $cm^{-2}$ and a wind velocity of less than 5 m $s^{-1}$ could be determined [17].

Using the radar altimeter and from measurements of the satellite's orbital perturbations, the shape of the geoid and the values of the Earth's gravitational field can be determined. Such data should improve the accuracies of ballistic trajectories resulting in smaller CEPs (circular error probables) for SLBMs.

Under the US Navy Satellite Oceanographic Research Program (SOREP) the development and demonstration of the effective use of satellites in oceanography are planned [16]. The sensors used on board the US Seasat 1 satellite were developed under SOREP.

## Early warning satellites

In the 1950s radars formed the basis of early warning of a surprise attack by ICBMs. The range of radars, however, was limited by the curvature of the Earth and so provided only about 15 minutes during which a response could be worked out. The use of artificial Earth satellites has extended this warning time to some 30 minutes by detecting the enemy missiles as soon as they are launched.

Most military targets emit infra-red radiation to a varying degree. Early warning satellites detect enemy missiles through the use of sensors sensitive to the infra-red radiation emitted by the hot plume of a rocket. With the rapid developments (at least in the USA) in thermal imaging sensor technology—such as focal plane infra-red mosaic devices and integrated micro-electronics—the mission of early warning satellites will possibly extend to the detection of aircraft and cruise missiles [18]. Such devices consist of a telescope to form an image of a scene in its focal plane, a mosaic of many small photo-sensitive sensors placed in the focal plane of the telescope and signal processing micro-electronic circuits.

When an image falls on the sensors, it is converted into electrical signals which in turn are processed by the micro-electronic circuits coupled mechanically or electrically with the sensors. For high resolution and sensitivity, it is desirable to have many sensors which are individually very small in size (see paper 1 of part II). Such sensors are known as charged couple devices (CCDs). A typical device might contain some 2 048 × 96 elements per chip with a spacing of 13$\mu$m between elements [19]. Such chips could be mounted in rows and placed in the focal plane of an optical system. The French SPOT satellite will be equipped with such a CCD scanning sensor.

As mentioned above, CCD sensors will be used to monitor not only missile launches but also such targets as aircraft. The infra-red signatures of aircraft, surface and sea targets have been determined under a US programme called the HI-CAMP (High Resolution Calibrated Airborne Measurement Program) [20]. This programme is related to the actual sensor to be used on an early warning satellite called the Teal Ruby. The experimental sensor will be launched into space on board the space shuttle in 1983. It is planned to launch a second-generation infra-red array sensor into synchronous orbit using the shuttle during 1987. This sensor will be used not only to detect such targets as aircraft but even to track them and determine their flight path [20].

The testing of the adaptive optics for the above device was completed in the USA under the High Altitude Large Optics (HALO) programme [18]. Under this programme, such items as light-weight optical components, detector arrays and signal processing electronics were developed and their feasibility demonstrated.

It should be noted here that the ground-based radars also form a part of the early warning system, used to cross-check observations made by the satellites. Moreover, sophisticated sensors are being developed so as not only to give information of the launch of a missile but even to determine the nature of the attack. This is in order to select the appropriate response [21].

For details of the Soviet early warning satellites see paper 2 of part II.

### Nuclear explosion detection satellites

In the USA, 12 nuclear explosion monitoring satellites, the Vela satellites, have been launched and orbited at an altitude of some 110 000 km (see figure 4.3), but only three of these are operational. The last of the 12 satellites launched to monitor nuclear explosions in the atmosphere and in outer space detected what may have been a low-yield nuclear explosion (see figure 4.12) over the sea in the South Atlantic near the South African coast on 22 September 1979. The characteristic light flash of an atmospheric nuclear explosion was detected by a sensor called a 'Bhangmeter' on board the satellite. The explosion was estimated to be 2–4 kt in yield. However, the satellite's sensors could not accurately locate the position of the explosion.

A group of experts was set up in the USA to determine whether or not any other phenomenon could have produced the signals in the satellite sensor [22]. It has been reported that the US group concluded that there was no malfunction and that the signal was due to a natural phenomenon occurring close to the satellite. However, this is not a satisfactory explanation, considering that the instruments are such that signals due to natural phenomena cannot be confused with those due to a nuclear explosion [23]. In 1980, reports from South Africa, based on US intelligence information, stated that another nuclear explosion might have been conducted on 16 December 1980, again in the South Atlantic region. This time it was a US early warning satellite which had registered a flash of light in this region, but it was reported that a large meteor entering the atmosphere caused this phenomenon [24]. It is interesting to note that when examining all references to the first explosion, no mention was found of data collected from early warning satellites. The mysterious flashes still remain a matter of dispute.

## *III. Communications satellites*

Space-based sensors for surveillance of the Earth, together with land-based surveillance systems, generate considerable amounts of data. The transmission of this and other data for military purposes needs highly reliable and secure communications systems. The security aspect is considered in paper 4 of part II. However, a brief description of the satellite communications systems and programmes is given here. This is an area of vital interest as some 80 per cent of military communications are carried out using artificial Earth satellites [25].

The US military communications programme is basically divided into three projects: (*a*) a satellite system for command, control and communications; (*b*) a high capacity communications system for the National Command Authorities, the military commanders stationed in major headquarters around the world and the intelligence community—the World-Wide Military

Command and Control System (WWMCCS) is part of this system; and (*c*) communications for mobile forces such as aircraft, ships and anti-submarine warfare aircraft [26].

For command, control and communications, the Air Force Satellite Communications System (AFSATCOM) was developed. Under this programme, AFSATCOM I, equipment for satellite communications is developed and procured for selected DoD host satellites. An important component of the AFSATCOM system is the Satellite Data System (SDS) which consists of multi-purpose communications satellites. The first of two SDS experimental satellites was launched by the USA in March 1971 and, so far, seven operational satellites have been launched into highly elliptical orbits at orbital inclinations of about 63° and orbital periods of about 12 hours. The last one was launched in April 1981 [27].

Under the second US project, the Defense Satellite Communications System (DSCS) has been developed. Satellites under this programme are launched in pairs and in geosynchronous orbit. The most recent pair was launched in November 1979; presently there are four DSCS II satellites in orbit forming part of the WWMCCS. The DSCS III series of satellites are nuclear-hardened with anti-jam capability. Furthermore, these satellites, a replacement for the DSCS II series, are supposed not to depend on foreign-based US ground stations [28]. The first of the DSCS III has not yet been launched although this had been planned for 1981 [28]. Besides relaying military communications, the DSCS II satellites also relay data, gathered by photographic and electronic reconnaissance satellites, for treaty monitoring on the size and activities of military forces, the deployment of weapons, early warning satellites and crisis management [21].

The third project uses mainly the US Navy's Fleet Satellite Communication (FLTSATCOM) system. Four FLTSATCOM satellites have been launched, the first one in February 1978 and the latest one, the last in a series of five, in August 1981. The satellites are in geosynchronous orbit. Once FLTSATCOM satellites begin to reach the end of their lives, SDS will form the backbone of the AFSATCOM system [21].

To meet the communications requirements of the US nuclear forces, satellites with improved survivability against electromagnetic radiation and physical impact are planned for under the US Air Force's Strategic Satellite System. Two experimental satellites, LES 8 and 9, were launched in 1976 to demonstrate the use of a high-power radioisotope thermoelectric power supply to increase a satellite's resistance to nuclear attack and to eliminate the large radar cross-section of solar panels.

While most of the global communications satellites either planned or in use by the USA have made use of geosynchronous orbits (see figure 4.9), most Soviet civilian and military communications satellites are placed in highly elliptical orbits with periods of about 12 hours. Soviet communications satellite programmes are carried out by the Molniya series and some satellites in the Cosmos series. More details of the Soviet satellites are given in paper 2 of part II.

**Figure 4.9. Some military satellites in geosynchronous orbit**

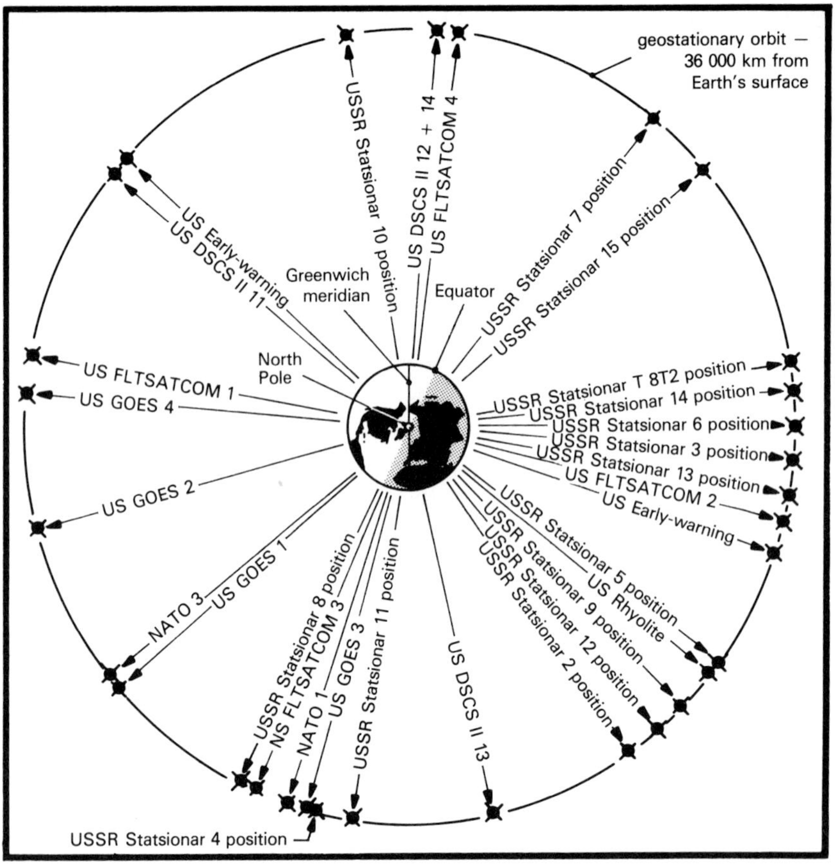


The first Soviet synchronous satellite was orbited in 1974 and it was not until the end of 1975 that a new series of such synchronous satellites was launched.

While the land-based ICBMs and the nuclear bomber force could be vulnerable to enemy attack, submarine-based ICBMs are relatively invulnerable at present. However, submarines have to surface for communications purposes, and a number of methods of communication are being investigated to avoid this. One such method is to use high-energy lasers and artificial Earth satellites. This is considered briefly below.

### Laser communication

The only electromagnetic waves which can travel significant distances in oceans are extremely low-frequency radiation and those corresponding to visible light. The blue-green lasers fall into the latter category. A laser, an acronym for Light Amplification by the Stimulated Emission of Radiation, usually emits light in the infra-red, visible or ultraviolet region of the spectrum of electromagnetic radiation. A ray of laser light differs from that of ordinary light in that the former is much more intense, directional, monochromatic and coherent. Thus it is made up of waves which are all of the same wavelength and are all in phase.

In the USA the Navy and the Defense Advanced Research Project Agency are investigating the possibility of using blue-green lasers for communications with submarines via satellites. An initial study was conducted by the US Navy in 1977. Two systems are being investigated under the Strategic Laser Communications (recently re-named Submarine Laser Communications) programme. In one, the transmitting laser would be placed on the ground. The signal-carrying laser beam would be aimed at a satellite in geosynchronous orbit carrying large mirrors to reflect the beam down to parts of the ocean where submarines might be travelling. In the other system, microwave signals would be transmitted from the ground to the satellite in a conventional way and a laser on board the satellite would retransmit the signals down to the submerged submarine [29]. The average power of the laser beam would be about one kilowatt. In either case the wavelength of the laser light would be in the range 460–530 nanometres (blue-green light). Xenon fluoride (XeF) is expected to be used in a ground-based system to initially produce an ultraviolet laser light. This light would then be passed through hydrogen to produce the blue-green laser light. The XeF is preferred because the same system would then have some weapon applications [30].

In the case of the ground-based laser transmitter, the need for clear skies means that many such transmitters would be needed so that at least one transmitter would be unhampered by cloud. In this case, a laser beam of an average power of a few megawatts would be needed [31]. Such a system should also have a high-precision optical system controlled by a sophisticated target acquisition, pointing and tracking system. Moreover, the optical system should be able to compensate for distortions of the laser beam as it travels through the atmosphere.

All of these problems are also common to other applications, such as the development of laser weapon technology. The target acquisition, pointing and tracking systems have already been under development for a decade or so and the level of precision (one microradian or less) required for the strategic optical communications system has been demonstrated [29]. Under the strategic reconnaissance data relay satellite programme (programme 405B, space communication system using laser) accurate acquisition, pointing and tracking capabilities over intercontinental distances have been demonstrated.

Almost two decades have been devoted to solving the problem of distortions introduced by the atmosphere when a laser beam travels through it. One way of overcoming this problem is to modify the optics of the laser instrument in such a way as to compensate for atmospheric distortions. For example, a large mirror (measuring a few metres square) for a high-energy laser device is made up of many small (10 cm) mirrors. These are mounted so that each of them can be moved slightly in order to distort the wavefront of the laser beam in such a way that after it has emerged through the atmosphere the distortions introduced by the atmosphere are exactly compensated for. It is claimed that this technology is now available [29]. In a recent laboratory experiment it was demonstrated that atmospheric distortions can be compensated for and atmospheric turbulence can be simulated [20]. In practice, considerable data obtained on meteorological and ocean conditions will be used [30].

Two-way communication with a submerged submarine has been demonstrated under the US Navy's Optical Ranging and Detection System and Optical Ranging, Identification, Communications System programmes [32]. Under the US Naval Air Development Center's Submarine Aircraft Optical Communications System and Optical Satellite Communications programmes, communication has also been made with a submerged receiver [32].

## *IV. Navigation satellites*

For many weapon systems it is important to know the exact position and velocity of the missile. This is particularly so for missiles launched from sea-based platforms. Satellites are beginning to fulfil these requirements. Naval surface ships as well as submarines, aircraft and missiles determine their positions and velocities using signals emitted continually by satellites. Recent US tests using navigation satellites and some estimates from simulation experiments suggest that, even in a civilian application, positions could be fixed with an accuracy of 200 m [33]. In a military system this is about 20 m when 18 satellites are used (for details see paper 3 of part II).

The principle of the use of satellites for navigation is illustrated in figure 4.10. A and B represent two satellites with the line OO bisecting a line joining the satellites. If two signals are transmitted simultaneously by the satellites, the times at which they arrive at a point X will be different as long as X is not on the plane through OO. If the pulse from A, for example, is coded then the difference between the times of arrival of the pulses at X can be measured. It would, in turn, determine which side of OO the point X lies. If this time difference is kept constant for various positions of X, a series of hyperboloids (the solid curves in figure 4.10) forming well-defined lanes could be obtained. With a third satellite at C, a similar set of hyperboloids could be obtained between B and C and A and C (dotted curves in figure 4.10). Thus a three-dimensional grid is formed in which the position of X can be accurately determined. For a world-wide navigation system and for greater accuracy a number of satellites would be needed. In order to avoid ambiguity, the transmitters on various satellites must be operated at different frequencies.

**Figure 4.10. Position-fixing by navigation satellites**

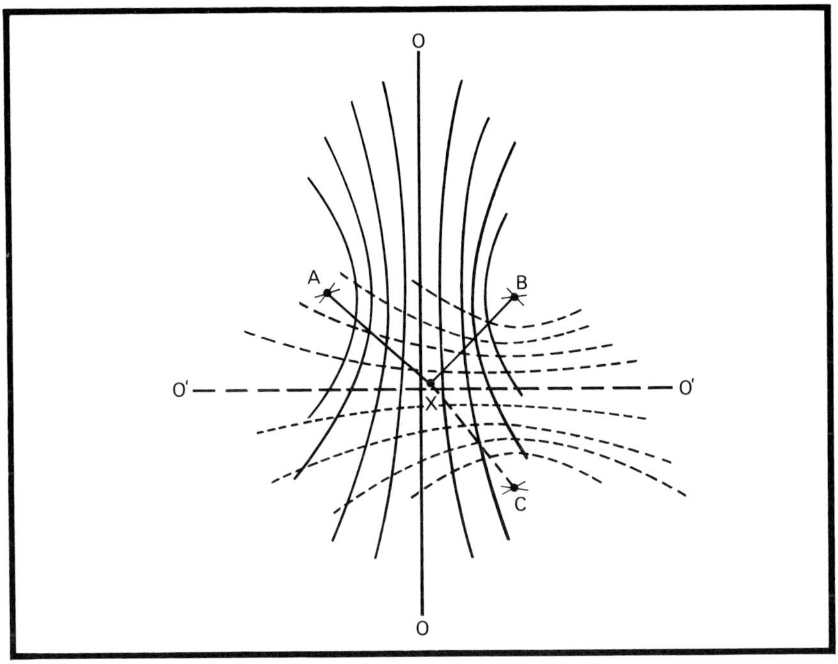


Both the USA and the USSR have developed such navigation systems. Paper 2 of part II describes the Soviet system in some detail while the US TRANSIT navigation satellites are dealt with in paper 3. The TRANSIT satellites were developed to provide position fixes in two dimensions particularly for the strategic ballistic missile submarines [26]. An improved TRANSIT satellite, Nova 1, was launched on 15 May 1981. The concept of a a second-generation system, the NAVSTAR Global Positioning System (GPS), had evolved by mid-1974 under a joint services development programme. Under the recently revised programme, 18 instead of the original 24 satellites in three orbital planes are planned for (see paper 3 of part II). The GPS is expected to contribute to improvement in the accuracy of ballistic missiles under the US Navy's Trident Improved Accuracy Program [34].

Although the USA has launched satellites specifically to detect nuclear explosions in the atmosphere and in outer space, it is now planned that NAVSTAR will carry sensors for this purpose under the Integrated Operational Nuclear Detection System (IONDS). The IONDS is intended to provide damage assessment both within one's own country and within enemy territories during and after a nuclear attack. This effort is in support of the new nuclear war doctrine which requires early warning of attack, information for assessing the size of the attack and data on the attacked target so that an appropriate response can be made [35].

## *V. Meteorological satellites*

Although military and civilian meteorological requirements are different and separate satellites have so far been launched, there is a recent trend in the USA to incorporate sensors for both types of mission on the same spacecraft, namely, Block 6 [36]. The current generation of military satellites are the Block 5D satellites, which are part of the Defense Meteorological Satellite Program, a joint services programme with the US Air Force in charge of management and funding for the system as a whole and the spacecraft, launch vehicle and tracking facilities in particular.

At any one time there are two Block 5D satellites, usually in polar sun-synchronous orbits, at an altitude of about 800 km. The orbits of the two satellites are such that if one satellite passes over a particular region at 0700 hours local time, then the other will pass over the same region some five hours later. Thus the two-satellite constellation provides global coverage four times a day [37].

These satellites carry a line-scanner radiometer sensitive in the visible and infra-red region of the spectrum, a sounder which measures infra-red radiation emitted from different heights within the atmosphere, a spectrometer which counts precipitating electrons at different energy levels and a density sounder which measures the oxygen and nitrogen density of the thermosphere. The temperature and moisture sounder allows temperature and water vapour determination at various altitudes. Therefore, the sensors aboard the satellite can detect and observe developing patterns of weather and can track existing weather systems. For a complete understanding and prediction of weather and climate formation, data collected from a very large number of ground stations are also used.

While the DoD does not at present have a military meteorological satellite in geosynchronous orbit, it has complete access to the Geostationary Operational Environmental Satellites (GOES) operated by the National Oceanic and Atmospheric Administration. There are four GOES in orbit (see figure 4.9). The Air Force Global Weather Central has its own terminal, for example, to receive data generated by the Tiros-N and other satellites.

The Soviet Union also uses satellites for meteorological observations. The routine monitoring of weather from space began in 1969. Mostly the Meteor series of satellites perform this mission. Presumably measurements similar to those made by the US satellites are also being made by the Soviet spacecraft.

While the knowledge of cloud formation and movements is of importance in the photography of targets of military interest by reconnaissance satellites and in planning bombing missions, the amount of data collected is more than sufficient. One reason for this could be that once man has understood the mechanics of weather and climate formation his military genius may be able to control these for hostile purposes.

Another not very widely known aspect of meteorological observations is their use in improving the accuracies of ICBMs. The accuracy of an ICBM depends on, among other factors, the knowledge of the Earth's gravitational field, of the relative positions of the target and the missile launching pad and the knowledge of the meteorological conditions along the missile's possible trajectory. Not only do the meteorological conditions determine the corrections made to missile trajectories but these conditions are also taken into account when predicting satellite orbital tracks [37]. Predictions of the other effects mentioned above are considered below.

## *VI. Geodetic satellites*

The knowledge of geophysics (the physics of the Earth and its environment, i.e., Earth, air and space) is continually being increased. This is, as seen from the above discussion on meteorology, partly because such knowledge is required for military use. Another very important aspect of geophysics is geodesy which includes determination of the size and shape of the Earth, its gravitational field, detailed maps and the location on the globe of cities, towns and villages as well as the precise positions of military targets. In the USA this is carried out by the Defense Mapping Agency.

With the development of digital data processing, considerable impact is expected on the performance of weapon systems, such as ICBMs and cruise missiles, the accuracy of which is largely dependent on geophysical knowledge in general and geodesy in particular. For example, the knowledge of target positions, the ability to correct the trajectory of missiles and determine accurately their launch orientation, and extensive knowledge of the values of the Earth's gravitational field throughout the globe are vital to the improved accuracies of delivery vehicles used for warheads. Satellites are exploited considerably to increase this knowledge [38].

## *VII. Military use of manned spaceflights*

The concept of man orbiting the Earth is not a new one. In the USA serious ideas of rocket-boosted glider spacecraft were introduced in 1950 and the USSR could not have been far behind since both powers launched men a decade later. By 1952 the US Air Force had designed a bomber-missile called Bomi.

However, it was not until the launch of the first artificial Earth satellite, Sputnik I, that ideas of man in space began to take firm shape. The US Air Force then began to design a spacecraft called Dyna-soar (dynamic soaring). This was a small one-manned winged spacecraft with manoeuvring capabilities, intended to land like an aircraft. However, in December 1963, the Dyna-soar programme was cancelled in favour of the Manned Orbiting Laboratory (MOL). Both the Dyna-soar and the MOL projects were conceived with military missions in mind.

The MOL spacecraft, to be launched using a Titan IIIC rocket, was to have been a modified Gemini spacecraft attached to a cylindrical laboratory. It was to have a total length of about 17 m and weigh 11 364 kg; some 2 300 kg were to have been due to reconnaissance equipment, including large optical instruments, equipment for tracking ground and space targets for ocean surveillance, instruments for electromagnetic intelligence and facilities to even test target post-strike assessments [39]. MOL was to be placed in a circular orbit inclined at between 85° and 92° at an altitude of about 480 km. The initial unmanned flight was to have taken place in 1967/68 followed by a manned flight in 1968/69 with a total cost of about $1.5 billion. However, for several reasons the MOL programme got delayed and many of the tasks that were going to be performed by men were, by 1969, being carried out using unmanned spacecraft. The MOL programme was cancelled in June 1969.

Besides the MOL, the USA had four other manned programmes, Mercury, Gemini, Apollo and the Skylab. In the meantime the USSR was vigorously pursuing its manned space programme and had a lead on the US programme; the first man to orbit the Earth was a Soviet cosmonaut, and the Soviet Union was also the first to launch a man into space. These programmes are dealt with in some detail in paper 9 of part II. For comparison, the US and the Soviet manned flights are summarized in table 4.2. While initially the Soviet Union led in terms of total man-hours in space, the USA quickly overtook the USSR and held the lead until 1975, after which the US manned space programme ceased. However, it was not until the middle of 1978 that the Soviet Union equalled the USA in total man-hours (see table 4.2). It must be noted here that while there was considerable publicity for all the successful manned flights, the deaths of three US astronauts and one Soviet in 1967 and three Soviets in 1971 seem to have been forgotten.

All of these above-mentioned manned space activities involved the use of expendable launchers. However, the successful launches in 1981 of the US reusable launcher, the Space Transportation System, or space shuttle (see table 4.2), have opened the door for further proliferation of manned military activities in outer space. The space shuttles were launched by the civilian agency NASA, but the US Air Force has suggested it should have more control over the future activities of the space shuttle [40]. It has also been suggested that under the Block II Shuttle Program, the capabilities of the space shuttle should be increased by enlarging the vehicle to accommodate a larger payload.

The second shuttle, Columbia 2, launched on12 November 1981, was a significant one. First, it showed that a space transportation vehicle could be used again and, second, it carried a number of test payloads which increased the mass lifted into orbit compared with the previous launch in April 1981. The test payloads included a manipulator arm to be used later for placing satellites into orbit and retrieving satellites from orbit for inspection and repair. It also included an image radar and a multispectral infra-red radiometer, both of which functioned successfully [41]. The ground resolution of the radar, the first side-looking radar to be orbited, is expected to be about 80 m with an instantaneous field of view (IFOV) of 40 m × 40 m [42]. Columbia 2 was placed in a circular orbit at an altitude of about 250 km with an orbital inclination of 38°. It orbited in an inverted position with its cargo bay doors facing the Earth and open.

The importance of the space shuttle system lies in the fact that it will enhance the effectiveness of military satellites. Compared with the existing expendable boosters, the shuttle will be able to launch greater payload weights and volumes. The shuttle system not only allows the recovery and refurbishing of satellites for re-use but also enables satellites to be repaired in orbit [26].

The emphasis put on the space shuttle by the military can be illustrated by the fact that the DoD is contributing about one-sixth of the cost of developing such a system, and some 40 per cent of the initial space shuttle's flights will be used by the DoD [43]. In fact, the US Air Force is responsible for the development of part of the US space shuttle system. It is developing the launch and landing facilities at the Vandenberg Air Force Base space complex and developing the inertial upper stage (IUS) for use on shuttle launchers to deliver DoD spacecraft to higher orbital altitudes and inclinations than the shuttle alone could provide. The IUS is a multi-stage vehicle which would use solid fuel as a propellant. The DoD version is a two-stage vehicle capable, when used on the shuttle, of delivering payloads weighing up to about 2 300 kg into geosynchronous orbit. During the early transition period, the IUS will also be used on the more conventional Titan 3 launcher [26].

The first operational flight is scheduled for November 1983 manned by four astronauts [44]. The first DoD payload is expected to be launched on this flight. Details of the DoD space shuttle payloads are not given but one of the first DoD payloads will be the Teal Ruby advanced test infra-red sensor.

The above brief review of the use of outer space has shown that not only is there a greater use of space for military purposes but that spacecraft are one of the important military targets. This is because the role of satellites in waging wars on Earth is increasing. Some of the anti-satellite (ASAT) systems are discussed below.

## *VIII. Anti-satellite systems*

Both the USSR and the USA are investigating a number of methods for disabling satellites in Earth orbit. These range from ground-based ASAT missiles to orbiting killer-satellites. Some of these systems have even been

**Table 4.2. Summary of the US and the Soviet manned flights between 1961**

USA

| Launch date | Vehicle | Astronauts | Duration of flight (hr:min) | Total man-hours in space (hr:min) | Number of orbits |
|---|---|---|---|---|---|
| 5 May 1961 | MR-3 | Shepard | :15 | :15 | |
| 21 Jul 1961 | MR-4 | Grissom | :16 | :31 | |
| 20 Feb 1962 | MA-6 | Glenn | 4:55 | 5:26 | 3 |
| 24 May 1962 | MA-7 | Carpenter | 4:56 | 10:22 | 3 |
| 3 Oct 1962 | MA-8 | Schirra | 9:13 | 19:35 | 6 |
| 15 May 1963 | MA-9 | Cooper | 34:20 | 53:55 | 23 |
| 12 Mar 1965 | GT-3 | Grissom (2)<br>Young | 4:53 | 63:41 | 3 |
| 3 Jun 1965 | GT-4 | McDivitt<br>White | 97:56 | 259:33 | 66 |
| 2 Aug 1965 | | Cooper (2)<br>Conrad | 190:55 | 641:23 | 128 |
| 4 Dec 1965 | GT-7 | Borman<br>Lovell | 330:35 | 1 302:33 | 220 |
| 15 Dec 1965 | GT-6 | Schirra (2)<br>Stafford | 25:51 | 1 354:15 | 17 |
| 16 Mar 1966 | GT-8 | Armstrong<br>Scott | 10:41 | 1 375:37 | 7 |
| 3 Jun 1966 | GT-9 | Stafford (2)<br>Cernan | 72:21 | 1 520:19 | 47 |
| 18 Jul 1966 | GT-10 | Young (2)<br>Collins | 70:47 | 1 661:53 | 46 |
| 12 Sep 1966 | GT-11 | Conrad (2)<br>Gordon | 71:17 | 1 804:27 | 47 |
| 11 Nov 1966 | GT-12 | Lovell (2)<br>Aldrin | 94:35 | 1 993:37 | 63 |
| 27 Jun 1967 | AS-204 | Grissom (2)<br>White (2)<br>Chaffee | Fatal fire on the launch pad killed all three astronauts | | |
| 11 Oct 1968 | Apollo 7 | Schirra (3)<br>Eisele<br>Cunningham | 260:09 | 2 774:04 | 174 |
| 21 Dec 1968 | Apollo 8 | Borman (2)<br>Lovell (3)<br>Anders | 147:01 | 3 215:07 | 2 (Earth)<br>10 (Moon) |
| 3 Mar 1969 | Apollo 9 | McDivitt (2)<br>Scott (2)<br>Schweickart | 241:01 | 3 938:10 | 162 |
| 15 May 1969 | Apollo 10 | Stafford (3)<br>Young (3)<br>Cernan (2) | 192:03 | 4 514:19 | 2 (Earth)<br>31 (Moon) |
| 16 Jul 1969 | Apollo 11 | Armstrong (2)<br>Collins (2)<br>Aldrin (2) | 195:19 | 5 100:16 | 2 (Earth)<br>30 (Moon) |

| USSR | | | | | |
|---|---|---|---|---|---|
| Launch date | Vehicle | Astronauts | Duration of flight (hr:min) | Total man-hours in space (hr:min) | Number of orbits |
| 12 Apr 1961 | Vostok 1 | Gagarin | 1:48 | 1:48 | 1 |
| 6 Aug 1961 | Vostok 2 | Titov | 25:18 | 27:06 | 17 |
| 11 Aug 1962 | Vostok 3 | Nikolayev | 94:25 | 121:31 | 64 |
| 12 Aug 1962 | Vostok 4 | Popovich | 70:59 | 192:30 | 48 |
| 14 Jun 1963 | Vostok 5 | Bykovskiy | 119:06 | 311:36 | 81 |
| 16 Jun 1963 | Vostok 6 | Tereshkova | 70:50 | 382:26 | 48 |
| 12 Oct 1964 | Voskhod 1 | Komarov, | 24:17 | 455:17 | 16 |
| | | Yegorov, | | | |
| | | Feoktistov | 26:02 | 507:21 | 17 |
| 18 Mar 1965 | Voskhod 2 | Belyayev | | | |
| | | Leonov | | | |

No Soviet manned flights between
18 March 1965 and 21 April 1967

| Launch date | Vehicle | Astronauts | Duration of flight (hr:min) | Total man-hours in space (hr:min) | Number of orbits |
|---|---|---|---|---|---|
| 21 Apr 1967 | Soyuz 1 | Komarov (2) (killed during return flight) | 26:37 | 533:58 | 18 |
| 26 Oct 1968 | Soyuz 3 | Beregovoy | 94:51 | 628:49 | 64 |
| 14 Jan 1969 | Soyuz 4 | Shatalov | 71:23[1] | 795:50 | 48 |
| | | (Yeliseyev) | | | |
| | | (Khrunov) | | | |
| 15 Jan 1969 | Soyuz 5 | Volynov | 72:56 | 868:46 | 49 |
| | | Yeliseyev | | | |
| | | Khrunov | | | |
| 11 Oct 1969 | Soyuz 6 | Shonin | 118:42 | 1 106:10 | 80 |
| | | Kubasov | | | |
| | | Filipchenko | | | |
| 12 Oct 1969 | Soyuz 7 | Gorbatko | 118:41 | 1 462:13 | 80 |
| | | Volkov | | | |
| 13 Oct 1969 | Soyuz 8 | Shatalov (2) | 118:50 | 1 699:53 | 80 |
| | | Yeliseyev (2) | | | |

| USA | | | | | |
|---|---|---|---|---|---|
| Launch Date | Vehicle | Astronauts | Duration of flight (hr:min) | Total man-hours in space (hr:min) | Number of orbits |
| 14 Nov 1969 | Apollo 12 | Conrad (3)<br>Gordon (2)<br>Bean | 244:36 | 5 834:04 | 2 (Earth)<br>45 (Moon) |
| 11 Apr 1970 | Apollo 13 | Lovell (4)<br>Haise<br>Swigert | 142:55 | 6 262:49 | 2 (Earth)<br>1 (Moon) |
| 13 Jan 1971 | Apollo 14 | Shepard (2)<br>Roosa<br>Mitchell | 216:02 | 6 910:55 | 2 (Earth)<br>34 (Moon) |
| 26 Jun 1971 | Apollo 15 | Scott (3)<br>Irwin<br>Worden | 295:12 | 7 796:31 | 2 (Earth)<br>74 (Moon) |
| 16 Apr 1972 | Apollo 17 | Young (4)<br>Duke<br>Mattingly | 265:51 | 8 594:04 | 2 (Earth)<br>65 (Moon) |
| 7 Dec 1972 | Apollo 16 | Cernan (3)<br>Schmitt<br>Evans | 301:52 | 9 499:40 | 3 (Earth)<br>75 (Moon) |
| 14 May 1973 | Skylab 1 | Occupied by successive crews | – | – | |
| 25 May 1973 | Skylab 2 | Conrad (4)<br>Kerwin<br>Weitz | 672:50 | 11 518:10 | 433 |
| 28 Jul 1973 | Skylab 3 | Bean (2)<br>Garriott<br>Lousma | 1 427:09 | 15 799:37 | 919 |
| 16 Nov 1973 | Skylab 4 | Carr<br>Gibson<br>Pogue | 2 017:16 | 21 851:25 | 1 298 |

No US manned flights between
16 November 1973 and 15 July 1975

USSR

| Launch date | Vehicle | Astronauts | Duration of flight (hr:min) | Total man-hours in space (hr:min) | Number of orbits |
|---|---|---|---|---|---|
| 1 Jun 1970 | Soyuz 9 | Nikolayev (2)<br>Sevastyanov | 424:59 | 2 549:51 | 286 |
| 19 Apr 1971 | Salyut | (Occupied by Soyuz 11 crew) | | | |
| 22 Apr 1971 | Soyuz 10 | Shatalov (3)<br>Yeliseyev (3)<br>Rukavishnikov | 47:46 | 2 693:09 | 32 |
| 6 Jun 1971 | Soyuz 11 | Dobrovolskiy[2]<br>Volkov (2)[2]<br>Patsayev[2] | 570:22 | 4 404:15 | 385 |
| 27 Sep 1973 | Soyuz 12 | Lazarev<br>Makarov | 47:16 | 4 498:47 | 32 |
| 18 Dec 1973 | Soyuz 13 | Klimuk<br>Lebedev | 188:55 | 4 876:37 | 127 |
| 24 Jun 1974 | Salyut 3 | (Occupied by Soyuz 14 crew) | | | |
| 3 Jul 1974 | Soyuz 14 | Popovich (2)<br>Artyukhin | 337:30 | 5 631:37 | 252 |
| 26 Aug 1974 | Soyuz 15 | Sarafanov<br>Demin | 48:12 | 5 728:01 | 32 |
| 2 Dec 1974 | Soyuz 16 | Filipchenko (2)<br>Rukavishnikov (2) | 142:24 | 6 012:49 | 96 |
| 26 Dec 1974 | Salyut 4 | (Occupied by Soyuz 17, 18 crews) | | | |
| 10 Jan 1975 | Soyuz 17 | Gubarev<br>Grechko | 709:20 | 7 431:29 | 466 |
| 5 Apr 1975 | | Lazarev (2)<br>Makarov (2) | :20 | 7 432:09 | |
| 24 May 1975 | Soyuz 18 | Klimuk (2)<br>Sevastyanov (2) | 1 511:20 | 10 454:49 | 993 |

USA

| Launch date | Vehicle | Astronauts | Duration of flight (hr:min) | Total man-hours in space (hr:min) | Number of orbits |
|---|---|---|---|---|---|
| 15 Jul 1975 | Apollo | Stafford (4)<br>Slayton<br>Brand | 217:28 | 22 503:49 | 145 |

No US manned flights between
15 July 1975 and 12 April 1981

USSR

| Launch date | | Vehicle | Astronauts | Duration of flight (hr:min) | Total man-hours in space (hr:min) | Number of orbits |
|---|---|---|---|---|---|---|
| 15 Jul | 1975 | Soyuz 19 | Leonov (2)<br>Kubasov (2) | 142:31 | 10 739:51 | 96 |
| 22 Jun | 1976 | Salyut 5 | (Occupied by Soyuz 21, 24 crews) | | | |
| 6 Jul | 1976 | Soyuz 21 | Volynov (2)<br>Zholobov | 1 182:24 | 13 104:39 | 790 |
| 15 Sep | 1976 | Soyuz 22 | Bykovskiy (2)<br>Aksenov | 189:54 | 13 484:27 | 127 |
| 14 Oct | 1976 | Soyuz 23 | Zudov<br>Rozhdestvenskiy | 48:06 | 13 580:39 | 32 |
| 7 Feb | 1977 | Soyuz 24 | Gorbatko (2)<br>Glazkov | 425:23 | 14 431:25 | 285 |
| 29 Sep | 1977 | Salyut 6 | (Occupied by Soyuz 26, 27, 28, 29, 30, 31, 32, 35, 36, 37, 38, T-2, T-3 crews) | | | |
| 9 Oct | 1977 | Soyuz 25 | Kovalenok<br>Ryumin | 48:46 | 14 528:57 | 32 |
| 10 Dec | 1977 | Soyuz 26 | Romanenko<br>Grechko (2) | 898:06 | 16 325:09 | 590 |
| 10 Jan | 1978 | Soyuz 27 | Dzhanibekov<br>Makarov (3) | 1 558:53 | 19 442:55 | 1 024 |
| 2 Mar | 1978 | Soyuz 28 | Gubarev (2)<br>Remek<br>(Czechoslovakia) | 190:17 | 19 823:29 | 125 |
| 15 Jun | 1978 | Soyuz 29 | Kovalenok (2)<br>Ivanchenkov | 1 911:23 | 23 646:15 | 125 |
| 23 Jun | 1978 | Soyuz 30 | Klimuk (3)<br>Hermaszewski<br>(Poland) | 190:04 | 24 026:23 | 125 |
| 26 Aug | 1978 | Soyuz 31 | Bykovskiy (3)<br>Jahn<br>(German D.R.) | 1 628:14 | 27 282:52 | 354 |
| 25 Feb | 1979 | Soyuz 32 | Lyakhov<br>Ryumin (2)<br>(Returned unmanned) | 2 596:24 | 32 475:39 | 1 711 |
| 10 Apr | 1979 | Soyuz 33 | Rukavishnikov (3)<br>Ivanov<br>(Bulgaria) | 47:01 | 32 569:41 | 31 |
| 6 Jun | 1979 | Soyuz 34 | Launched unmanned<br>(Returned Soyuz 32 crew) | 1 770:17 | 35 778:05 | 1 153 |
| 9 Apr | 1980 | Soyuz 35 | Popov<br>Ryumin (3) | 1 321:29 | 38 421:03 | 868 |
| 26 May | 1980 | Soyuz 36 | Kubasov (3)<br>Farkas<br>(Hungary) | 1 580:54 | 41 582:51 | 355 |
| 5 Jun | 1980 | Soyuz T-2 | Malyshev<br>Aksenov (2) | 94:41 | 41 772:13 | 62 |
| 23 Jul | 1980 | Soyuz 37 | Gorbatko (3)<br>Pham Tuan<br>(Viet Nam) | 1 911:17 | 45 594:47 | 1 257 |

| USA | | | | | |
|---|---|---|---|---|---|
| Launch date | Vehicle | Astronauts | Duration of flight (hr:min) | Total man-hours in space (hr:min) | Number of orbits |
| 12 Apr 1981 | STS-1 Columbia | Young (5) Crippen | 54:29 | 22 558:18 | 36 |
| 12 Nov 1981 | STS-2 Columbia | Engle Truly | 54:14 | 22 612:32 | 36 |

MR: Mercury Redstone
MA: Mercury Atlas
GT: Gemini Target
STS: Space Transportation System

[1] Crew duration for two transferring 47:49 each.
[2] All three died of asphyxiation after separation from Salyut. Soyuz 11 landed with the dead crew.

tested and in some cases deployed. An ASAT weapon may not necessarily be deployed against a spacecraft. It could be aimed at satellite command, control and space surveillance systems which are vital to the efficient operation of satellites. In the continuing development of ASAT systems it has to be taken into account that satellites are meanwhile being hardened. Some of the developments in these fields are reviewed briefly below.

**Space surveillance, command, control and data acquisition centres**

In the USA these are mainly located at: (*a*) the US Air Force's Eastern Test Range (ETR) facility, (*b*) the Western Test Range managed by Space and Missile Test Center (SAMTEC), (*c*) the US Army's White Sands Missile Range (WSMR), (*d*) Satellite Control Facility (SCF), and (*e*) the US Navy's Pacific Missile Test Center (PMTC).

Of these, the ETR, SAMTEC and WSMR provide both data acquisition as well as satellite-tracking facilities. The SCF has nine world-wide satellite-tracking stations while the Navy's PMTC provides a data acquisition facility. The US Air Force's Space Detection and Tracking System is primarily a data-processing and cataloguing operation whereas the Department of

| USSR | | | | | |
|---|---|---|---|---|---|
| Launch date | Vehicle | Astronauts | Duration of flight (hr:min) | Total man-hours in space (hr:min) | Number of orbits |
| 18 Sep 1980 | Soyuz 38 | Romanenko (2) | | | |
| | | Tamayo Mendez | 188:43 | 45 972:13 | 124 |
| | (Cuba) | | | | |
| 27 Nov 1980 | Soyuz T-3 | Kizim | | | |
| | | Makarov (4) | | | |
| | | Strekalov | 307:08 | 46 893:37 | 204 |
| 12 Mar 1981 | Soyuz T-4 | Kovalenok (3) | | | |
| | | Savinikh | | | |
| 22 Mar 1981 | Soyuz 39 | Dzanibekhov | 188:53 | 47 082:30 | 127 |
| | | Gurragchaa | | | |
| | | (Mongolia) | | | |
| 14 May 1981 | Soyuz 40 | Popov (2) | 188:38 | 47 271:08 | 127 |
| | | Prunariu | | | |
| | | (Romania) | | | |

Defense's Space Surveillance System uses radars to illuminate all satellites crossing over the USA. This is maintained by the Navy.

Only a few of the US tracking and data acquisition facilities are listed above. In fact, the Department of Defense maintains many radars, Backer-Nunn cameras, interferometers and other equipment all over the World. The information collected by all these facilities is passed on to NORAD, the North American Air Defense Command.

Five additional systems are being developed to be placed at various locations on the Earth under the Ground Electro-Optical Deep Space Surveillance System [45]. DARPA is also developing a mosaic staring sensor under the High Altitude Large Optics programme for the detection and warning of satellite attack.

Details of the Soviet satellite-tracking systems are scanty, but it is known that the Soviet Union has a network of satellite-tracking and data acquisition stations. Within the Soviet Union there are seven known facilities at Yevpatoriya, Tbilisi, Dzhusaly, Kolpashevo, Ulan-Ude, Ussuriyisk and Petropavlosk. However, the USSR does not have as extensive a satellite-tracking system outside its borders as does the USA. Because of the limited number of foreign-based tracking stations, the Soviet Union has developed several sea-based satellite-tracking stations. These consist of a number of ships operating in the mid-Pacific, in the tropical Atlantic and in the Mediterranean.

### Anti-satellite weapons

With the development of weapon systems for attacking satellites the dangers of warfare have now been extended into outer space. An ASAT system may be based on Earth or else carried by a satellite. The Earth-based ASAT weapons are of two types—missiles and high-energy laser weapons. Concepts of space-based ASAT systems involve the use of satellites as weapons, conventional explosives or lasers carried on board satellites and possibly charged-particle beam weapons. However, before considering these, it is important to realize that the space surveillance and spacecraft command and control systems mentioned above are the most sensitive components of military space technology. The ground receiving stations linked to satellites and the space surveillance network are also vulnerable to attack. This aspect is dealt with first.

*An EMP bomb?*

The vast amount of energy released from a nuclear weapon originates from the nuclei of the nuclear explosive material. The rate at which nuclear processes take place is so great that the entire energy is released in approximately one millionth of a second (or about one microsecond). This energy is in the form of thermal radiation, blast and shock waves and nuclear (initial and residual) radiations consisting of gamma rays and neutrons. The distribution of energy from a fission weapon exploded in the atmosphere at an altitude of about 30 km is shown in figure 4.11. These proportions can be varied with different designs of the weapon.

It is possible to reduce the blast and heat effect and enhance only the nuclear radiation produced from the weapon. The significance of this lies in the fact that the gamma rays (or prompt gamma rays) emitted from the nuclear explosion and those produced by neutron interactions with the remaining material of the weapon or the surrounding medium interact with air atoms and molecules to produce a sharp and relatively short but high intensity pulse of electromagnetic radiation called an electromagnetic pulse (EMP). The duration of an EMP is shown in figure 4.12; the time sequences and duration of gamma rays and neutrons are also indicated. The processes which result in the production of an EMP are complex. Essentially, on absorption of gamma rays, electrons are removed from the atoms of the medium surrounding the explosion producing charged atoms or ions. The electrons move away from the ions and the rapid motion of these charged particles creates intense electric and magnetic fields propagating outwards from the point of explosion at the velocity of light.

The characteristics of the EMP depend on the explosive yield and the height at which the explosion occurs. The EMP from a nuclear explosion 40 km above the Earth, for example, will affect an area as large as the

**Figure 4.11. Distribution of energy from a fission weapon exploded at an altitude of 30 km in the atmosphere**

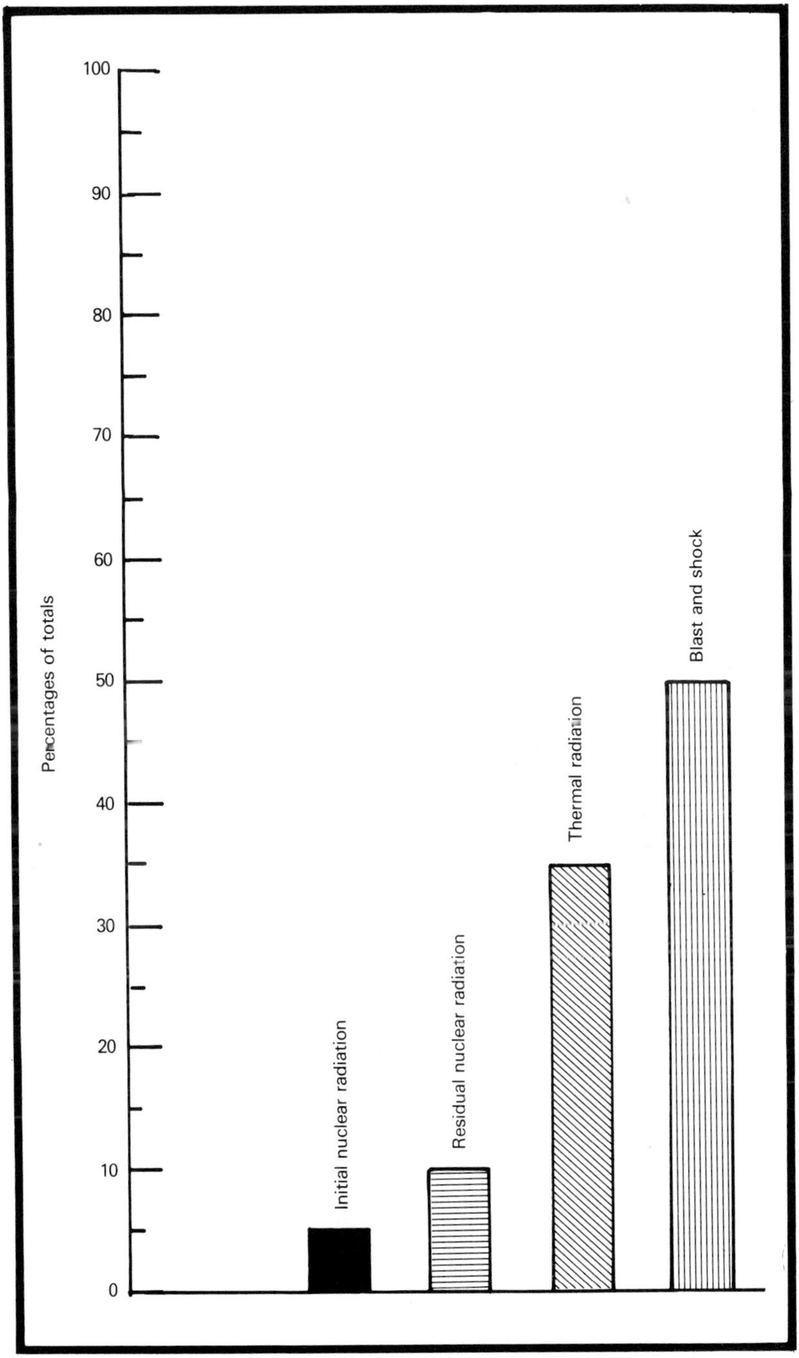

European continent while that from a surface burst is mainly localized. Moreover, an exo-atmospheric explosion may affect spacecraft as well, particularly because of the nuclear radiation.

*Nuclear radiation.* This consists mainly of gamma rays and neutrons. A pulse of high-energy gamma rays lasts for several thousand millionths of a second (see figure 4.12). Such radiation affects semi-conductor devices, which can be destroyed unless protective shielding is used. This can vary from a few centimetres to several tens of centimetres thick depending on the shielding material used. The neutron pulse which follows the gamma rays lasts for a few thousandths of a second (see figure 4.12). Neutron pulses can change permanently the electrical properties of semi-conductor devices. Moreover, it can generate radioactivity in materials around electric equipment which can last for a long period of time. The most effective shielding against such radiation is paraffin wax or water.

**Figure 4.12. Production of gamma ray and neutron pulses, thermal and blast effects and the EMP in time in a nuclear explosion**

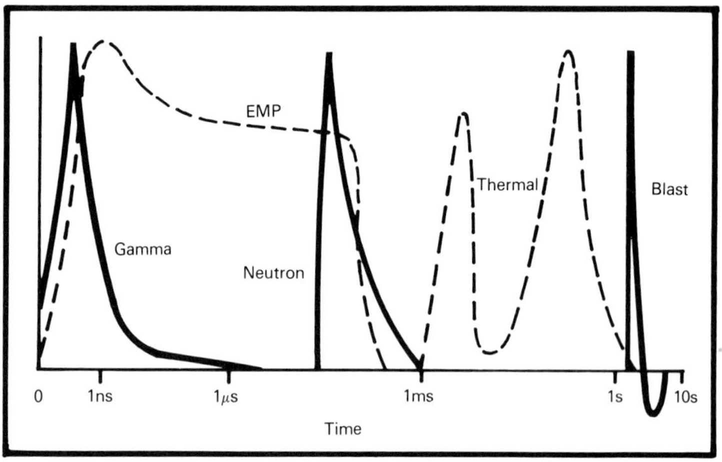


*Source:* Reference [47]

The energy in the form of an EMP, though only about 0.01 per cent of the total energy released from a nuclear explosion, is significant since the energy released from a nuclear weapon is very large. The field strength of an EMP, for example, could be as much as 50 000 volts per metre [46]. Such field strength can be picked up by antennae or cables leading to sensitive electronic equipment thus causing considerable damage to it. Considerable effort is being devoted to hardening both the ground and the space segment of the $C^3I$ [35, 47].

The implications of the use of nuclear weapons as EMP weapons are discussed in some detail in paper 10 of part II.

*Conventional ASAT weapons*
While the feasibility of the use of high-energy beams as ASAT weapons is still being investigated, conventional weapons have been tested and deployed. Basically two types of system exist. A missile carrying an explosive warhead can be aimed at a satellite from the ground or from space, or a satellite can itself act as a weapon by ramming against its prey. In the following sections the US and Soviet conventional ASAT systems are briefly reviewed.

*US Conventional ASAT Weapons.* Between 1964 and 1975 the USA had an operational anti-satellite system which was based on Johnston Island in the Pacific Ocean. The system, known as the direct ascent system, was tested in 1963 using a missile against a US booster which was orbiting the Earth at the time. The missile came within the calculated kill range of its simulated nuclear warhead. The US Air Force investigated, under the direct ascent anti-satellite programme, the possibility of such a system in which the kill range would be improved and a conventional warhead used. Four direct-ascent anti-satellite vehicles guided by infra-red homing devices were built. Two were launched but neither was successful. These and other programmes, for example, the US Army's Zeus and Nike X missiles and Project SAINT (Satellite Inspector Technique), were either cancelled or suspended.

However, the interest in anti-satellite weaponry has been revived. The current US programme contains three separate projects, each investigating different techniques. Under the US Air Force programme two types of anti-satellite systems are being investigated. In one, a direct-ascent system, a cylinder, containing a non-nuclear warhead, would be launched from a high speed F-15 aircraft flying at high altitude. The warhead would be separated close to the target and guided to it by an infra-red homing device. Although this programme dates back to the 1960s, further impetus was given in 1975. The system could become operational in the second half of the 1980s [48, 49].

Another approach being developed does not make use of any explosives. Instead, a small manoeuvrable vehicle carrying an infra-red sensor for guidance, would ram and destroy the target. A number of such vehicles would be carried into Earth orbit by a rocket which would dispense them near a target satellite. Initial testing of this system, called the Miniature Homing Intercept Vehicle (MHIV), has already begun on the ground [26, 50]. A full-size model of the missile for launching MHIV from an altitude of 20 km into an orbit has been tested on an F-15 aircraft [51]. Since this system is not confined to any specific launch pad on Earth, it will have a considerable advantage over the Soviet system. The system can be used as a direct-ascent system against satellites in any orbital inclination. This flexibility together with the small size of the MHIV increases its survivability.

The third US anti-satellite project is similar to that developed by the Soviet Union. Either an anti-satellite satellite would be put into the same orbit as the target and then manoeuvred towards it, or else it would be put into a different orbit but given a fast approach to the target. Three such satellites will be built initially. The anti-satellite satellites would be guided by radar to the target and then exploded to destroy it.

During the fiscal year 1981 alone $179.9 million was allocated to US conventional ASAT activities—this includes the development and operation of ASAT systems, space surveillance technology and a spacecraft-tracking system. Moreover, considerable sums of money are being spent on making satellites less vulnerable to detection and nuclear attack [52]. It is interesting to note the statement made by former head of R&D for the US Air Force and former astronaut, Thomas P. Stafford. Testifying before the House Space Subcommittee, he said, "arms control agreements may contribute to stability by limiting the numbers and kinds of ASATs, but will probably not result in abolition of ASATs" [53].

*Soviet conventional ASAT weapons.* The Soviet anti-satellite programme may have begun as early as 1963 but rigorous testing with targets and interceptors began in 1967. Until 1970, both the targets and the interceptors were launched from Tyuratam. However, since 1971 all the targets have been launched from Plesetsk and the interceptors from Tyuratam. It is interesting to note here that, at least as far as the Soviet photographic reconnaissance satellites are concerned, Plesetsk appears to be used as an operational launching site while the experimental or specialized flights take place from Tyuratam.

During tests the interceptions have been made basically in three ways. In a so-called fast flyby, the interceptor, usually launched at an orbital inclination of 62°, makes several orbital manoeuvres to achieve an eccentric orbit. The interceptor then passes close to the target spacecraft near the perigee of the former's orbit. Soon after the pass is made, the interceptor is exploded, presumably to test the kill mechanism for the destruction of the target. The second manner is when the interceptor makes a slower approach to the target. The interceptor and target are usually launched at an orbital inclination of 65° and in the same orbital plane. A third method has been tested in which the interceptor ascends close to the target. The interception is made before the interceptor completes a full orbit and it is commanded back to Earth after the interception has been made. The interceptor enters the Earth's atmosphere and disintegrates. For a detailed description of a more recent test, three satellites launched in 1981, see paper 2 of part II.

It has recently been reported that the Soviet Union has deployed the US MHIV type of infra-red homing interceptor on an anti-satellite battle station in low Earth orbit [54] but US official sources have denied this possibility [55].

*ASAT weapons based on high-energy beams*

Besides the above types of conventional weapon, more exotic weapons—the so-called directed-energy or high-energy beam weapons—are being investigated or developed not only for space warfare but also for ballistic missile defence (BMD) and other applications. These weapons, high-energy lasers, charged and neutral particle beams and electromagnetic pulse weapons, could have drastic effects on the stability of the strategic balance because they offer virtually no warning of attack (see paper 7 of part II).

*Laser.* A beam of laser is a ray of light which is coherent (the waves are in step with one another), usually monochromatic (with a single wavelength), highly focused and concentrated. It can deliver a vast amount of destructive or damaging energy on a target almost instantaneously since the destructive energy moves at the speed of light (300 000 km $s^{-1}$). This means that a laser weapon does not have to lead a target (see paper 7 of part II).

Essential elements of a laser weapon are the high-energy laser-producing system and the beam control system which aims the laser beam at the target and focuses the laser on it. Three types of lasers are thought to be suitable for weapon applications: a chemical laser using hydrogen fluoride as the lasing medium; a gas dynamic laser using carbon dioxide; and an electron-discharge laser which uses a beam of electrons passed through a gas lasing medium.

For industrial use 20 kW lasers are available, while the current plans for military use have lasers with power output of between 2.5 and 5 MW [56]. A 10 kW laser can cut through about 2.5 cm of steel plate in a fraction of a minute and a 2.5 MW laser would destroy a fast-moving aircraft and an air-to-air missile. The US Air Force has an airborne laser laboratory equipped with a 400 kW gas dynamic laser. In early February 1981, the US Air Force tested its aircraft-mounted laser weapon at full power on the ground [57]. Later, on 1 June, the airborne laser weapon was tested against an air-launched AIM-9L Sidewinder air-to-air missile. While the beam hit the target, it did not destroy it [58]. Two days later a second test was carried out against an AIM-9L missile. It was reported that this second test had achieved better results because the beam was able to lock onto the target for a long period [59]. The US Navy, under a programme called the Sea Lite, is investigating a more powerful chemical laser. DARPA is also investigating the use of a ground-based chemical laser device (ALPHA) to demonstrate the feasibility of a laser suitable for deployment in outer space [20].

Under the Talon Gold programme, the USA is investigating the target acquisition, tracking and precision pointing techniques. The Talon Gold system includes a laser radar and improved inertial reference platforms [20]. It is planned to orbit the pointing and tracking equipment for satellite high-energy defence applications using the space shuttle. Coupled with this is the Large Optics Demonstration Experiment (LODE). The purpose of the LODE programme is to establish the feasibility of large aperture beam control for space-based lasers. Under the Large Optics Technology programme the feasibility of ultra-light-weight glass mirrors has been shown [20].

*Particle beam.* The second type of potential beam weapon consists of a stream of highly accelerated atomic or subatomic particles such as electrons, protons, neutrons or heavy ions. These are accelerated to close to the speed of light using particle accelerators, more colourfully known as atom smashers, and destroy or damage a target either by penetrating the target or through secondary radiation produced by their interactions with the target material or the surrounding medium. Particle beams could detonate high explosives in nuclear warheads. Since charged particles would be deflected by the Earth's magnetic field, a neutral particle beam has been envisaged for ASAT applications over long distances. Such a beam is produced by first accelerating negatively charged hydrogen atoms and then neutralizing them by passing the beam through a charge-exchange device which strips off the extra electron on the hydrogen atom. In the USA, methods of neutralizing accelerated charged particles are being investigated under the White Horse project at Los Alamos Laboratory [60].

The first generation of such beam weapons may be used as tactical weapons over a range of several kilometres. In the USA, under the DARPA Particle-Beam Technology Program, an Advanced Test Accelerator (ATA) facility is being constructed (see also paper 7 of part II). When it is completed at the end of 1982, studies on propagation properties of an electron beam in the atmosphere will begin [20]. It has been suggested that some of the first potential applications of particle-beam weapons may be for the defence of large ships against cruise missiles and for terminal ballistic missile defence at the missile silos and national command authorities [61]. However, another application which has not been discussed much is for defence against tanks in battlefields. When a beam of electrons is moving through the atmosphere, much of its energy lost is converted into high-energy gamma rays. This may damage or disrupt electronic equipment on weapon systems as well as kill humans over a range of about a kilometre. It may even replace the enhanced radiation weapon (the neutron bomb). The narrow beam of electrons is converted into a much wider beam of gamma rays, a few metres in diameter. Thus the beam does not have to score a pin-point hit on the target.

*EMP.* EMP weapons discussed earlier are based on nuclear weapons and therefore have no directional property. However, if a burst of radio-frequency or microwave radiation were to be produced using a very high peak-power microwave generator, then such an energy could be directed by an antenna system [62]. This type of radiation could disrupt or even destroy, for example, electronic components of communications networks and computers. Relatively little is known at present about this technology.

The above brief review indicates the kind of effort being devoted to the investigations of beam weapons. This discussion is mainly confined to US activities simply because comparatively more data is available. Scanty reports on Soviet efforts in the field, however, suggest that similar efforts are being put into the investigation of such weapons.

A number of problems need to be solved before such systems can be used in practicable weapons. For example, considerable effort is being devoted to solving the problem of producing high enough energies in a compact weapon system and large resources have been devoted to studying the nature of beam propagation. In the case of lasers, beam propagation is reasonably well understood [63]. The problem of tracking and aiming at a fast-moving target have to some extent been solved but much work still needs to be done.

These are problems of a technological nature; no clear evidence has been found which indicates that beam weapons are clearly impossible. Since a sound scientific basis for such systems exists, there is every chance that high-energy beam weapons may be developed and even deployed. Moreover, the impetus for the development of such weapons will be derived from the fact that the technology is very similar to that needed and in some cases even used for other applications. For example, it has been proposed that high-energy beam weapons should be developed for BMD application. Furthermore, high-energy beams are used for measuring the physical properties of materials at high temperatures and pressures, providing useful information for modelling some aspects of nuclear weapon design and performance. Very high pressures, densities and temperatures are produced in nuclear explosions and in inertial confinement fusion (the basis for a hydrogen bomb). To study the phenomenon connected with such conditions, it is necessary to know the equations of state—that is, the relationship between pressures, density and internal energy of materials. Such knowledge facilitates understanding the behaviour of, for example, fusion pellets in a thermonuclear weapon. Until now such studies have been carried out in underground nuclear explosions, but these have obvious limitations. Moreover, with the use of high-energy lasers and particle beams, it would be possible to simulate some of the effects of nuclear weapons to study the vulnerability of, for example, missile warheads and artificial Earth satellites [64]. The use of high-energy lasers for communications with submarines has already been discussed above. Finally, high-energy lasers have been developed for enriching uranium-239 in naturally occurring uranium.

For these reasons, more efficient and higher-energy lasers and particle beams than are available today will no doubt be developed. The issues connected with high-energy beam technology are complex. In recent years much has been written on the subject discussing arguments for and against the use of the technology for weapon purposes [65–88]. Many have argued that these weapons are not cost-effective, that there are other weapons which can do the same damage more cheaply and that there are countermeasures to high-energy beam weapons. High-energy beam technology has, as pointed out above, several military applications and, of course, there are a number of civilian applications as well. For many of these applications the technology has already been developed, and even used, so that any assessment of cost-effectiveness must take these into account. The history of weapon systems indicates that man has never been contented; military technology has always striven to develop better weapons. High-energy beam weapons clearly present many advantages over most of the weapons available today.

Not nearly enough data are available to support the view that high-energy beam weapons are useless. Nor is it possible to predict what solutions future technology will come up with. But it is clear and even urgent that a better objective assessment should be made of this new technology before a new round of the arms race of a totally different quality starts.

## Satellite system survivability

In any major war involving the two great powers, there is little doubt that artificial Earth satellites and related ground systems will play a vital role and will therefore themselves be targets. As such, the ground-based satellite systems are probably more vulnerable than the satellites since the former tend to be large and stationary. These systems—command, control and communications sytems, data receiving stations and satellite-tracking and surveillance stations—can be destroyed relatively easily by more conventional techniques, while the destruction of satellites requires sophisticated methods.

In order to overcome this problem in the USA (and no doubt in the USSR also), there is a considerable effort to develop methods of making satellites more autonomous so that they function, as much as possible, independent of the ground systems. Efforts to make satellites less vulnerable include hardening satellites, particularly from the point of view of an attack from high-energy lasers and EMP bombs. Methods are being developed to warn satellites of the approach of an unfriendly spacecraft as well as to make the communications systems free of interference. Another concept involves the design of satellites which are not easily observable. These are the so-called 'dark satellites' orbited at altitudes beyond the range of surveillance sensors. A dark satellite would remain inactive in the depths of space until a signal were sent to it to move into an operational orbit and begin its required mission. Yet another method being investigated is to use large numbers of satellites as decoys. Above all, satellites of the future will be designed with considerable manoeuvrability so as to evade an attacking interceptor [34].

## References


1. 'Military race in space', *The Defense Monitor*, Vol. 9, No. 9, 1980.
2. 'Target list', *Department of Defense Authorization for Appropriations for Fiscal Year 1980*, Part 6, Hearings before the Committee on Armed Services, United States Senate, March–April 1979 (US Government Printing Office Washington, D.C., 1979), pp. 3037–3038.
3. 'Space reconnaissance dwindles', *Aviation Week & Space Technology*, Vol. 113, No. 14, 6 October 1980, p. 18.
4. 'CIA spied on anti-war demonstrations', *Center for National Security Studies News*, 16 July 1979.

5. 'China's Earth satellite returns', *Peking Review*, No. 51, 17 December 1976, p. 9.
6. *Le Monde*, 18 January 1973.
7. 'French debate dropping triad concept', *Aviation Week & Space Technology*, Vol. 109, No. 10, 4 September 1978, p. 193.
8. *World Armaments and Disarmament, SIPRI Yearbook 1980* (Taylor & Francis Ltd., London, 1980, Stockholm International Peace Research Institute), p. 194.
9. 'Military satellite', *Interavia Airletter*, No. 9600, 6 October 1980, p. 6.
10. 'Etudes de définition du satellite française de reconnaissance SAMRO', *Air et Cosmos*, No. 845, 31 January 1981, p. 55.
11. 'Décision attendue en 1982 sur le satellite-SAMRO', *Air et Cosmos*, No. 866, 27 June 1981, p. 65.
12. 'Japan plans launch of earth resources satellite in 85' *Defense Daily*, Vol. 11, No. 13, 21 July 1980, p. 93.
13. 'Satellite militaire Japonais', *Air et Cosmos*, No. 889, 9 January 1982, p. 31.
14. Melton, W.C., 'Time of arrival measurements possible with the use of Global Positioning System', *Defense Electronics*, Vol. 13, No. 12, December 1981, pp. 90–98.
15. Jasani, B., 'Ocean surveillance by earth satellites', in *Ocean Yearbook 2*, eds. E.M. Borgese and N. Ginburg (The University of Chicago Press, Chicago, Illinois, 1980), pp. 250–269.
16. Augustine, F.J. and Lujetic, V.J., 'The Navy Satellite Oceanographic Research Program (SOREP) Plan', *Ramcor Report*, No. U-RC149 A-002, September 1978.
17. Chelton, D.B., Hussey, K.J. and Parke, M.E., 'Global satellite measurements of water vapour, wind speed and wave height', *Nature*, Vol. 294, No. 5841, 10 December 1981, pp. 529–32.
18. Fossum, R.R., *Department of Defense Authorization for Appropriation for Fiscal Year 1981*, Part 4, Statement to the 96th Congress, Second Session, Hearings on Military Posture, Committee on Armed Services, US House of Representatives, February–March 1980 (US Government Printing Office, Washington, D.C., 1980), pp. 578–606).
19. 'CCD sensor sensitivity increased 100 fold', *Defense Electronics*, Vol. 13, No. 11, November 1981, p. 86.
20. Fossum, R.R., *Research and Development, Title II*, Part 4, Statement to the 97th Congress, First Session, Hearings on Military Posture, Committee on Armed Services, US House of Representatives, March–April 1981 (US Government Printing Office, Washington, D.C., 1981), pp. 82–170.
21. Burke, K.H., *FY 1982 Budget Estimates, Research, Development, Test & Evaluation, United States Air Force*, Part 4, Statement to the 97th Congress, First Session, Hearings on Military Posture, Committee on Armed Services, US House of Representatives, March–April 1981 (US Government Printing Office, Washington, D.C., 1981), pp. 385–415.
22. UN document A/34/674, Annex II, Appendix, 12 November 1979.
23. Barasch, G.E., *Light Flash Produced by an Atmospheric Nuclear Explosion*, Report LASL-79-84, (Los Alamos Scientific Laboratory, Los Alamos, Calif., November 1979.)
24. 'Early-warning satellite picked up flash over S. Atlantic', *Defense Daily*, Vol. 114, No. 34, 23 February 1981, p. 271.

25. Baker, D., 'Defense and space equation', *Jane's Defense Review*, No. 1, 1980, pp. 73–77.
26. Perry, W.J., Under Secretary of Defense for Research and Engineering, *NASA Authorization for Fiscal Year 1980*, Part 3, Statement to the 96th Congress, First Session, Hearings before the Committee on Commerce, Science and Transportation, US Department of Defense Space Activities, US Senate, March, May and June 1979 (US Government Printing Office, Washington, D.C., 1979), pp. 1715–57.
27. Martin, J.J., *Department of Defense Authorization for Appropriations for Fiscal Year 1980*, Part 5, Statement to the 96th Congress, First Session, Hearings before the Committee on Armed Services, US Senate, April 1979 (US Government Printing Office, Washington, D.C., 1979), p. 2647.
28. 'Nuclear hardened, autonomous satellites readied for launch', *Defense Electronics*, Vol. 13, No. 9, September 1981, pp. 78–85.
29. Wood, L., *Department of Defense Authorization for Appropriations for Fiscal Year 1980*, Part 6, Statement to the 96th Congress, First Session, Hearings before the Committee on Armed Services, US Senate, March–April 1979 (US Government Printing Office, Washington, D.C., 1979), pp. 3326–88.
30. 'Ground-based laser has the edge for sub communications', *Laser Focus*, Vol. 18, No. 1, January 1982, pp. 38–40.
31. 'Blue-green laser link to subs', *Laser Focus*, Vol. 16, No. 4, April 1980.
32. Wiener, T.F., 'Strategic laser communications', *Signal*, Vol. 35, No. 1, September 1980, pp. 43–48.
33. 'Navstar still too accurate', *Defense Electronics*, Vol. 12, No. 8, August 1980, p. 20.
34. Perry, W.J., Under Secretary of Defense for Research and Engineering, *Department of Defense Authorization for Fiscal Year 1981*, Part 4, Statement to the 96th Congress, Second Session, Hearing on Military Posture, Committee on Armed Services, US House of Representatives, February–March 1980 (US Government Printing Office, Washington, D.C., 1980), Research and Development, pp. 7–301.
35. *Strategic Command, Control and Communications: Alternative Approaches for Modernization, A CBO Study* (Congressional Budget Office, Washington, D.C., October 1981).
36. 'Larger meteorological satellite with greater sensor capability will be in orbit by 1984', *Defense Electronics*, Vol. 11, No. 11, November 1979, pp. 29–30.
37. Doubleday, V.C., 'Air Force weather programs', *Department of Defense Appropriations for 1977*, Part 5, Hearings Before the Subcommittee on Appropriations, House of Representatives, 94th Congress, Second Session (US Government Printing Office, Washington, D.C., 1976), pp. 1441–44.
38. Nicholson, W.L., III, *Statement before the Committee on Armed Services*, Part 4, Department of Defense Authorization for Appropriations for Fiscal Years 1982 and 1981, 97th Congress, First Session, Research and Development, March –April 1981 (US Government Printing Office, Washington, D.C., 1981), pp. 1174–1176.
39 Peable, C., 'The manned orbiting laboratory', *Spaceflight*, Vol. 22, No. 4, Part I, 4 April 1980, pp. 155–160.
40. 'USAF needs space command for shuttle', *Flight International*, Vol. 120, No. 3787, 5 December 1981, p. 1674.

41. 'Second shuttle mission was a success', *Interavia Air Letter*, No. 9896, 10 December 1981, pp. 6–7.
42. Taranik, J.V. and Settle, M., 'Space shuttle: a new era in terrestrial remote sensing', *Science*, Vol. 214, No. 4521, 6 November 1981, pp. 619–626.
43. Bell, T.E., 'America's other space program', *The Sciences*, December 1979.
44. 'NASA sets manifest for 72 space shuttle flights', *Defense Daily*, Vol. 119, No. 32, 18 December 1981, pp. 251–53.
45. *Department of Defense Appropriations for Fiscal Year 1980, Hearings, Committee on Armed Services*, Part 6, US Senate, March–April 1979 (US Government Printing Office, Washington, D.C., 1979), p. 3022.
46. Broad, W.J., 'Nuclear pulse (I): awakening to the chaos factor', *Science*, Vol. 212, No. 4498, 29 May 1981, pp. 1009–1012.
47. Lukeman, T.J. and Williamson, R., 'Nuclear hardening of equipment', *International Defense Review*, Vol. 14, No. 11, 1981, pp. 1444–1448.
48. Klass, P.J., 'Ballistic missile defense tests set', *Aviation Week & Space Technology*, Vol. 112, No. 24, 16 June 1980, pp. 13–18.
49. 'Antiballistic missile flight test nears', *Defense Electronics*, Vol. 12, No. 12, December 1980, pp. 22–24.
50. Smith, B.A., 'Vought test small antisatellite system', *Aviation Week & Space Technology*, Vol. 115, No. 19, 9 November 1981, pp. 24–25.
51. 'Launch pylon for antisatellite system tested', *Aviation Week & Space Technology*, Vol. 116, No. 3, 18 January 1982, p. 19.
52. 'Reagan's $180 billion strategic package', *Defense Daily*, Vol. 118, No. 19, October 1981, p. 146.
53. 'Stafford proposes Air Force takeover of shuttle', *Defense Daily*, Vol. 118, No. 12, 24 September 1981, p. 92.
54. 'Killer satellites', *Aviation Week & Space Technology*, Vol. 115, No. 17, 26 October 1981, p. 15.
55. 'Pentagon denies Soviet orbital battle station', *Flight International*, Vol. 120, No. 3783, 7 November 1981, p. 1363.
56. 'Lasers light up the battlefield', *High Technology*, Vol. 1, No. 2, November–December 1981, pp. 76–81.
57. 'USAF tests high-energy laser weapon', *Flight International*, Vol. 119, No. 3744, 7 February 1981, p. 334.
58. 'Laser fails to destroy missile', *Aviation Week & Space Technology*, Vol. 114, No. 23, 8 June 1981, p. 63.
59. 'Second laser laboratory test', *Interavia Air Letter*, No. 9778, 26 June 1981, p. 8.
60. Lemer, E.J., 'Killer particle beams', *IEEE Spectrum*, September 1981, pp. 44–47.
61. 'Generating intense electron beams for military applications', *Energy and Technology Review*, Lawrence Livermore National Laboratory, Report UCRL-52000-81-12, December 1981.
62. 'Soviets study potential of electromagnetic pulse weapon', *Aerospace Daily*, Vol. 100, 28 November 1979, p. 129.
63. 'DARPA Chief assesses laser weapon program', *Laser Focus*, Vol. 17, No. 12, December 1981, pp. 34–38.
64. 'Inertial confinement fusion', *Fiscal Year 1981 Arms Control Impact Statements*, May 1980 (US Government Printing Office, Washington, D.C., 1980), pp. 460–75.

65. Klass, P.J., 'Research nears applications level', *Aviation Week & Space Technology*, Vol. 97, No. 7, 14 August 1972, pp. 12–15.
66. Klass, P.J., 'Power boost key to feasibility', *Aviation Week & Space Technology*, Vol. 97, No. 8, 21 August 1972, pp. 32–40.
67. Klass, P.J., 'Advanced weaponry research intensifies', *Aviation Week & Space Technology*, Vol. 103, No. 7, 18 August 1975, pp. 34–39.
68. Klass, P.J., 'Pentagon seeks to channel research', *Aviation Week & Space Technology*, Vol. 103, No. 9, 1 September 1975, pp. 50–56.
69. Klass, P.J., 'Current systems still more cost effective', *Aviation Week & Space Technology*, Vol. 103, No. 10, 8 September 1975, pp. 53–59.
70. Klass, P.J., 'Progress made in high energy lasers', *Aviation Week & Space Technology*, Vol. 106, No. 10, 7 March 1977, pp. 16–17.
71. Robinson, C.A., Jr., 'Soviets push for beam weapons', *Aviation Week & Space Technology*, Vol. 106, No. 18, 2 May 1977, pp. 16–23.
72. Robinson, C.A., Jr., 'US pushes development of beam weapons', *Aviation Week & Space Technology*, Vol. 109, No. 41, 2 October 1978, pp. 14–22.
73. Robinson, C.A., Jr., 'Key beam weapons tests slated', *Aviation Week & Space Technology*, Vol. 109, No. 15, 9 October 1978, pp. 42–53.
74. Robinson, C.A., Jr., 'Army pushes new weapons effort', *Aviation Week & Space Technology*, Vol. 109, No. 16, 16 October 1978, pp. 42–52.
75. Robinson, C.A., Jr., 'Air Force emphasizes laser weapons', *Aviation Week & Space Technology*, Vol. 109, No. 18, 30 October 1978, pp. 51–55.
76. Robinson, C.A., Jr., 'Power generation key to beam weapons development', *Aviation Week & Space Technology*, Vol. 109, No. 19, 6 November 1978, pp. 50–57.
77. Robinson, C.A., Jr., 'Beam pointing technology explored', *Aviation Week & Space Technology*, Vol. 109, No. 19, 6 November 1978, pp. 59–71.
78. Robinson, C.A., Jr., 'Soviet test beam technology', *Aviation Week & Space Technology*, Vol. 109, No. 20, 13 November 1978, pp. 14–20.
79. Robinson, C.A., Jr., 'Beam weapons effort to grow', *Aviation Week & Space Technology*, Vol. 110, No. 14, 2 April 1979, pp. 12–16.
80. Robinson, C.A., Jr., 'Technical survey, particle beams, laser weapons–1', *Aviation Week & Space Technology*, Vol. 113, No. 4, 28 July 1980, pp. 32–66.
81. Robinson, C.A., Jr., 'Technical survey, particle beams, laser weapons–2', *Aviation Week & Space Technology*, Vol. 113, No. 5, 4 August 1980, pp. 46–68.
82. Robinson, C.A., Jr., 'Laser technology demonstration proposed', *Aviation Week & Space Technology*, 16 February 1981, pp. 16–19.
83. Robinson, C.A., Jr., 'Technical survey', *Aviation Week & Space Technology*, Vol. 114, No. 21, 25 May 1981, pp. 40–71.
84. Tanimoto, D.H., 'Opportunities for space applications of directed-energy technologies', *Signal*, Vol. 35, No. 6, February 1981, pp. 37–42.
85. Wright, W.E., 'Charged particle beam weapons: should we? could we?', *US Naval Institute Proceedings*, Vol. 105/11/921, November 1979, pp. 28–35.
86. Beane, W.J., 'The Navy and directed energy weapons', *US Naval Institute Proceedings*, Vol. 107/11/945, November 1981, pp. 48–52.
87. Bekefi, G., Feld, B.T., Permentol, J. and Tsipis, K., 'Particle beam weapons', *Program in Science and Technology for International Security, Report No. 4*, (Department of Physics, Massachusetts Institute of Technology, Cambridge, MA, December 1978).

88. Cullaham, M. and Tsipis, K., 'High energy laser weapons—a technical assessment', *Program in Science and Technology for International Security*, *Report No. 6*, (Department of Physics, Massachusetts Institute of Technology, Cambridge, MA, November 1980).

# Chapter 5. Contribution of space technology to the arms race

> 'Tis a principle of war that when you can use the lightning 'tis better than cannon.
>
> Napoleon I

## *I. Nuclear war doctrines*

As advances are made in military technology, improvements in weapons occur. This in turn may refine war-fighting tactics as well as give rise to new ones. Over the past two decades or so nuclear war-fighting tactics have undergone a change from the doctrine of mutual assured destruction (MAD) to those of limited strategic options, 'flexible response' and counterforce, to mention a few.

According to the MAD concept, tremendous destructive power would be unleashed against enemy cities and industrial centres in the event of being attacked; this was a policy of deterrence, the assumption being that no side would attack first for fear of its own cities being destroyed. But with the sophistication of war-fighting policies and the interlinked technological developments this is no longer a safe assumption. For these policies and developments seek to make it possible to actually fight and win a nuclear war, thus making such a war more likely.

The counterforce strategy, for example, emphasizes the necessity to strike an enemy's military targets accurately, minimizing damage to civilian population and property; among such a policy's requirements are high weapon accuracies and precise knowledge of targets. The flexible response strategy asserts that aggression should be appropriately matched, facilitating smooth adjustment of response when situations change, and requiring a wide range of options in response to potential action by the opposite side. Among requirements for this are counterforce ability and the availability of secure communication, command, control and intelligence ($C^3I$) systems. Such developments could lead, eventually, to a side imagining that it can actually win a nuclear war by destroying, in a first strike, the enemy's ability to counteract, or at least, it could lead to a situation in which a side might attempt a first strike merely to prevent its own strategic arsenal being destroyed in one massive strike by the other side's devastatingly accurate weapons. Another danger—one likely to emerge at an even earlier stage—is that counterforce and flexible response policies, in particular, could lead to a perception that the effects of nuclear war can be confined to a small area and to military targets, that is, a limited nuclear war could be fought in which civilians would be largely unaffected, and in which one side could emerge as winner.

Recently a more unified concept seems to be emerging, at least in the USA. This concept, a countervailing strategy, requires the maintenance of MAD capability as well as the ability "for flexible, controlled use of strategic weapons against appropriate targets for any attack at any level of conflict" [1]. Moreover, the countervailing strategy seems to provide for a whole range of military options from fighting a conventional war to a general nuclear war.

A number of technological developments either contribute or will contribute in the near future towards fulfilment of the aims of the countervailing strategy. The two main areas which have direct relevance are new developments in nuclear weapons and the advances made in military space technology. The latter includes improvements in space-based sensors for surveillance; in communications, command and control systems; and in the space-based navigation technology which enhances the accuracies of the delivery systems for both conventional and nuclear weapons. Herein lies an important link between the arms race and the militarization of outer space, the other link being the fact that in nuclear weapon states, spacecraft are put into orbit using launchers based on missiles developed to carry nuclear warheads (see table 5.1). It is interesting to note that there is a new trend in which some of the countries outside the nuclear weapon states have developed launchers primarily for orbiting satellites.

Another development in space technology worth mentioning in connection with war doctrines is that of high-energy beam weapons. Not only would large-scale deployment present the threat of space warfare, but it is difficult to imagine current strategies surviving such a deployment. The reason for this is the dramatic change it would bring in the time-scale of warfare (the speed of the destructive energy delivered to the target is about 300 000 km $s^{-1}$), and there is the threat, seen by some experts, that high-energy beam weapons could render nuclear weapons obsolete (see papers 7 and 10 of part II).

## *II. Contribution of space technology to the countervailing strategy*

The term 'countervailing strategy' surfaced only recently. However, as far back as 1975, the then Secretary of Defense James Schlesinger's statement

**Table 5.1. Some milestones in the development of strategic weapon delivery vehicles and various space systems**

| Type of vehicle | Date of deployment: USA | Date of deployment: USSR |
|---|---|---|
| Intercontinental strategic bomber | 1948 | 1954–55 |
| Carrier-based aircraft with nuclear weapons | 1950 | – |
| Forward land-based aircraft with nuclear weapons | early 1950s | – |
| Submarine cruise missile | 1954 (not operational) | 1962 |
| Low penetration bomber | 1959 | – |
| ICBM | 1959 | 1957 |
| Reconnaissance satellite | 1959 | 1962 |
| Forward land-based missiles | 1959–60 (not operational) | – |
| Nuclear-powered strategic submarine | 1960 | 1968 |
| Supersonic bomber | 1960 | – |
| Heavy ICBM | 1960 | 1960 |
| Navigation satellite | 1960 | 1970 |
| Forward bases for strategic submarines | 1961 | – |
| Solid-propellant ICBMs | 1962 | 1968 |
| MRV | 1964 | 1973 |
| Communications satellite in synchronous orbit | 1966 | 1974 |
| Early warning satellite in synchronous orbit | 1966 | 1975 |
| ICBM MIRV | 1970 | – |
| SLBM MIRV | 1971 | – |

that with "a reserve capability for threatening urban-industrial targets, with offensive systems capable of increased flexibility and discrimination in targeting, and with concomitant improvements in sensors, surveillance, and command-control, we could implement response options that cause far less civilian damage than would now be the case" [2] contained most of the ingredients of the countervailing strategy. Another element of the strategy is that the weapons and weapon systems should survive enemy attacks in order to be able to respond to the attacks. To this end considerable efforts are being made to develop new methods of ballistic missile defence (BMD) based in space. The above statement also indicates some of the requirements to support the strategy. For example, increased flexibility and discrimination in targeting could be obtained from accurate knowledge of the targets and their locations as well as through all target information being instantly available at headquarters. In a flexible response, which would not involve weapons with massive explosive yields, the damage to the target needs to be rapidly assessed. Moreover, a greater co-ordination of battle operations is necessary. In all of these tasks communications and reconnaissance satellites play a vital role. The importance of navigation satellites in a flexible response is that they increase the accuracy of missiles, minimizing collateral damage. Other satellites also play a role in improving accuracy.

## Table 5.2. Summary of possible military satellites by type of mission

| | Photographic reconnaissance satellites | | | Electronic reconnaissance satellites | | US MIDAS and Vela satellites | | Early-warning satellites | | Ocean-surveillance satellites | | Navigation satellites | |
|---|---|---|---|---|---|---|---|---|---|---|---|---|---|
| Year | USA | USSR | China | USA | USSR | MIDAS | Vela | USA | USSR | USA | USSR | USA | USSR |
| 1958 | – | – | – | – | – | – | – | – | – | – | – | – | – |
| 1959 | 6 | – | – | – | – | – | – | – | – | – | – | 1 | – |
| 1960 | 6 | – | – | – | – | 2 | – | – | – | – | – | 2 | – |
| 1961 | 13 | – | – | – | – | 3 | – | – | – | – | – | 3 | – |
| 1962 | 26 | 5 | – | 4 | – | 1 | – | – | – | – | – | 1 | – |
| 1963 | 17 | 7 | – | 7 | – | 2 | 2 | – | – | – | – | 3 | – |
| 1964 | 24 | 12 | – | 8 | – | – | 2 | – | – | – | – | 3 | – |
| 1965 | 21 | 17 | – | 5 | – | – | 2 | – | – | – | – | 4 | – |
| 1966 | 23 | 21 | – | 10 | – | 2 | – | 1 | – | – | – | 4 | – |
| 1967 | 18 | 22 | – | 8 | 5 | – | 2 | – | 1 | – | 1 | 3 | – |
| 1968 | 16 | 29 | – | 7 | 7 | – | – | 1 | 1 | – | 1 | 1 | – |
| 1969 | 12 | 32 | – | 6 | 11 | – | 2 | 1 | – | – | – | – | – |
| 1970 | 9 | 29 | – | 7 | 10 | – | 2 | 3 | – | – | 1 | 1 | 1 |
| 1971 | 7 | 28 | – | 3 | 15 | – | – | 1 | – | 4 | 2 | – | 2 |
| 1972 | 8 | 30 | – | 3 | 7 | – | – | 2 | 1 | – | 1 | 1 | 3 |
| 1973 | 5 | 35 | – | 2 | 12 | – | – | 2 | 1 | – | 1 | 1 | 3 |
| 1974 | 5 | 28 | – | 3 | 10 | – | – | – | 1 | – | 2 | 1 | 4 |
| 1975 | 4 | 34 | 1 | 2 | 8 | – | – | 2 | 2 | 1 | 3 | 1 | 4 |
| 1976 | 4 | 34 | 1 | 1 | 11 | – | – | 1 | 1 | 4 | 2 | 1 | 8 |
| 1977 | 3 | 33 | – | – | 8 | – | – | 2 | 3 | 4 | 3 | 1 | 8 |
| 1978 | 2 | 35 | 1 | 1 | 6 | – | – | 2 | 2 | 1 | – | 4 | 8 |
| 1979 | 2 | 35 | – | 1 | 5 | – | – | 2 | 2 | – | 3 | – | 6 |
| 1980 | 2 | 35 | – | 1 | 6 | – | – | – | 5 | 4 | 4 | 2 | 6 |
| 1981 | 2 | 37 | – | – | 4 | – | – | 2 | 5 | – | 8 | 1 | 5 |
| **Total by country** | **235** | **538** | **3** | **79** | **125** | **10** | **12** | **22** | **25** | **18** | **32** | **39** | **58** |
| **Total by mission** | | **776** | | **204** | | **22** | | **47** | | **50** | | **97** | |

[a] Fractional orbital bombardment system.

| Communications satellites | | | | | Meteorological satellites | | | | Geodetic satellites | | | FOBSs[a] | Interceptor/ destructor satellites | Yearly total | Cumulative total |
|---|---|---|---|---|---|---|---|---|---|---|---|---|---|---|---|
| USA | USSR | NATO | UK | France | USA | USSR | France | UK | USA | USSR | France | USSR | USSR | | |
| 1 | – | – | – | – | – | – | – | – | – | – | – | – | – | 1 | 1 |
| – | – | – | – | – | – | – | – | – | – | – | – | – | – | 7 | 8 |
| 2 | – | – | – | – | 2 | – | – | – | – | – | – | – | – | 14 | 22 |
| 2 | – | – | – | – | 1 | – | – | – | – | – | – | – | – | 22 | 44 |
| 3 | – | – | – | – | 4 | – | – | – | 1 | – | – | – | – | 45 | 89 |
| 4 | – | – | – | – | 3 | 2 | – | – | – | – | – | – | – | 47 | 136 |
| 3 | 3 | – | – | – | 3 | 2 | – | – | 2 | – | – | – | – | 62 | 198 |
| 7 | 8 | – | – | – | 6 | 4 | – | – | 6 | – | – | – | – | 80 | 278 |
| 11 | 2 | – | – | – | 6 | 2 | – | – | 4 | – | 1 | 2 | – | 89 | 367 |
| 17 | 5 | – | – | – | 6 | 4 | – | – | 1 | – | 2 | 9 | 1 | 105 | 472 |
| 11 | 4 | – | – | – | 4 | 2 | – | – | 1 | 2 | – | 2 | 4 | 93 | 565 |
| 5 | 2 | – | 1 | – | 3 | 2 | – | – | 1 | 2 | – | 1 | 2 | 83 | 648 |
| 3 | 14 | 1 | 1 | – | 5 | 6 | – | – | 1 | – | 1 | 2 | 3 | 100 | 748 |
| 5 | 21 | 1 | – | – | 2 | 4 | 1 | 1 | – | 2 | – | 1 | 6 | 106 | 854 |
| 3 | 24 | – | – | – | 4 | 5 | – | – | – | 2 | – | – | – | 94 | 948 |
| 4 | 33 | – | – | – | 2 | 3 | – | – | – | 1 | – | – | – | 105 | 1 053 |
| 3 | 24 | – | 2 | 1 | 4 | 6 | – | – | – | 2 | – | – | – | 96 | 1 149 |
| 5 | 37 | – | – | 1 | 3 | 5 | – | – | 1 | 2 | 1 | – | – | 117 | 1 266 |
| 11 | 29 | 1 | – | – | 3 | 3 | – | – | 1 | 1 | – | – | 1 | 118 | 1 384 |
| 4 | 16 | 1 | – | – | 2 | 3 | – | – | – | 1 | – | – | 7 | 99 | 1 483 |
| 6 | 42 | 1 | – | – | 4 | – | – | – | – | 1 | – | – | 1 | 117 | 1 600 |
| 3 | 27 | – | – | – | 2 | 4 | – | – | – | – | – | – | 2 | 94 | 1 694 |
| 3 | 36 | – | – | – | 2 | 2 | – | – | – | – | – | – | 3 | 111 | 1 805 |
| 2 | 39 | – | – | – | 2 | 2 | – | – | – | – | – | – | 3 | 112 | 1 917 |
| **118** | **366** | **5** | **4** | **2** | **73** | **61** | **1** | **1** | **19** | **16** | **5** | **17** | **33** | **1 917** | |
| | | **495** | | | | | **134** | | | **40** | | **17** | **33** | **1 917** | |

## Accurate delivery of weapons to their targets

An intercontinental ballistic missile (ICBM) or other vehicle carrying its lethal payload is subjected to a number of forces which affect the final accuracy with which it reaches the target. The effects of these can be minimized using, for example, the considerable amount of information collected on atmospheric conditions from space by meteorological satellites.

Other information that can enhance the accuracy with which a nuclear warhead is delivered to its target concerns the Earth's gravitational field and some geodetic parameters, such as the shape of the Earth. Knowledge of these parameters can be increased using geodetic satellites.

Finally, one of the most important items of information needed to accurately deliver a weapon to its target is the recognition and location of the target. This task is performed using photographic and electronic reconnaissance satellites for the land-based targets and ocean surveillance satellites for naval targets. The latter, together with oceanographic satellites, are used to improve methods for the detection of submarines.

The extensive use of various types of satellite is indicated in table 5.2. It can be seen that a quarter of all military satellites are communications satellites. Some 40 per cent of all the military satellites are launched to carry out photographic reconnaissance. If the other types of reconnaissance satellite, such as the electronic, early warning and ocean surveillance satellites, are included then the proportion is 55 per cent. The extent of observation time obtained by photographic reconnaissance satellites launched during 1981 is shown in figure 5.1. It can be seen that both the USA and the USSR almost continuously observe the Earth throughout the year, the only difference being the very long life of the US satellites (over three years) compared with a few days (sometimes 30 days) for the Soviet satellites. This explains why so few US photographic reconnaissance satellites are launched (about two a year).

With the ability to see an object only 30 cm in diameter and the detailed geodetic knowledge of the Earth, it is not surprising that most targets can be described and located accurately relative to sites from where missiles could be launched against them. Such a missile is then guided, using various guidance systems, to its target. Guidance systems fall into two groups, those which operate from within the vehicle, such as an inertial guidance system, and those which need instruments both on board the vehicle as well as in a ground station or in a satellite (radio guidance and navigation systems).

For long flights, a small error in the flight at the beginning may become quite large by the time the missile or long-range bomber nears its destination. With most inertial guidance systems, such errors are introduced because of the so-called gyroscope drifts. Radio guidance techniques do not suffer from such problems so that they are well suited for applications where long flights are involved. However, whereas an inertial system can use the same hardening that protects the missile itself, it is difficult to harden the antennae of radio-command guidance systems. Modern methods, therefore, use both

**Figure 5.1. Coverage by Soviet and US photographic reconnaissance satellites during 1981**

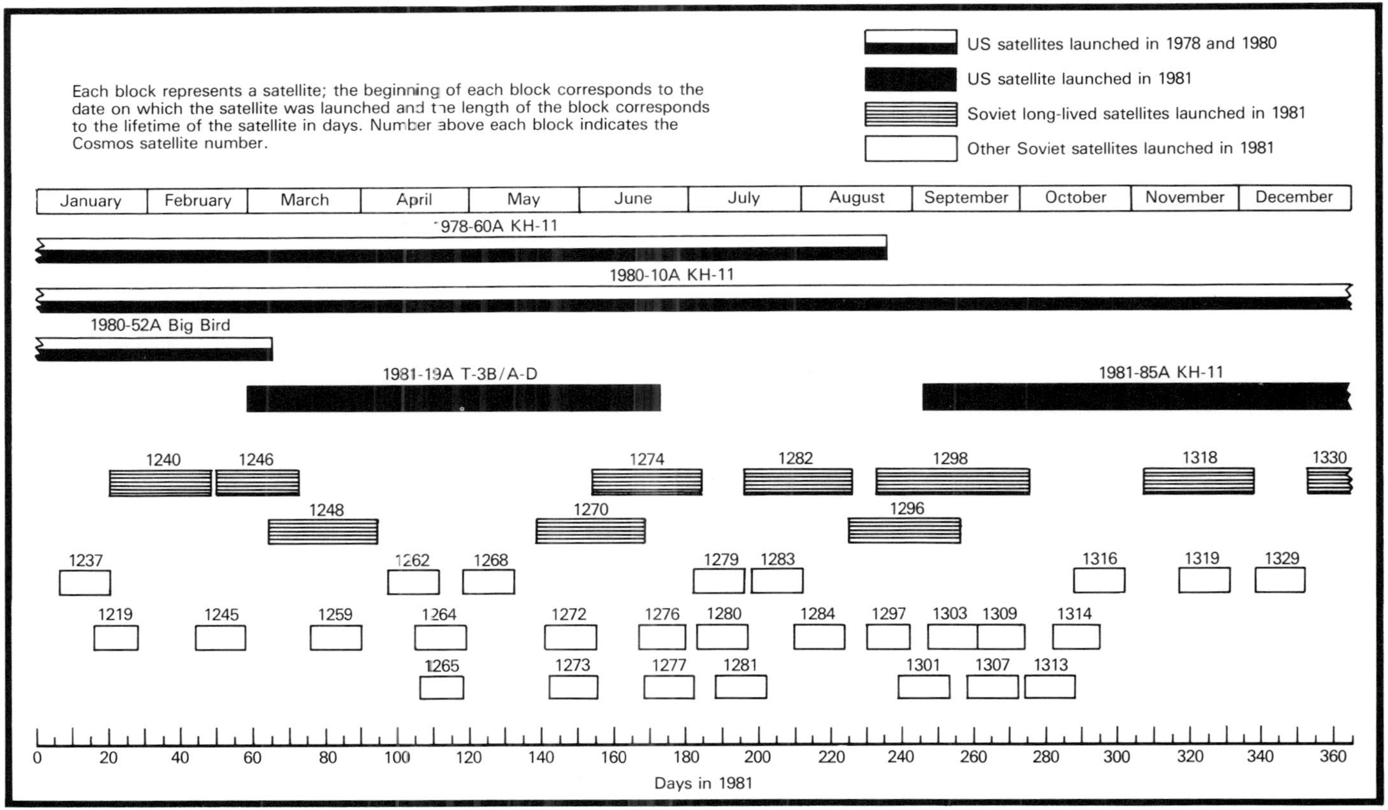

**Figure 5.2. Improvement in CEP of various Soviet and US strategic delivery vehicles**

Where precise values of CEPs are not available, a range of possible values is indicated by the vertical lines through the mean CEP for a particular missile.

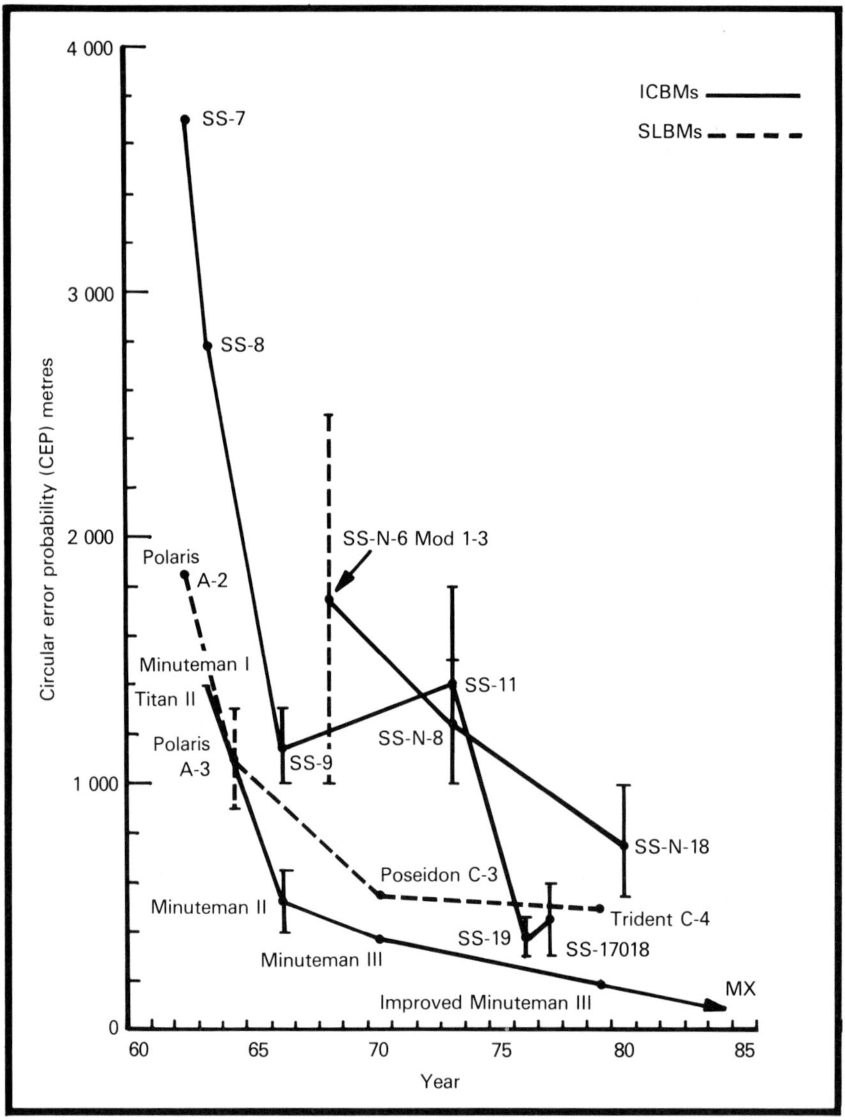


systems to some extent. For example, a missile would be guided most of its way to the target by an inertial guidance system but would up-date guidance information from, for example, satellites. Naval surface ships as well as submarines, aircraft and missiles determine their positions and velocities using signals emitted continually by satellites.

The probability of destroying a target, for example, a silo containing an ICBM, depends on several factors. Some of these are the yield of the warhead, the accuracy with which it can be delivered to the target, the hardness of the silo and the reliability of the missile carrying the warhead and the re-entry vehicle. The accuracy, known as the circular error probable (CEP) of a missile, is defined as the radius of a circle within which half of the missiles fired at the centre of the circle would fall. In figure 5.2, the improvements in CEP for various US and Soviet strategic missiles over the last decade or so are shown. It can be seen that land-based ICBMs with CEPs of between 200 m and 400 m are available today while the values of CEPs for submarine-launched ballistic missiles (SLBMs) range from 400 m to over 1 000 m.

The development of more accurate nuclear weapons makes it more difficult to protect land-based weapons from an attack and encourages a first-strike option. There are limitations to hardening silos so that new methods for countermeasures are now being investigated. One trend, for example, is investigation into, and in some cases even testing of, new types of BMD systems.

**Ballistic missile defence**

In the debate in the late 1960s and early 1970s on strategic nuclear arms stability between the USA and the USSR, it was generally assumed that mutual vulnerability was a requirement for such stability. This was apparently one of the important reasons for concluding the Anti-Ballistic Missile (ABM) Treaty of 1972 between the two powers. The objectives of the Treaty were to stop the strategic defensive arms race by limiting the deployment of anti-ballistic missile systems designed to counter strategic missiles, but with possibilities left open in the ABM Treaty for further development and deployment of new BMD systems, it is not surprising that both the USSR and the USA are developing new systems that would limit damage to their strategic weaponry. Since 1972 the concept of and the approach to the ABM have changed considerably. Outstanding progress in a wide spectrum of BMD research and development activities resulted in a major new thrust for the BMD programme during the late 1970s and particularly from 1980 [3]. It is therefore useful to review some of the new developments in BMD systems, particularly the use of high-energy lasers.

Essential elements of any BMD system are a target-detection, -recognition, -tracking and -destruction system. At present these tasks are being performed by ground-based radar sensors and by target interceptors armed with nuclear warheads. Such a system suffers from two main disadvantages. First, with a radar sensor, it is not possible to discriminate effectively between re-entry vehicles released halfway during an ICBM's trajectory and fragmented boosters and decoys. Although high resolution radars, such as phased-array radars, have been developed, they need very large antennae to achieve the required high resolution and high beam energy for detecting targets at long ranges. Thus these systems are very large and vulnerable to enemy attack. Second, electromagnetic radiation resulting from a nuclear explosion near enemy missiles, as well as clouds of chaff released by enemy re-entry vehicles, could, at least temporarily, blind ground-based radars. Moreover, nuclear BMD interceptors could not be tested under realistic conditions because the Partial Test Ban Treaty prohibits nuclear explosions in the atmosphere as well as in outer space.

Because of these limitations, considerable efforts have been devoted to R&D on new concepts of target-detection, -recognition and -tracking systems and on non-nuclear interceptors.

The USA is developing a so-called layered defence system (see figure 5.3). Three types of 'non-nuclear kill' (NNK) interceptors as well as infra-red and advanced radar sensors are being investigated. For example, under a programme called Homing Overlay Experiment (HOE), an NNK interceptor is to be tested in conjunction with an infra-red telescope against a MIRVed ICBM. During the tests, which are expected to begin in 1982–83, the Minuteman ICBM will be launched first, followed by the infra-red telescope and then the NNK interceptor to 'kill' the re-entry vehicles ejected from the Minuteman missile above the atmosphere. The infra-red telescope and its onboard data processor would replace large ground-based radars and computers for acquisition, assessment, tracking and discrimination of the target. Once the telescope is above the atmosphere, the long-wavelength infra-red sensor searches for the target against the cold background of outer space. Information is relayed from the telescope to a homing infra-red sensor and data processor on board the interceptor. The latter is then aimed towards the assigned target, which is destroyed by collision between it and the interceptor [4]. In another NNK interceptor, metal pellets are ejected in a controlled sequence so as to place them in concentric circles in the path of the incoming warheads. These pellets would then destroy them on impact [4]. Under operational conditions, the infra-red telescope and the NNK interceptors are launched on information received from an early warning satellite about the launch of an enemy missile. Some tests of the infra-red telescope and NNK interceptor have already been carried out.

Despite the fact that infra-red sensors operating above the atmosphere provide higher resolution and better discrimination of targets from other objects than those obtained by ground radars, the development of the latter continues. In fact, the current US BMD system envisages the development and deployment of both the exo-atmospheric and endo-atmospheric systems. Under the above-mentioned layered defence system (see figure 5.3), the use of radar sensors is envisaged for detection and tracking of targets that escape the exo-atmospheric infra-red sensors [3]. There would be a system of phased-array radars known as the Low Altitude Defense (LoAD) system. These need not be high power radars, and they would be operated at high frequencies so as to reduce the overall size of the system. The LoAD system is designed to be used in conjunction with MX advanced ballistic missiles [3].

During a HOE test in September 1980, the ground-based phased-array

Figure 5.3. Drawing of a concept of the layered BMD system

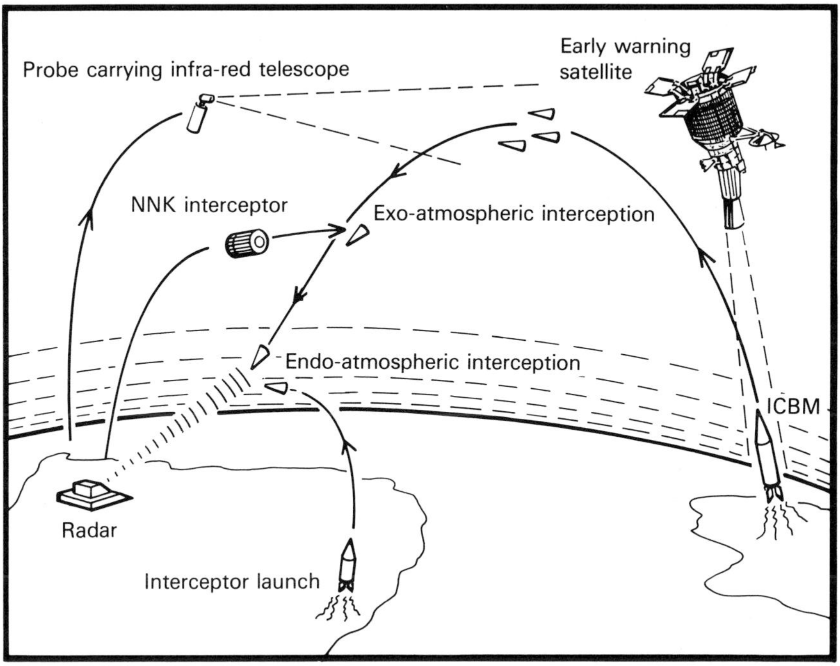


radars, together with the infra-red telescope and onboard data processing system, were also tested. The US Army is now planning to build a prototype demonstration LoAD system for terminal defence [5].

As ground-based radars form important sensors for an endo-atmospheric BMD system, considerable effort has been devoted to improving the radar system. For example, some progress has been made in reducing the effects of nuclear blasts and radiation on phased-array antennae of radars and in the development of millimetre-wave radars [4].

Another field in radar technology which is receiving much attention from the point of view of BMD is that of laser radar. Low-energy laser beams have been used for tracking satellites and in some cases even for determining the type of optics photographic reconnaissance satellites carry.

The US NNK BMD-interceptor programme dates back to the 1960s when the system was conceived as an ASAT weapon; this ASAT system could become operational by the second half of the 1980s [4, 6]. Now a second ASAT system, one which uses high-energy lasers, is thought to be applicable as a BMD system, and is approaching the stage when realistic tests are planned for the next six years [7]. In use in a BMD system, a high-energy laser beam is most likely to cause damage to an ICBM during the boost phase of the missile. During this phase an ICBM presents itself to the beam as a large, slow-moving target and has a large amount of fuel under great pressure. Moreover, the structure is under great stress and is also heated by the atmosphere, making the ICBM both most easily detected and most vulnerable to attack in this phase. Several tests of high-energy lasers against military targets have already been carried out.

The Soviet Union also has a BMD programme and there are BMD systems based around Moscow. Much effort is also being devoted to research and development on new systems. It has been suggested that the Soviet Union may be developing a rapidly deployable BMD system which would consist of a phased-array radar, a missile-tracking radar and an interceptor with a high altitude intercept capability [8]. Considerable research is also being carried out in the field of high-energy lasers for application as an ASAT weapon which could also be used as a BMD system.

BMD forms an essential component of the countervailing strategy because it enables a low level of response, with the remainder of the arsenal protected. In addition, BMD, on a large scale, would undermine the concept of deterrence because it would encourage attacks on cities instead of military targets.

## *III. Future contributions?*

Is there a possibility of using the newly found space technology for a really peaceful application? The next chapter discusses one such possibility—an international arms control monitoring agency. However, a different kind of international agency is discussed in paper 8 of part II—a space patrol guarding world peace. The paper suggests that such an activity could be carried out by the USA and the USSR, but that it might be more useful if the international community were also involved.

### References

1. 'Countervailing strategy demands revision of strategic force acquisition plans', Report No. MASA D-81-35 (United States General Accounting Office, Washington, D.C., 5 August 1981).
2. *Annual Defense Department Report, FY 1975*, Report of the Secretary of Defense James R. Schlesinger to the Congress on the FY 1975 Defense Budget and FY 1975–79 Defense Program (Department of Defense, Washington, D.C., 4 March 1975).
3. 'Status of the ballistic missile defense research and development program, 25 February 1980, Statement by G.D. Tate', *Department of Defense Authorization for Appropriations for Fiscal Year 1981, Hearings on Military Posture*, Part 4, Book 1, Committee on Armed Services, House of Representatives, February–March 1980 (US Government Printing Office, Washington, D.C., 1980), pp. 920–28.

4. Klass, P.J., 'Ballistic missile defense tests set', *Aviation Week & Space Technology*, Vol. 112, No. 24, 16 June 1980, pp. 213–18.
5. 'Technology milestone met in missile defense testing', *Aviation Week and Space Technology*, Vol. 113, No. 13, 29 September 1980, pp. 25–26.
6. 'Antisatellite missile flight test nears', *Defense Electronics*, Vol. 12, No. 12, December 1980, pp. 22–24.
7. 'Statement by W.J. Perry', *Department of Defense Authorization for Appropriations for Fiscal Year 1980, Hearings before the Committee on Armed Services*, Part 1, US Senate, January–February 1979 (US Government Printing Office, Washington, D.C., 1979), pp. 384–85.
8. 'The ballistic missile defense research and development program, Statement by S.C. Meyer', *Department of Defense Authorization for Appropriations for Fiscal Year 1981, Hearings before the Committee on Armed Services*, Part 1, US Senate, January–February 1979 (US Government Printing Office, Washington, D.C., 1979), pp. 309–15.

# Chapter 6. Satellites for crisis and arms control monitoring

> There is still too much uncertainty. Too many verification needs are unfulfilled and many more have to be satisfied if and when desirable disarmament agreements are to be reached. The international responsibility for verification of disarmament therefore must have an organized form. [1].
>
> Alva Myrdal

## *I. Satellites for crisis monitoring*

A study of satellite launches during the past eight years or so suggests that in addition to verifying the implementation of the SALT I and II agreements, US and Soviet reconnaissance satellites have monitored most of the major armed conflicts in the world during this time.

Some of the most striking examples of this are as follows: In December 1971, two manoeuvrable satellites were launched by the USSR to observe the conflict in East Pakistan (which subsequently became Bangla Desh) [2]. During the periods immediately before and after the 1973 Middle East War, the USSR orbited a series of Cosmos photographic reconnaissance satellites [3]. In 1974, both the USA and the USSR orbited their reconnaissance satellites over Cyprus, Greece and Turkey to observe the army coup of Cyprus on 15 July and the subsequent Turkish invasion of Cyprus on 20–22 July (see reference 1 in chapter 1). For further details on these examples of crisis monitoring, see paper 11 of part II.

More recently, the conflict which began on 22 September 1980 between Iran and Iraq was also observed. During this time the Soviet Cosmos 1210 satellite, launched on 19 September, was already in orbit, Cosmos 1211 was launched on 23 September, a day after the conflict began, and Cosmos 1212 was launched on 26 September. Of these three, Cosmos 1210 is the one that most probably made detailed observations of the conflict.

In figure 6.1 the ground tracks of Cosmos 1210 are plotted. The heights of the satellite over the region at different times were computed and it was found that the perigee of the satellite's orbit was over the area. It can be seen that the ground tracks are closely spaced. However, from the 158th orbit, the spacing between two consecutive tracks becomes greater. The closeness of the tracks would indicate detailed observation of the area beneath the satellite. It must be noted here that such close spacing of the tracks and the position of the perigee also occur elsewhere over the Earth's surface, so that the areas of observation in figure 6.1 are not the only ones. Other regions of particular interest for the USSR are the USA and the People's Republic of China. The satellite ground tracks considered here also pass over these countries—at about midnight local time over the USA and in the late afternoon over China. However, several of the orbital changes made for this satellite appear just before the satellite passes over Iran and Iraq. It was found that the perigee moved across Iran and Iraq and was over the border of the two countries on 24 and 25 September, over Khorramshahr and Abadan, respectively. The local time during the passes was about 1115 hours.

**Figure 6.1. Ground tracks over Iran and Iraq of Cosmos 1210, launched on 19 September 1980**

The date and orbit numbers are indicated for each ground track.

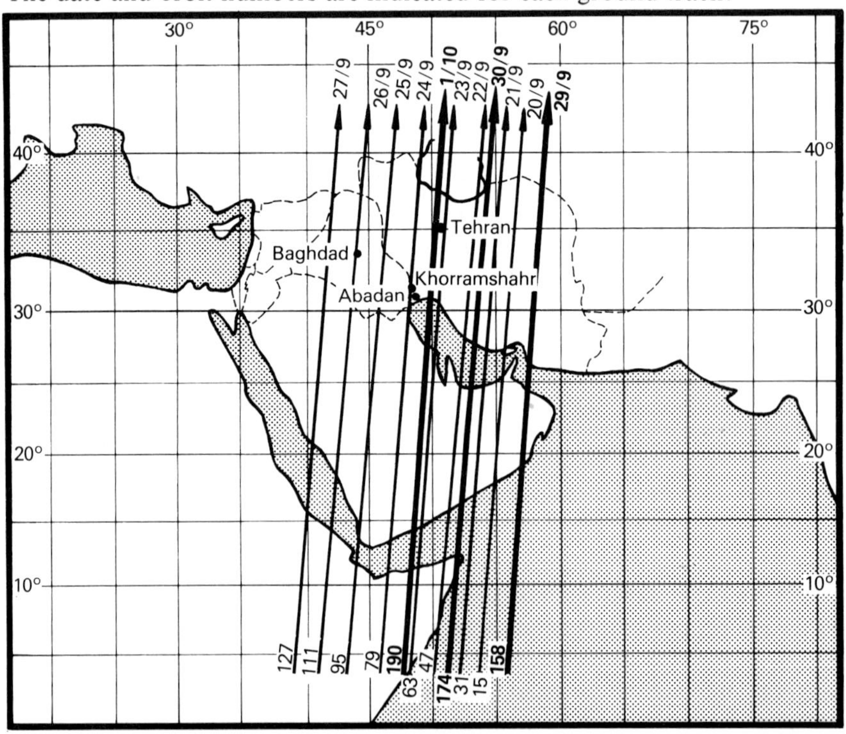


The above shows that spacecraft are used to monitor the development and progress of conflicts. The technical requirements for a crisis monitoring system are discussed in detail in paper 11 of part II. The usefulness of such a mission is clear provided it is performed under international control. At present the capability and the will to carry out verification and crisis monitoring is in the hands of only two nations, and there is considerable reluctance on the part of the USA and the USSR to relinquish this monopoly.

During the first UN Special Session on Disarmament, France proposed the establishment of an international satellite monitoring agency to observe the implementation of international disarmament and security agreements. The proposal thus suggested the use of spacecraft for the purpose of settling disputes between states [4]. Moreover, such methods could also assist the United Nations in its peace-keeping role. It is interesting to note here that moves have already been made to help man fight natural disasters. A joint feasibility study has been carried out by Canada, France, Norway, the USA and the USSR on a Search and Rescue Satellite System (SARSAT) [5]. Canada is to build the ground station for the SARSAT which will be used to locate marine and aviation distress signals anywhere in the world [6]. It is desirable that some serious consideration also be given to man-made disasters such as war and conflict.

## *II. International verification of disarmament agreements*

One of the important and controversial issues in the history of disarmament and arms control negotiation has been the verification of compliance with the terms of negotiated treaties. The lack of real progress in the disarmament field is likely to continue if no special efforts are made to resolve the issue, particularly by the international community, and the arms race will continue even more fiercely. The public at large and even some governments are getting increasingly impatient with this absence of progress.

More and more treaties have relied for their verification on the so-called 'national technical means' which are possessed by only a few states and in some cases by only two—the USA and the USSR. This often prevents states from participating in treaties. Ways and means have to be found so that all nations are fully able to take part in the disarmament process. The fact that some disarmament measures may require more intrusive methods of verification complicates negotiations even more because a considerable amount of co-operation based on trust and good faith is needed.

The best method of resolving these dilemmas is to create an international verification agency, an idea which is by no means a new one. Apart from several official proposals, many prominent personalities and scholars have suggested the creation of such an agency [7–12]. A recent proposal made by France has attracted much interest [4]. Moreover, considerable progress has been made in the technology of verification, particularly in space technology, and this would facilitate the establishment of such an agency.

Though there are many countries which will not be able to acquire on their own the technical capabilities that can be used for verification purposes certain industrialized and some industrializing countries are obtaining the required space technology (see paper 12 of part II). For example, the University of Tokai in Japan has developed an image processing technique which

Figure 6.2a. A Landsat image over Japan

Figure 6.2b. Enlargement of a section of figure 6.2a in which aircraft can clearly be seen on Haneda Airport's runway in Tokyo

can bring out a considerable amount of detail using an image obtained from the US Landsat satellite orbiting at an altitude of 900 km. In figure 6.2a such an image is shown over Japan. In an enlarged section of the photograph (figure 6.2b) aircraft on the runway of Haneda Airport in Tokyo can be seen. In the next 20 or 30 years many nations will possess space launchers and space systems and some will acquire deep-water vehicles, systems for ground surveillance and large computers and data processing techniques. All of these could be used for verification purposes.

However, if this process is allowed to develop in an unorganized way, it may create capabilities for military purposes only. These would include, for example, better targeting, damage assessment capabilities and military command, control and communications systems. Another consequence of such a random development is that it will be slow and possibly technically inferior and less cost-effective, a situation which would be intolerable from the point of view of verification. The most important factor is that verification must be carried out by well-calibrated equipment if data from observations made by individual parties to an agreement are to be consistent. Contradictory data may result in unwarranted accusations and suspicions which would undermine the agreement in question.

The verification procedure is best performed if many mutually supporting methods are used. The discovery of violations is more certain and concealment of such actions more difficult. Many disarmament agreements may require similar verification methods so that a single agency would be able to perform the task of verification more efficiently and in a more comprehensive and accurate manner. Above all, if an agency were to be established solely for the purpose of verifying disarmament agreements then, for the first time in the history of disarmament, the acquisition of verification capability would not have been the result of military technology. The verification capability would be less prone to military application not only because it would have been acquired with a peaceful purpose in mind but also because of its international character.

The recent report of the group of governmental experts on the implications of establishing an international satellite monitoring agency emphasized the usefulness of present civilian space technology [13]. Table 6.1 shows various arms control/disarmament treaties for which satellites—in some cases even civilian ones—could be used to verify compliance with the provisions of the treaties. The possible usefulness of civilian satellites in verification is well illustrated in figures 6.2a and b.

Although the problems associated with the establishment of an international verification agency are considerable (as indicated in paper 12 of part II), the capabilities of civilian space technology for observing the Earth's surface are beginning to approach those of military technology in many respects (as seen in figures 6.2a and b). Moreover, not only is satellite technology spreading to many more countries (see table 6.2) but launcher technology has reached countries not otherwise considered to be advanced in this field (see table 3.9). More importantly many more countries are also

**Table 6.1. Contribution from observation satellites to verification process of arms control agreements**[1]

| Treaty | Type of satellite; sensors needed for observation | Type of activities that need be observed | Comments |
| --- | --- | --- | --- |
| Geneva Protocol | | | See CW Convention below |
| Antarctic Treaty | Military; synthetic aperture radar (SAR), optical cameras, electronic surveillance equipment | Establishment of military bases, military manoeuvres and testing of weapons | Some of the activities prohibited could be observed by some civilian satellites; more details could be observed if sensors on civilian satellites were orbited at lower altitudes |
| Treaty banning nuclear weapon tests in the atmosphere, in outer space and under water (PTBT) | Military; nuclear radiation detectors, X-ray and optical sensors | Preparatory activities for a test and subsequent radiation could be observed when a nuclear test is carried out | Optical and heat sensors on some civilian satellites (e.g., weather satellites) could observe atmospheric and outer space nuclear explosions |
| Outer Space Treaty | | | |
| The Tlatelolco Treaty | Military; optical and SAR | Construction of unknown nuclear reactors or enrichment plants, roads and railway tracks; preparatory activities for a nuclear test | Optical and radar sensors on civilian satellites (e.g., weather satellites) could observe construction of clandestine nuclear facilities, particularly if they are orbited at lower altitudes |
| Treaty on the non-proliferation of nuclear weapons (NPT) | Military; optical and SAR | Construction of unknown nuclear reactors or enrichment plants, roads and railway tracks; preparatory activities for a nuclear test | Optical and radar sensors on civilian satellites could observe constructions of clandestine nuclear facilities, particularly if they are orbited at lower altitudes |
| Sea-Bed Treaty | | | Satellites cannot contribute to verification of this treaty |
| Biological Weapons Convention | Military; optical sensors in visible and IR range, SAR | Manoeuvres connected with testing and training drills; decontamination facilities and segregation barriers; railway lines and specially-designed tanker-wagons | Again optical sensors and SAR on some civilian satellites could detect some of the BW activities if the sensors are orbited at lower altitudes |

| | | | |
|---|---|---|---|
| Environmental Modification Convention | Military and civilian weather satellites; visible, IR and microwave radiometers; optical sensors | Information on cloud conditions, temperature on Earth's surface and top of clouds, temperature and water vapour determination at various altitudes, density sounder measures oxygen and nitrogen content of the thermosphere | Although satellites may detect any unusual developing weather pattern, it would be difficult to know who might have caused such a change |
| Limitation of Anti-Ballistic Missile Systems (ABM Treaty) | Military; optical and electronic sensors | ABM systems normally consist of launchers, interceptor missiles and radars which can be easily observed by military satellites; high-energy beams of ABM radars are detected by electronic reconnaissance satellites | Civilian satellites are not equipped with sensors to detect electronic signals but their optical sensors may be able to detect large radar installations from low altitudes |
| Strategic Arms Limitation Talks (SALT I) | Military; optical and electronic sensors | Missile tests during which telemetry could be monitored; construction of missile silos, observation of launch and impact areas to get range of missiles, shipyards to observe construction of submarines | Some of the activities, e.g., construction of silos, may be detected by some civilian satellites but the sensors do not have high enough resolution to observe details |
| Threshhold Test Ban Treaty (TTBT) | – | – | Satellites cannot contribute to verification; in future sensors to measure ground motion and atmospheric overpressure due to underground nuclear explosion may be developed |
| Peaceful Nuclear Explosion Treaty (PNET) | – | – | Satellites cannot contribute to verification; in future sensors to measure ground motion and atmospheric overpressure due to underground nuclear explosion may be developed |

[1] Satellite observations could contribute considerably to the monitoring of peace treaties such as that between Egypt and Israel.

acquiring the technology for image processing, essential for the interpretation of data from space (see table 6.3). Undoubtedly these trends will continue and space technology will spread beyond the industrial nations. Once many states are able to observe the Earth from space, the fear of releasing sensitive data – one of the most serious objections to establishing an international verification agency—may no longer be relevant.

**Table 6.2. Satellite launches as of the end of 1981**

| | Number of satellites | | | | |
|---|---|---|---|---|---|
| Country | Satellites | Manned flights | Deep-space probes | Moon flights | Total excluding manned flights |
| Australia | 2[a] | – | – | – | – |
| Canada | 9[a] | – | – | – | – |
| China, People's Republic of | 11 | – | – | – | – |
| Czechoslovakia | 1[b] | – | – | – | – |
| Countries of the European Space Agency | 11[c] | – | – | – | – |
| France | 25[d] | – | – | – | – |
| FR Germany | 5[e] | – | 1[a] | – | – |
| India | 5[f] | – | – | – | – |
| Italy | 5[g] | – | – | – | – |
| Japan | 23[h] | – | – | – | – |
| NATO | 5[a] | – | – | – | – |
| Netherlands | 1[a] | – | – | – | – |
| Spain | 1[a] | – | – | – | – |
| UK | 13[i] | – | – | – | – |
| Sub-total | **117** | – | **1** | – | **118** |
| USA | 859 | 37 | 22 | 26 | 944 |
| USSR | 1 559 | 55 | 22 | 27 | 1 663 |
| Total | **2 535** | **92** | **45** | **53** | **2 725** |

[a] Launched by USA
[b] Launched by USSR
[c] One launched by Ariane, the rest by USA
[d] Seven launched by USSR, one by USA
[e] One launched by France and the rest by USA
[f] One launched by USSR and one launched by France
[g] Launched using US launcher
[h] Two launched by USA
[i] One launched by UK

Among the aspects to be considered before such an international verification agency could be established is that verification could not be carried out from space alone; data from other sources would be necessary. This could be obtained by involving a number of existing international organizations (table 6.4) in the verification of a specific arms control/disarmament treaty.

Difficult questions concerning the modalities of data acquisition and dissemination of direct relevance to the sensitive security considerations of states must be dealt with. However, solutions to such problems will not be found unless discussions continue. These discussions may even consider the

**Table 6.3. Landsat ground stations**

| Country | Date agreement signed | Date of data reception | Date of data processing |
|---|---|---|---|
| Argentina | 1976 | 1980 | 1980 |
| Australia | 1979 | 1980 | 1980 |
| Brazil | 1973 | 1973 | 1974 |
| Canada, Prince Albert | 1972 | 1972 | 1972 |
| Canada, Shoe Cove | 1976 | 1977 | 1977 |
| Chile | 1975 | – | – |
| China, People's Republic of | 1980 | – | – |
| India | 1978 | 1979 | 1980 |
| Iran[a] | 1974 | – | – |
| Italy (ESA) | 1974 | 1975 | 1976 |
| Japan | 1979 | 1979 | 1979 |
| Sweden (ESA) | 1978 | 1978 | 1979 |
| Thailand | 1979 | – | – |
| Zaire | 1975 | – | – |
| Upper Volta | (1977) | – | – |

[a] The Landsat station in Iran was largely completed and began receiving some test data in late 1978 but the station ceased operation in early 1979.

possibilities of a verification agency on a multinational or regional basis. For example, once the Ariane launcher becomes routinely available, the ESA could contribute to the verification of any arms control measures that may be worked out in Europe. In this context the Intercosmos could also have a vital role to play.

## *III. An ASAT treaty*

In March 1977, the USA suggested that the two great powers should agree to ban the arming and destruction of each other's Earth satellites. In fact, both the parties met several times between 1978 and 1979. The discussions do not seem to have been very productive and no further meetings have taken place. These discussions, present treaties protecting satellites and other issues arising from the ASAT activities of the two powers are dealt with in paper 14 of part II.

However, since these talks, the Soviet Union, in 1981, proposed to the United Nations a new treaty banning ASAT weapons [14]. This proposal is discussed in paper 15 of part II. One of the significant aspects of this proposal is the fact that the ASAT discussions were, for the first time, removed from a bilateral forum to a multilateral one.

One of the problems with the ASAT measures proposed so far is that they would help eliminate only part of the arms race in space and might not

**Table 6.4. Some international organizations potentially relevant to arms control agreements**

| Organization | Date of establishment | UN affiliation | Number of parties | Organs with control functions | Function of the organ |
|---|---|---|---|---|---|
| Inter-governmental Maritime Consultative Organization (IMCO) | 1959 | UN agency | 118 | | Fulfilment of the convention on the International Maritime Satellite Organization |
| International Civil Aviation Organization (ICAO) | 1947 | Co-operation | 146 | | |
| International Telecommunications Union (ITU) | 1865 | UN agency | 154 | International Frequency Registration Board, International Radio Consultative Committee | Assignment of radio frequencies, investigation of harmful interference, monitoring of spectrum utilization |
| World Health Organization (WHO) | 1948 | Co-operation | 155 | | Surveillance of communicable diseases, collecting data, reports on drugs, pollutants |
| World Meteorological Organization (WMO) | 1951 | UN agency | 152 | World Weather Watch Programme, Global Data Processing System, Global Telecommunications System, Executive Committee Panel of Experts on Satellites, Weather Modification Programme, Environment Programme, Operational Hydrology Programmes | Observations, co-ordination of satellite operations, transmitting meteorological data, co-ordination of weather modification research, standardization of hydrological data |
| International Atomic Energy Agency (IAEA) | 1957 | Co-operation | 110 | | Dissemination of information, safeguarding nuclear operations in accordance with safeguards agreements with some 100 states |
| OECD Nuclear Energy Agency (NEA) | 1958 | | 17 | NEA Data Bank with Computer Programme Library, Neutron Data Compilation Centre | Continuous survey of uranium resources, exchange of information |
| Western European Union (WEU) | 1955 | | 7 | Agency for the Control of Armaments | Control of non-manufacture of certain types of arms, control of level of stocks of armaments of member states |

| | | | | | |
|---|---|---|---|---|---|
| International Council of Scientific Unions (ICSU) | 1919 Cospar 1958 | | 69 (104) | Committee on Antarctic Research, Committee on Oceanic Research, Committee on Space Research, Committee on Genetic Experimentation | Co-ordination of international scientific research in various fields |
| European Space Agency (ESA) | 1975 | | 11 | Space Research and Technology Centre, Space Operations Centre, Space Documentation Centre | Co-operation in space research and technological developments |
| International Academy of Astronautics (IAA) | 1960 | | 556 of 31 states | | Research in space rescue and safety, etc. |
| International Astronautical Federation (IAF) | 1950 | | 58 societies of 36 states | International Academy of Astronautics, International Institute of Space Law | Fostering development of astronautics for peaceful purposes |
| UN Committee on the Peaceful Uses of Outer Space (COPUOS) | 1958 | UN organ | 37 | | Establishment of Thumba sounding rocket range |
| Eurospace | 1961 | | 11 states, 96 organizations, 31 banks | | |
| International Telecommunications Satellite Organization (INTELSAT) | 1964 | | 105 | | |
| InterCosmos | 1970 | | 9 | | Data interpretation, tracking assistance |
| Interamerican Nuclear Energy Commission (IANEC) | 1959 | | 28 | | |
| International Seismological Datacentre (ISD), Swedish | 1979 | | | | |

affect the arms race on Earth. It would, in fact, be a partial measure, as are most arms control measures. There are a number of other examples of partial measures related to outer space.

The 1967 Outer Space Treaty has not helped to check either the militarization of space or the development and deployment of ASAT systems. The treaty only prohibits the placing in orbit around the Earth any objects carrying nuclear weapons or any other kind of weapon of mass destruction. This and other treaties are more fully discussed in paper 13 of part II.

There are other treaties which have complicated matters by enshrining the use of military satellites.

The three US–Soviet strategic arms agreements (SALT I, ABM and SALT II) prohibit interference with "national technical means of verification" which include reconnaissance satellites but do not indicate which types of reconnaissance satellite.

An international telecommunications agreement forbids interference with communications. However, this includes communications satellites implicitly and therefore does not prohibit the military use of such spacecraft.

The recent Soviet proposal has two merits. One is, according to the proposal, that the participating nations pledge themselves "not to put objects carrying any types of weapon into orbit around the earth, not to install such weapons on celestial bodies and not to deploy such weapons in outer space in any other way . . . ". Secondly, the proposed treaty calls for negotiations on an international level and not just on a bilateral basis.

Unfortunately, the Soviet proposal does not seem to cover ground-based or air-based ASAT weapons, such as, the anti-satellite missiles launched from high-altitude aircraft and high-energy beam weapons. However, the proposed treaty not only bans interference with spacecraft of member states but also bans interference with the flight trajectory of spacecraft.

A significant drawback of this proposal may be seen in its verification clause. It states that "each participating nation shall use the national technical control facilities at its disposal" to assure compliance with the provisions of the treaty. Only two nations today possess the technology needed to perform this task. Thus it is difficult to visualize many nations becoming parties to the treaty unless an international verification agency is created—of the kind proposed by France in 1978.

While a ban on ASAT weapons would be a step in the right direction from an arms control point of view, it nevertheless would be regarded as another measure to protect vital elements of the military. In fact, the element that would be protected actually enhances the continuing nuclear and conventional arms race.

## References

1. Myrdal, A., *The Game of Disarmament* (Pantheon Books, New York, 1976), p. 304.

2. Perry, G.E., 'Cosmos coverage of the Indo-Pakistani war', *Spaceflight*, September 1972, p. 350.
3. Perry, G.E., 'Looking down on the Middle East war', *Flight International*, Vol. 105, No. 3389, 21 February 1974, pp. 240–41, 245.
4. UN Document A/S-10/AC.1/7, 1 June 1978.
5. Langereux, P., 'Démonstration du sauvetage par satellite durant la Transat', *Air et Cosmos*, No. 869, 18 July 1981, p. 54.
6. 'Canadian astronautics awarded government contract', *Interavia Air Letter*, No. 9523, 17 June 1980, p. 6.
7. Feld, B.T., 'Inspection techniques of arms control', in *Arms Control, Disarmament and National Security*, ed. D.G. Brennan (George Braziller, New York, 1961), p. 318.
8. Davies, M.E. and Murray, B.C., 'A role for the UN', in *The View from Space* (Columbia University Press, New York and Landa, 1971), pp. 23–33.
9. SIPRI, *World Armaments and Disarmament, SIPRI Yearbook 1973* (Almqvist & Wiksell, Stockholm, 1973), pp. 73–74.
10. 'Stevenson calls for international arms monitoring Satcom net', *Defense/Space Daily*, 23 July 1979, p. 38.
11. Toyoda, T., 'Technical feasibility of international satellite monitoring system for strategic arms', *Proceedings of the 28th Pugwash Conference on Science and World Affairs—Global Aspects of Disarmament and Security*, Varna, Bulgaria, 1–5 September 1978, p. 238.
12. Myrdal, A., 'The international control of disarmament', *Scientific American*, Vol. 231, No. 44, October 1974, pp. 21–33.
13. UN report A/AC.206/14, 6 August 1981.
14. UN Document A/36/192, 20 August 1981.

# Chapter 7. Conclusions

> Non-violence . . . is a rule of conduct for society if it is to live consistently with human dignity and make progress towards the attainment of peace . . .
>
> Mahatma Gandhi

During the time it takes to read this sentence the USA will spend some $2 000 on its military space programme. Assuming that the Soviet budget is the same, then the amount spent every 10 seconds amounts to over $4 000. The military space effort includes the launch of one military satellite every third day, the prime aim of these being to increase the fighting efficiency of the military forces on Earth.

Since the first military satellite was orbited more than 20 years ago, nearly 2 000 spacecraft, or some three quarters of all spacecraft launched, have circled the Earth for military purposes—reconnaissance, communications, navigation, weather forecasting and early warning of an attack by enemy missiles. Meanwhile, the romantic image of spacecraft unravelling the mysteries of the unknown has diverted people's attention from their military function. Few people realize, for example, that developments are already underway that could result in weapons orbiting the Earth in the foreseeable future. Space-based weapons are being investigated that would knock down enemy missiles early in their flight trajectory, before they have released their MIRVs towards their targets. In addition, weapons to obliterate Earth-orbiting satellites have been developed and exotic weapons such as anti-satellite high-energy laser and particle beam weapons are vigorously being investigated.

With the rapid development of space technology, apparent changes in nuclear war-fighting doctrines are surfacing which emphasize limited nuclear war-fighting capabilities at various levels. This emphasis results from concepts such as flexible response, an outcome of the ability to monitor from space the effects of nuclear strikes; real-time assessment of a nuclear attack anywhere in the world can be provided by the space-based surveillance and communications systems. The need for precise target information and the need to strike the target accurately in the current countervailing doctrine for fighting a limited nuclear war are also fulfilled by space systems. In fact, space technology has become an important element in the strategic doctrines of both the USSR and the USA.

The dependence on space systems results in a desire to have them continuously available but, at the same time, to deny the enemy use of its own spacecraft. Military satellites have, therefore, become essential military targets. Thus the research and development begun in the 1970s on advanced sensor technology for identifying and tracking from space military targets both in orbit and on Earth will continue, as will the development of systems for the destruction of such targets.

Many of the new space systems being investigated need large structures. Thus their deployment will no doubt be facilitated by the space-shuttle-type transportation systems under development by the USA and the USSR.

Among the large-structure systems being developed are advanced imaging and tracking radar, electronic surveillance and communications systems; the antennae for such systems range from 30 m to 200 m in diameter. Advances in the technology of large structures in outer space will, for example, allow an individual soldier to communicate from the battlefield to his command post by enabling the placement of multi-beam phased-array antennae in geosynchronous orbit [1]. Large-structure technology will also enhance the capabilities of microwave sensors and of sensors using advanced concepts of adaptive optics with lenses of over 30 m in diameter. The 1980s will also see space-based systems for the detection and tracking of aircraft and cruise missiles. Such a device, the US Teal Ruby, will be launched into a 750-km polar orbit by the space shuttle.

Another focal point for the military use of outer space during the 1980s is expected to be the increasing presence of man in space with the view to eventually establishing a permanent manned space station. At present the Soviet Union holds the record for the longest manned space flight (over 108 days), while the USA is expected to initiate the development of a permanent manned space station [2]. The cost of such a venture would be approximately $5–7 billion. The Soviet Union is reported to have begun testing a modular space station.

However, the most important implication of the development in large-structure technology is for ASAT systems. For example, the contribution of the space shuttle system and large-structure technology to the deployment of space-based high-energy beam weapons would be considerable. Of the two types of beam weapon envisaged, laser and particle beam, the former is likely to be deployed first.

Besides the development of offensive weapons—conventional or beam weapons—the 1980s will see considerable effort being devoted to protecting and improving the survivability of military satellites in orbit. This will be achieved by hardening the satellite surfaces and by making their communications systems resistant to interference. Moreover, surveillance of outer space will be increased and improved.

Apart from technological developments expected for protecting military satellites, a political approach for achieving the same result was initiated in 1978 when the USA and the Soviet Union began talks on the control of their ASAT activities. These did not progress but in August 1981 the USSR proposed to the United Nations a new treaty banning the placement of any kind of weapon in Earth orbit.

Arms control and disarmament issues have reached a stage when a decision has to be made between a controlled arms race or disarmament. Most existing treaties either ban weapons which are not weapons at all or weapons which cannot be used. This, by and large, has become the basis for arms control efforts and the concept of disarmament seems to be slowly disappearing. Outer space has proved no exception, but it may, nevertheless, provide a way to begin the process of disarmament.

The effectiveness with which nuclear and in some cases even conventional weapons can be delivered to their targets, along with the quality of these weapons, is improving dramatically. In this process, the artificial Earth satellites play a major role. Thus it would seem natural to suggest that the military use of satellites be banned so that the qualitative nuclear arms race could be checked [3]. This would give some meaning to the pledge given by the nuclear weapon powers in Article VI of the 1968 Treaty on the Non-Proliferation of Nuclear Weapons (NPT) that they will work towards nuclear disarmament. Very little, if anything, has been done so far towards the fulfilment of this pledge. In fact, the opposite seems to be true since considerable efforts, from the point of view of technology as well as from a political standpoint, are made to protect the very spacecraft which enhance the arms race on Earth.

It is, however, recognized that space technology could contribute considerably to building confidence and security among nations, for example, by verifying some arms control agreements and by monitoring crisis areas. Although the ability to perform such tasks is at present confined to only two powers, this monopoly will eventually be broken. There is considerable impetus to internationalize the verification of arms control agreements from space since, firstly, this will strengthen international security and confidence and, secondly, space technology is spreading to nations other than the USA and the USSR. Moreover, as missile forces, particularly those carrying nuclear weapons, of other nations increase, more countries will find it necessary to deploy early warning satellites. An increase in security and confidence among nations would again be given as the reason for such deployment.

While the political climate is being created for such changes in the use of outer space, there is an urgent need for some measures to slow down and even to stop the militarization of this environment. In the absence of a total ban on the use of satellites for military purposes, the following are suggested:

1. Much more effort is needed to inform the public at large on the extent of the militarization of outer space and the disastrous consequences of allowing space technology to add impetus to the arms race. In this connection, setting up a United Nations Centre for Outer Space (similar to the UN Centre for Disarmament) would help.

2. While all the satellites launched are registered with the United

Nations in conformity with General Assembly Resolution 1721 B (XVI), the information given on most spacecraft is scanty. Important details, such as, the launch time and site, physical characteristics of satellites, degree of manoeuvrability, satellite lifetime and often the orbital characteristics of satellites, should be given. At present, the purpose of a spacecraft is often difficult to determine.

3. Since disarmament is a concern for all, it is essential that discussions on the concept of an international satellite monitoring agency are kept alive.

4. Much effort is now devoted to making satellite ground stations invulnerable and to improve the survivability of satellites in orbit. It might be useful if such survivability measures did not take place, or were at least limited, just as has been agreed upon in the case of strategic weaponry—the 1972 ABM Treaty limits defence of ICBMs, for example.

5. To ensure availability of satellites during a time of crisis, both the USSR and the USA intend to station spare satellites in orbit. Moreover, for a continuous world-wide availability of accurate navigation capability, it is necessary to have a number of spacecraft in a space-based navigation system. Similarly, a world-wide communications system needs at least three satellites in orbit. It might be suggested that the number of military satellites launched per year by a country should be limited so that, at least, the performance of some weapon systems on Earth could be degraded and even the efficiency of performance and effectiveness of military forces could be reduced.

6. The minimum that the two major powers could do to bring at least some military activities in outer space under control is to agree to an ASAT treaty. However, an ASAT treaty would protect military spacecraft. While such a treaty would improve the image of the two great powers, protection of military satellites would not help to control the nuclear arms race.

The Second Special Session of the General Assembly devoted to disarmament in June-July 1982 and the Second United Nations Conference on the Exploration and Peaceful Uses of Outer Space in August 1982 are of great importance. It is crucial that emphasis be placed on the military uses of outer space with a view to clearly defining which of the military satellites are to be permitted and for what purposes. While there are still only two countries which are able to enhance the performance of their nuclear arsenals using space technology, a truly comprehensive outer space treaty containing real disarmament measures must be worked out. Another opportunity will be missed if greater control over the militarization of space is not brought about now.

## References

1. 'SAMSO charts future in space', *Defense Electronics*, Vol. 11, No. 11, November 1979, pp. 43–50.

2. 'Air Force looking at military space station needs at NASA's request', *Flight International*, Vol. 119, No. 30, 16 December 1981, p. 233.
3. Jasani, B., 'Why we should ban the military use of satellites', *Nature*, Vol. 273, No. 5662, 6 June 1978, p. 416.

# PART II

# Space technology, crisis monitoring and arms control

# Paper 1. Reconnaissance satellites

**B.G. Blair***

*The Brookings Institution, Foreign Policy Studies, 1775 Massachusetts Avenue, N.W., Washington, D.C. 20036, USA*

## I. Introduction

Reconnaissance satellites serve a wide range of purposes including arms control verification, crisis monitoring, early warning of attack, weapons targeting, and execution of current strategic employment policies. This versatility can carry benign or adverse implications for national security and international stability.

Consideration of the implications begins with an understanding of the key technical aspects and capabilities of the spacecraft and sensors. This paper attempts, therefore, to sketch the curve of technological evolution in satellite reconnaissance. Reliance on public sources of information necessarily limits the discussion, particularly as regards Soviet programmes. For this reason, the discussion focuses mainly on US early warning and photo-reconnaissance spacecraft.

## II. Early warning satellites

Successful launches of early warning satellites during the early 1970s led quickly to a dramatic improvement in the ability of the United States to detect a nuclear missile attack. The first satellite to become operational had been launched in 1971 and 'parked' over the Indian Ocean at an altitude of about 36 000 km. Its purpose was to monitor missile launches in the Eastern Hemisphere. Because of its high altitude, geographic coverage was extensive. And because of its geosynchronous orbit, the area under continuous coverage did not vary. As figure 1.1 shows, almost all of the Soviet Union,

---

* The opinions expressed in this paper are those of the author and do not necessarily reflect the official policy of his professional affiliation.

China and other potential missile launch and test areas came within view.

A second satellite, launched in 1972, was positioned over Panama to monitor areas of the Atlantic, Pacific, and Caribbean in which Soviet missile submarines patrolled. Figure 1.1 demarcates the area within line of sight of the satellite.

**Figure 1.1. Coverage by early warning satellites**

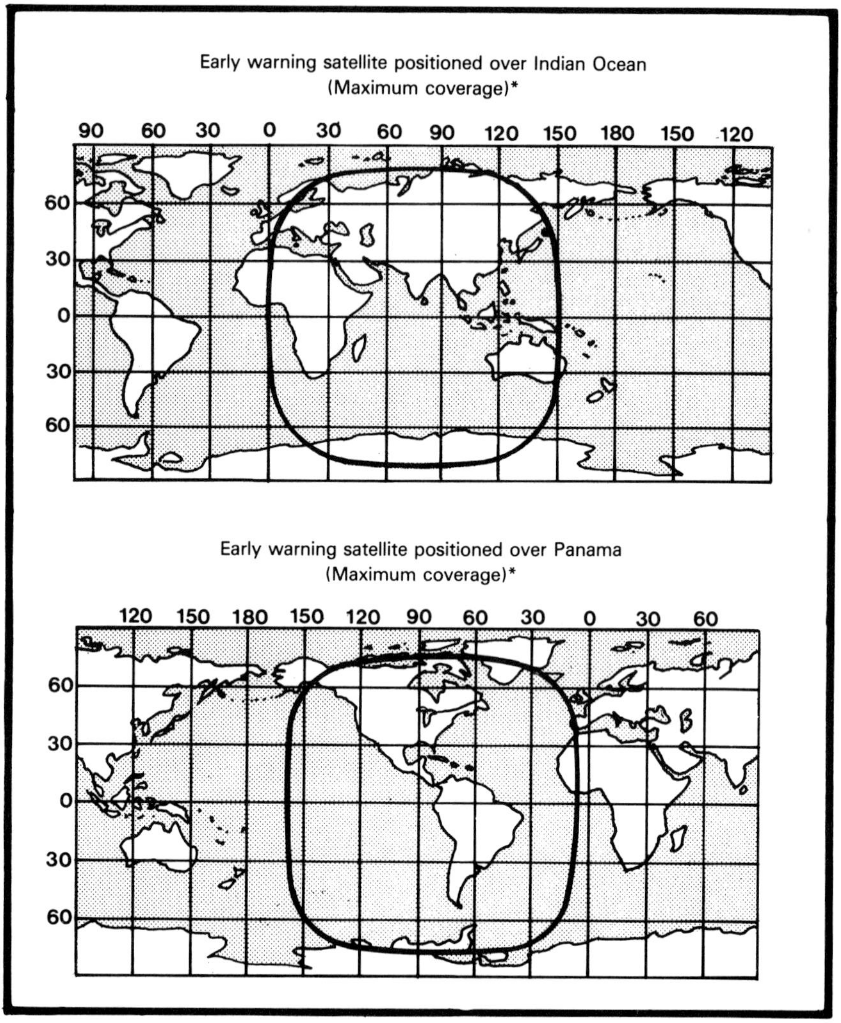


* *Note:* Line-of-sight limits coverage. Referring to diagram below, the distance to the horizon ($D_h$) for a satellite (S) at a given altitude ($H$) is equal to:

$$D_h = \sqrt{[(E_r + H)^2 - (E_r)^2]}$$

where $E_r$ is the Earth's radius (7 341 km). Since $\sin a = D_h/(H + E_r)$, we can compute angle $a$ in degrees using a table of trigonometric functions. In this example, $H$ = 22 300; $D_h$ = 25 963; $\sin a$ = 0.9885; and $a$ = 81° (as projected on above maps). Ground distance from centre of circle to edges = $a$(69), or 10 351 km.

These early warning satellites have been credited with the capability to detect a missile within a few tens of seconds after lift-off. That would mean that they could provide 30 minutes of advance warning of land missile attack and nearly as much warning of submarine-launched missile attacks, assuming that Soviet submarines fired from their normal patrol areas in the Eastern Atlantic. The warning time would be less, of course, if missile submarines had moved closer to US shores prior to launch.

The detection capability of the satellites derives from the thermal infra-red sensors that they carry. They were designed to sense the radiated energy of the hot plume of a missile during its relatively slow climb to altitude. This exhaust plume reaches a temperature of about 2 000° K, radiating peak energy at about 1.5 μm (according to Wien's displacement law[1]). The sensitivity of thermal infra-red devices on early warning satellites of US vintage is not centred exactly on this energy wavelength. The devices are designed instead to have maximum sensitivity at about 3 μm. They are so designed because the atmosphere absorbs most of the thermal radiation that is emitted at wavelengths below 3 μm. The devices thus exploit the existence of an 'atmospheric window' located at 3.0 to 5.0 μm of the thermal infra-red band. Though the amount of energy that a rocket emits at 3 μm is considerably less than that emitted at 1.5 μm, it is enough for the sensors to pick up. Since 1971, onboard sensors have detected well over 1 000 missile test launches by the Soviets, Chinese and French.

Despite this impressive record of performance, technical limitations and deficiencies have been and still are numerous. For example, sun glint and glare from clouds and ocean surface occasionally trigger false alarms or create blind spots that degrade coverage of certain areas. The time of the occurrences and the affected area are fairly predictable, as they tend to be correlated with seasonal changes. And the problem has been alleviated somewhat by the addition of a third satellite to complement the one positioned over Panama. The coverage of the two satellites overlap extensively, but they are spaced far enough apart that blind spots normally would not coincide. If one satellite experiences solar or other problems, the other satellite could monitor the trouble area from a different position and look angle.

Not predictable are computer malfunctions and communications outages at satellite receiving stations on the ground. Read-out stations associated with intercontinental ballistic missile (ICBM) and submarine-launched ballistic missile (SLBM) detection are situated near Woomera, Australia, and Buckley Airfield, Colorado, respectively. Problems with the communications and data processing equipment at those locations have resulted in intermittent breakdowns in service as well as false alarms.

Another significant technical limitation concerns the instantaneous

---

[1] Wien's displacement law relates temperature and the wavelength of peak energy as follows:

$$W_m T = 2\,898$$

where $W_m$ is the wavelength (μm) and $T$ is the temperature (° K) field of view of the satellite sensors. Figure 1.1 demarcates the total surface area within line-of-sight range of a particular satellite, not the specific area actually under surveillance at any given instant in time. Only a minute fraction of the total area of potential surveillance, and only a very small fraction of the area from which Soviet missiles could be launched against US territory, can be monitored at any moment in time. For instance, a satellite positioned over Panama cannot simultaneously monitor both oceans or all possible launch points within any given body of water. A Soviet 'Yankee' submarine armed with an SS-N-6 missile can hit Washington from an area of the Atlantic that is roughly 2.5 million km. The satellite over Panama probably has less than one per cent of this area in view at any time[2]. Full coverage of the area thus requires the satellite to scan portions of the surface methodically, bringing one part of the ocean into view as another part leaves it. There is obviously some chance of a delay in detection, and probably some chance that an event of short duration would be missed completely, because (*a*) this scanning technique can capture only a fraction of the total energy that a missile emits during the burn phase of its trajectory, and (*b*) repeated scans are required in order to minimize the risk of false alarms.

The basic physical limitation imposed by Earth curvature also proves to be a significant constraint on detection capability now that the Soviets have deployed 'Delta'-class submarines armed with very long-range SS-N-8 missiles. With a range in excess of 7 964 km, this missile can be launched from points in the North Sea and Barents Sea that are well beyond the line-of-sight range of early warning satellites in geosynchronous orbit. There is speculation, but no solid evidence, that other US satellites in polar orbits fill in the gap in coverage.

In the foreseeable future, non-scanning or staring sensors will remove many of the technical constraints that limit performance and reliability. The next generation of satellites (1990s) will probably incorporate several new technologies—notably, charge-couple devices (CCDs), mosaic sensors, and advanced cryogenic coolers—which together promise an improvement of several orders of magnitude in resolution and signal-to-noise ratio.

CCDs are a variation on large-scale integrated circuit technology. In remote sensing applications their versatility is such that CCDs can process and store signals from a detector that is functionally separate, function as an integral component of the detector itself, or perform both functions in an integrated fashion. With mosaic sensors, infra-red detectors are integrated into a layer of silicon containing CCDs. Thus, the CCDs form a part of the detector array itself. Charge coupling refers to the technique of shifting an electrical charge from one detector element to the next. A chain of signals representing a projected infra-red pattern is transferred to a readout device as a time sequence of electrical pulses ready for transmission. To achieve

---

[2] This estimate is based on reports that the satellite sensor contains 2 000 infra-red elements, each capable of subtending an angle corresponding to less than five square kilometres of the Earth's surface.

maximum sensitivity of the monolithic, staring mosaic sensor, the sensing element will require a powered cryogenic cooling system.

The factors driving investment in this combination of technologies are several. First, the US Department of Defense (DoD) desires to eliminate the foreign basing of readout stations. With CCD technology, a great amount of the signal processing now performed on the ground can be performed on board the satellite. As a consequence, it will be possible to scale down the ground-based operation to a point where it might be dispensed with. Second, the DoD is concerned about the vulnerability of current readout sites. These facilities are large and fixed because the satellites themselves cannot perform the necessary signal processing. CCD technology will allow for deployment of smaller, simpler and mobile versions of those facilities. That means that exposure to attack can be greatly lessened. Third, the starting sensor will provide continuous coverage of potential launch points, providing additional warning time and reducing the risk of false alarms. This improvement is seen as doubly significant in light of recent policy shifts towards increased reliance on launch-under-attack as a means of ensuring the survival of US ICBM forces. Fourth, an early warning capability of this sort would better serve a strategy of flexible nuclear war-fighting.

The system would not only provide more precise information about the character of ICBM and SLBM attacks, but would also allow for the detection and tracking of 'colder' targets, such as tactical aircraft, cruise missiles, theatre nuclear missiles, post-burnout vehicles (e.g., multiple independently targetable re-entry vehicle 'MIRV' busses), and spacecraft (e.g., 'killer satellites')

To monitor moving targets in the atmosphere, the satellites will probably be equipped with multispectral thermal infra-red sensors which combine infra-red bands in the region between 2.5 and 8.0 μm. The heat from a jet engine runs at about 1 000°K, corresponding to peak energy at 3.5 μm. Because the bandwidth of the sensor would overlap this region, as well as the region around 4.3 μm where absorption of infra-red radiation by the atmosphere is unusually low, it should be capable of detecting the engine emission from aircraft (and small missiles).

Reconnaissance satellites designed to track very cold objects, such as MIRV busses, re-entry vehicles and satellites, would have to be equipped with sensors whose sensitivity is centred at about 10.0 μm. However, the Earth naturally emits radiation at approximately the same wavelength, and hence a satellite looking down at the Earth might be saturated with this 'noise', rendering it incapable of distinguishing the target from its background. The satellite could be placed in a low Earth orbit instead of a geosynchronous orbit, in order for it to 'look up' at objects against the cold, uncluttered background of outer space.

To achieve the high resolution needed to detect and trace 'dim' objects from geosynchronous altitudes, the DoD envisions eventual deployment of very large optical structures that incorporate adaptive optics as well as the technologies discussed earlier. Conventional glass optics will not suffice, because size determines weight. Precision glass optics would be too heavy to lift, even with the space shuttle. Lightweight modular mirrors, however, offer a practical alternative. Adaptive optics could electronically tune the mirrors to compensate for system imperfections, vibration and atmospheric turbulence. In effect, one has a 'rubber' or deformable mirror which can compensate in real time for mechanical or thermal distortions of optical systems. Consequently, size would not be the constraint it is today. Astronauts, using a manned space station or the shuttle, might even erect the optical structures in space.

Data collected by early warning satellites during missile test firings have contributed to arms control verification. Photo-reconnaissance satellites, however, are the principal means by which the two great powers check each other's compliance with arms control agreements. We next turn to examine the technical features of these reconnaissance tools.

## *III. Photo-reconnaissance satellites*

While thermal infra-red sensors provide round-the-clock, all-weather coverage of certain events, the sensors used for photo-reconnaissance can perform well only during daylight and clear weather. When the conditions are obtained, cameras carried by photo-reconnaissance satellites can take remarkably detailed pictures of objects ('close-look') or survey large areas of the Earth ('search-and-find') searching for those objects that deserve a closer look.

'Close-look' cameras carried by US Big Bird and KH-11 satellites are generally believed to be capable of achieving such high resolution that objects smaller than 30 cm can be distinguished from an altitude of 185 km or more. Better resolution is possible when the target has linear features (e.g., runways), a characteristic that heightens the contrast between the target and its background. To illustrate, a non-military satellite used for remote sensing has distinguished linear objects as small as 14 m in width, though its normal resolution was limited to 82 m. Contrast, and hence resolution, is furthermore amenable to enhancement by computer techniques of the sort that brought us the startling clear pictures during the 'Pioneer' and 'Voyager' explorations.

'Close-look' photography doubtless makes possible the detection and identification of most military targets on the ground. Table 4.1 in Part I contains estimates of the resolutions required for various interpretive tasks.

Resolution alone, however, does not determine the interpretability of satellite imagery. Indeed, there are many circumstances in which relatively low resolution, when accompanied by a broad view, gives a more interpretable image than a high resolution photo of a more limited area. 'Search-and-find' or area surveillance cameras are employed under such circumstances.

Area surveillance, as opposed to 'close-look', operations increasingly use multispectral scanning devices which combine the visible light and photographic infra-red regions of the spectrum. A device used on one of NASA's Landsat satellites is probably very similar, at least in principle, to devices designed for military use. The NASA scanner operates in three different bands of the visible light spectrum (0.5–0.6 μm, 0.6–0.7 μm, and 0.7–0.8 μm) and one band in the infra-red (0.8–1.0 μm). After an oscillating mirror causes the energy to cross the focus of a small telescope, the energy passes through spectral filters to detectors that convert it into electrical signals. A continuous strip photograph of the passing ground results, and it is stored as electronic imagery until transmitted in digital form to receivers on the ground.

Digital data acquired from multispectral scanners allow highly detailed colour photos to be produced by assigning a different colour to each spectral band and then superimposing the individual spectral images. Such 'false-colour' photographs have many advantages over conventional colour photography. Reconstruction of photographs using the false-colour approach can eliminate colours produced by a combination of different colour spectra, making it easier to understand and interpret the photographs. Camouflaged objects are more readily detected. For example, while the human eye cannot readily perceive, either by direct observation or by looking at conventional colour photos, the difference between ordinary green paint and a background formed by deciduous leaves, false-colour infra-red film renders the foliage bright red and the paint blue.

Multispectral images expressed in digital form contain a far more subtle store of data than can be found in standard photography. Computers can manipulate this data to bring out detail that would otherwise be lost. The results are exceedingly sharp images. Algorithms developed by the US Geological Survey's EROS centre, for instance, correct for geometric distortion, extremes of contrast, and the scattering effects of the Earth's atmosphere. Thus, it is possible to obtain imagery resolution beyond the limits imposed by the atmosphere. Algorithms also help reconstruct more clearly defined boundaries and edges. The development of such algorithms, and computer processing of digital data, generally, is a relatively recent achievement. Much progress has been made over the past five years, and further progress can be expected. For example, experiments have shown that low resolution satellite imagery may contain as much useful information as high resolution imagery. Using such statistical techniques as principal components analysis and canonical correlation, computers can extract an enormous amount of 'unseen' information.

The resolution of multispectral scanners used in military reconnaissance must be at least as good as existing NASA scanners, which themselves could attain a resolution of less than 6 m if orbited at the lower altitude flown by the military satellites. The KH-11 military satellites, thought to carry the most advanced sensor of this kind, probably attain resolution of the order of

2 to 3 m. That degree of resolution is sufficient to meet most of the detection requirements and many of the more demanding interpretive tasks.

Return beam vidicon sensors similar to those used by NASA's Landsat vehicles also have great potential for use in military area surveillance. The latest NASA version provides resolution of the order of 37 m from an altitude of 926 km. That converts to 8-m resolution at a lower altitude (185 km). A more capable sensor is easily within the state of the art. These television-like sensors are destined to become primary sources of real-time military video-intelligence.

The Soviet Union frequently uses multispectral photography. In September 1976, for example, the crew of the Soyuz 22 spacecraft photographed the Earth with a six-band multispectral camera having an estimated resolution of 6-20 m. Cosmos 912 and 948, both officially described as Earth resources missions, also carried multispectral cameras.

I expect that the trend in satellite photo-reconnaissance will be towards greater reliance on multispectral photography, partly because computer extraction of information from the digital data is very effective, and partly because the data can be collected and transmitted by radio link in real time.

## *IV. Ocean surveillance and ferret satellites*

Cloud cover and darkness are the bane of photo-reconnaissance. Microwave and longer wavelengths, however, are not affected by these conditions. For this reason, some observers have speculated, or better assumed, that US reconnaissance spacecraft use active radar to lift the veil of cloud cover and darkness. Although the Soviet Union uses active radar for ocean surveillance, the United States lacks a radar imaging capability and that situation is not likely to change in the immediate future. Or so it seems, based on the weight of evidence in the unclassified literature.

If NASA forecasts are reasonably accurate, however, the next 25 years could witness the development of space-borne radar that matches the performance of aircraft radar in use today. At the longer radar wavelengths, Apollo 17 has already demonstrated the remarkable resolution that can be obtained with a technique known as side-looking synthetic-aperture radar (SAR). Released from military security classification in 1970, SAR systems are just beginning to find civil satellite application.

SAR is important because conventional antennae for satellites cause images to have far coarser resolution along the orbital path than across it. SAR solves this asymmetry by taking advantage of the motion of the satellite. A short antenna can be made to perform like a very long antenna. Using this technique, Apollo 17 produced radar images of the Moon's surface from a distance of 111 km and with a ground resolution of 9 m. Preserved on film, they resembled ordinary high altitude photos. And, of course, they can be obtained during the day or night, almost irrespective of the weather.

NASA equipped Skylab with two microwave channels, and tested newer radar systems when Seasat A—an ocean dynamics monitoring satellite—was launched into a 926-km orbit in 1978.

Seasat A carried an imaging SAR having a swathwidth coverage of 232 km and a resolution of 23 m. The SAR processor employed CCD signal processors (as opposed to the CCD detector arrays discussed previously) for real-time reduction of radar data to images. Until CCD signal processing became available, radar to image conversion took several hours of high speed processing for each 10 minutes of observed data. A military version of a SAR system, placed in orbit at low altitude, could provide good resolution, all-weather effectiveness, and real-time performance over land or sea. The implications for verification are manifold. No such version is programmed for deployment, as best one can judge the prospects at the moment.

The US Air Force has proposed putting a large, cloud-peering active radar on medium altitude satellites. To meet antenna and power demand, it envisages use of an antenna measuring 183 by 274 m (larger by more than a factor of 10 than antennae on existing satellites), requiring 50 kW of power, or roughly 100 times greater than the wattage of existing high power systems. The space shuttle might be used to assemble large antennae in space, and to orbit a 25-kW power station for other spacecraft, possibly including surveillance satellites. Onboard nuclear reactors or radioisotope-powered thermoelectric generators (RTGs) offer two alternatives to an orbiting power station, as does the new technology known as modified bi-static radar.

The US Navy's only fling with radar ocean surveillance has ended with the cancellation of its Clipper Bow project. Under that programme, the Navy had expected to deploy a pair of satellites that would have provided Navy commanders at sea with all-weather, day and night coverage of the oceans.

The Soviet Union, on the other hand, has been conducting an active ocean surveillance programme for the past six years. Soviet spacecraft carry active radar in a low altitude orbit. Conflicting assessments of their performance abound. It has been reported that they can make prompt and accurate interpretation of changes in the disposition of US naval forces. Contrariwise, it has also been reported that the Soviet Union is still striving to achieve a satellite radar and electronic intelligence (ELINT) capability sufficient to detect, locate to the required degree of precision, and identify an ocean target.

Ferret satellites are used in an ELINT role to lock onto signals emanating from ship radar. Such satellites generally orbit at medium altitudes. Intercepted data, including military communications as well as ship radar signals, can be relayed via radio to friendly surface vessels and/or attack submarines.

US use of ferret satellites does not appear to be devoted to any specific type of eavesdropping mission such as ocean surveillance. But it is known that the current generation of satellites, known as 'Rhyolite', has been used to intercept telemetry broadcast by Soviet missiles under test. A new ferret satellite equipped with a long antenna tailored for telemetry interception is reportedly under development.

# Paper 2. Identification of military components within the Soviet space programme

**G.E. Perry**

*The Kettering Group, 101 Northampton Road, Kettering, Northamptonshire NN15 7JY, UK*

## *I. Introduction*

On 16 March 1962, the Soviet Union initiated the Cosmos programme "to continue the study of outer space". Ever since that date, launches in this programme have been announced with a standard wording, for example,

> The Soviet Union launched an artificial satellite, Cosmos . . . , on . . . The onboard scientific equipment is designated for continuing space exploration. The satellite was placed in orbit with the following parameters: the initial period of revolution is . . . , the minimum distance is . . . , the inclination of the orbit is . . . The onboard equipment is functioning normally.

When it was announced that Cosmos 4 had been recovered from orbit after a flight of only four days, it was clear that the name Cosmos was a blanket-term covering a variety of missions. A systematic study of the orbits, ground-tracks, locations of perigee, launch sites, flight durations, and routine monitoring of shortwave transmissions by The Kettering Group since 1962 has revealed distinct sub-sets, the largest of which contains those satellites which disappear from orbit when their orbital period is still high enough to rule out natural decay.

Table 2.1 shows the annual breakdown of recoverable launches in the Cosmos series. It can be seen that just under half of the Cosmos launches fall into this category with between 30 and 40 launches annually over the past six years. In practice this has ensured that there has been at least one such satellite in orbit for more than 80 per cent of the time with sometimes two or more in orbit simultaneously. Clearly a large proportion of the Soviet space budget is devoted to this class of mission which is conducted on a routine operational basis.

Initially all such launches were made from Tyuratam, but since the first launch from Plesetsk, in 1966, the majority of these launches have been

**Table 2.1. Annual breakdown of recoverable Cosmos launches**

| Year | 1962 | 1963 | 1964 | 1965 | 1966 | 1967 | 1968 | 1969 | 1970 | 1971 |
|---|---|---|---|---|---|---|---|---|---|---|
| TT recoverable | 5 | 7 | 12 | 17 | 15 | 8 | 14 | 13 | 13 | 13 |
| PL recoverable | – | – | – | – | 6 | 14 | 15 | 18 | 16 | 15 |
| Recoverable total | **5** | **7** | **12** | **17** | **21** | **22** | **29** | **31** | **29** | **28** |
| Cosmos total | **12** | **12** | **23** | **36** | **36*** | **61** | **64** | **55** | **65** | **67** |
| Annual % recoverable | 41.7 | 58.3 | 52.2 | 47.2 | 58.3 | 36.1 | 45.3 | 56.4 | 44.6 | 41.8 |
| Cumulative % | 41.7 | 50.0 | 51.1 | 49.4 | 52.1 | 46.7 | 46.3 | 48.2 | 47.5 | 46.6 |
| **Year** | **1972** | **1973** | **1974** | **1975** | **1976** | **1977** | **1978** | **1979** | **1980** | **1981†** |
| TT recoverable | 11 | 8 | 9 | 12 | 13 | 10 | 8 | 2 | 3 | 12 |
| PL recoverable | 18 | 27 | 19 | 21 | 21 | 23 | 27 | 34 | 32 | 18 |
| Recoverable total | **29** | **35** | **28** | **33** | **34** | **33** | **35** | **36** | **35** | **30** |
| Cosmos total | **58** | **64** | **60** | **64** | **79** | **79** | **67** | **64** | **67** | **62** |
| Annual % recoverable | 50.0 | 54.7 | 46.7 | 51.6 | 43.0 | 41.8 | 52.2 | 56.3 | 52.2 | 48.4 |
| Cumulative % | 47.0 | 47.9 | 47.8 | 48.2 | 47.6 | 47.1 | 47.5 | 48.0 | 48.3 | 48.3 |

Notes: PL = Plesetsk launched, TT = Tyuratam launched.
* includes two unannounced missions related to FOBS or interceptor tests.
† up to and including 30 September.

made from there. In 1979 and 1980 there were only 2 and 3, respectively, of these launches from Tyuratam, but 1981 has seen an increased launch activity at Tyuratam with 12 of the 30 recoverable Cosmos satellites launched in the first three quarters of the year being made from there.

## *II. Photographic reconnaissance*

A prime requirement for a wide-area search-and-find photographic reconnaissance mission is that complete coverage of the target area is achieved within the duration of the flight. This can be ensured by the selection of an orbital period giving a small daily displacement in the longitude of the ground-track until, by the end of the mission, it coincides with the position occupied by the following ground-track at the start of the survey.

Complete photographic coverage for a non-manoeuvring mission is obtained with orbital periods given (in minutes) by

$$T = \frac{518\,400n}{(16n \pm 1)\,(360.98 + 8.95 \cos i)}$$

when $n$ = length of mission in days; $i$ = inclination of orbit to the equator; and the sign is taken as negative for orbits with a westward daily drift and positive for those greater than 81° which drift eastward.

This leads, in general, to longer orbital periods for flights at higher inclinations; for 12-day missions at 62.8° and 72.8° the periods evaluate as

89.23 and 89.58 min, respectively. Reference to the tables produced in the SIPRI Yearbooks or the Royal Aircraft Establishment's Table of Earth Satellites, 1957–1980, [1, 2] will produce confirmation of this. The formula given above is quasi-empirical in so far as the value of 8.95 is concerned. The motion of the Earth in its orbit around the sun and the precession of the orbital plane about the Earth's axis due to the equatorial bulge are both contributory factors.

Over the years the duration of the missions has tended to increase from 8 days initially to the current 14 days. The trend towards longer duration is a direct corollary of a general improvement in photographic resolution. Obtaining increased resolution on the same film format by use of longer focal lengths results in a smaller ground area being photographed. The more closely spaced ground-tracks and the consequent shorter period are obtained by lowering the apogee. Perigee heights have always been as low as practicable consistent with keeping air-drag to an acceptable minimum.

Examples of ground-tracks giving complete photographic coverage have been provided elsewhere [3, 4a, 5].

Although analyses of orbital parameters, ground-tracks, and flight durations contribute to the classification of such missions, it has been the monitoring of the in-flight radio transmissions and those from the recovery beacons which has made it possible to distinguish four generations of reconnaissance spacecraft.

## First generation

Shortwave transmissions from these satellites on 19.995 MHz were frequency-shift keyed (f.s.k.) with the 'off' periods being transmitted on the second frequency. The telemetry frame format, like that of Korabl Sputnik 1, consisted of a characteristic 'purr' due to a train of rapid synchronizing pulses followed by 15 words transmitted at a rate of approximately one word per second. These words were pulse-duration modulated (PDM), the length of each word being a measure of some parameter onboard the satellite [6].

It was found that the first word of the frame following the synchronizing train existed in several modes [7]. During the main part of the mission it could be thought of as being either long or short, or mode 1 and mode 2 as they became known in the West. Other modes were observed immediately prior to recovery when long changed to very long and short changed to medium, which became known as modes 4 and 3 according to the order in which they were discovered. The change of mode occurred at the instant of recovery capsule separation from the instrument package. Even during the main part of the mission it was sometimes found that mode 1 would change to mode 2 in mid-transmission when the satellite had risen well above the horizon of a principal ground-station on Soviet territory, suggesting that the change was due to line-of-sight rapid play-back of stored intelligence on another frequency.

Three types of transmission were observed: mode 1-only, mode 2-only and mode-changers. The first of these came to be associated with the low resolution missions whereas the others were thought to be high resolution.

In 1966 the Group first noticed a sequence of Morse code groups consisting of the letters TK which appeared some six minutes after the loss of the f.s.k. transmission when the instrument package burned up on entering the dense layers of the atmosphere [8]. As time went by it became clear that these signals originated from the recovery capsule and were intended to assist recovery teams in locating the payload. The Morse code transmission begins at the instant the parachute is deployed and an abrupt decrease in signal strength some six to eight minutes later indicates the time of touch-down. The length of time for which the signals persist beyond this is a measure of the precision of the recovery.

As time went by, TGs were found on a frequency of 20.005 MHz, a frequency sometimes used simultaneously with 19.995 MHz for TK transmissions. These TGs were exclusive to the mode 1-only satellites. Thus there was an additional characteristic by which the missions could be classified—TKs on only one frequency, TKs on two frequencies simultaneously, or TGs. Pen-recordings showed that the dash of the T had twice the duration of the dashes in the K and the G. T is the letter common to all recovery beacons so far observed.

During the R&D period of 1962 flight durations varied between 4 and 6 days, but the operational missions were extended to 8 days. Initially they were launched by the A-1 but later the A-2 was used to launch them and is still in use today for the third and fourth generation spacecraft. The next development was to an extended duration of 12 days giving increased resolution. This coincided with a repositioning of a very long word within the telemetry frame from 11 to 7. On occasions, word 7 was not of very long duration initially but lengthened suddenly towards the end of the mission. It was shown that this coincided with the ejection of a pick-a-back scientific payload [9]. The fact that scientific missions were eventually reported with their results only for these particular missions strongly suggests that all other missions were solely military in nature.

Four flights in 1978, one of which produced a scientific pick-a-back payload, were the last examples of these first generation recoverable spacecraft.

**Second generation**

The introduction of a manoeuvring capability with Cosmos 251 in 1968 coincided with the appearance on 19.150 MHz of a new type of transmission in the form of Morse code groups. Twelve groups of three Morse code characters were transmitted repetitively at a rate well within the reading capability of a competent radio amateur. With one exception the groups were formed of characters chosen to give a total of seven pulses per group in a 2–3–2 configuration. The seven pulses of the remaining group had a 3–1–3 configuration and served to denote the beginning of the telemetry frame. It was shown that these represented 7-bit pulse code modulation (PCM) binary code in which the first digit rather than the last took the value 2°—the digits were read in the reverse direction to that used conventionally [6].

The sixth group was invariably MWI throughout the main part of the mission but changed to MRI on the day prior to recovery at the same time that an additional object was catalogued in orbit. This additional object was once thought to be the manoeuvring engine but inspection of the biological Cosmos satellites displayed in Soviet exhibitions in recent years has cast some doubt on that theory as it is not immediately apparent why it should have been necessary to discard those engines at all as their presence would not impede the action of the retro-engine in the configuration.

The steady increase in the numerical value of the 7th word in the PCM frame and the 13th word in the pulse duration modulation (PDM) frame indicated the use of some consumable. The choice between attitude-stabilization gas and photographic film was resolved in favour of the latter when the Cosmos 463 was observed to exhaust this consumable at twice the normal rate in its half the normal duration mission at the time of the Bangladesh war [46].

At times when two PDM satellites or two PCM satellites were in orbit simultaneously it became the practice for the earlier satellite to change its transmission frequency in flight prior to the launch of the new satellite, from 19.994 to 19.989 MHz or from 19.150 to 19.300 MHz. The recovery beacon remained on 19.995 and/or 20.005 MHz. This in-flight change of frequency often provided advanced warning of a new launch [10]. 19.150 and 19.300 MHz have not been employed since the second generation satellites were phased out in 1974.

### Third generation

Another type of f.s.k. transmission was discovered during the flight of Cosmos 364 in 1970 although it is possible that such a transmission was used by Cosmos 317 at the end of the previous year. In this type there appears to be no telemetry and the two-tone signal on 19.989 MHz must serve as a tracking beacon. Each tone is transmitted for equal durations and the time for one complete cycle varies between 1.5 and 1.75 s. The variation in cycle duration, pointed out by Peter Wakelin of The Kettering Group, serves to distinguish between different payloads simultaneously in orbit since the frequency-changing procedure of the first generation satellites has long been abandoned. Sven Grahn, also of The Kettering Group, suggested that there appeared to be some correlation between cycle duration and orbital period, but incontrovertible evidence of this has not been found and neither has the consequent change in cycle duration which would be expected following a major manoeuvre.

A new type of complete photographic coverage was introduced with

Cosmos 867 at the end of 1976. One day after launch the satellite is manoeuvred into an orbit with perigee in excess of 300 km and with a period of approximately 92.3 min. Photography from such a height would provide wide-area medium resolution imagery. Ground-tracks on consecutive days are widely separated by some rather more than one-half of the track separation giving rise to what has been termed interlaced coverage. Figure 2.1 shows such coverage as obtained during the early phases of the Iraqi–Iranian war from Cosmos 1216.

**Figure 2.1. Ground tracks of Cosmos 1216 over the Middle East, September 1980, showing interlaced coverage**

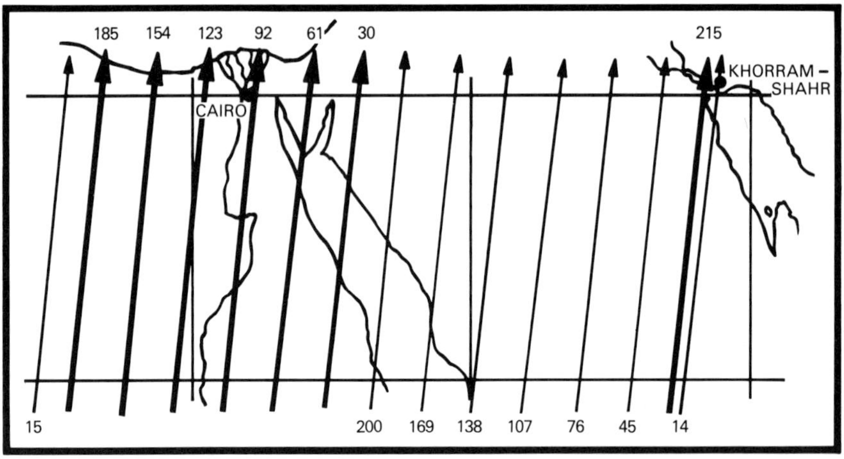


Consecutive ground-tracks for revs 14 and 15 on 17 September 1980 are shown as thin lines at the extremes of the figure. Between these, other thin lines depict the ground-tracks for revs 45, 76, 107, 138, 169 and 200 at two-day intervals thereafter. It will be seen that the ground-track for rev 200, on the day prior to recovery, falls almost exactly midway between the two extremes. Ground-tracks for revs 30, 61, 92, 123, 154 and 185 on intermediate dates, shown by thicker lines, fill the remaining space between the midway point and the ground-track for rev 15, completing the coverage and rev 215, the recovery orbit, falls just within the space between revs 14 and 45.

A variant of this interlacing which might be considered as triple-interlacing was exhibited by Cosmos 1283, the first of the high-perigee flights at 82.3°. The orbital period of 91.6 min ensured complete coverage with a three-day sub-cycle of 47 revs.

Figure 2.2 is a schematic representation of the pattern of the triple-interlacing with day-date of July 1981 and corresponding rev number plotted against ascending node longitude. This pattern was also adopted for the following Cosmos launch and may prove to be characteristic of high-perigee flights at 82.3° inclination.

A special sub-set of satellites exists within these third generation

**Figure 2.2. Ascending node with new numbers for Cosmos 1283 showing triple-interlacing of ground tracks**

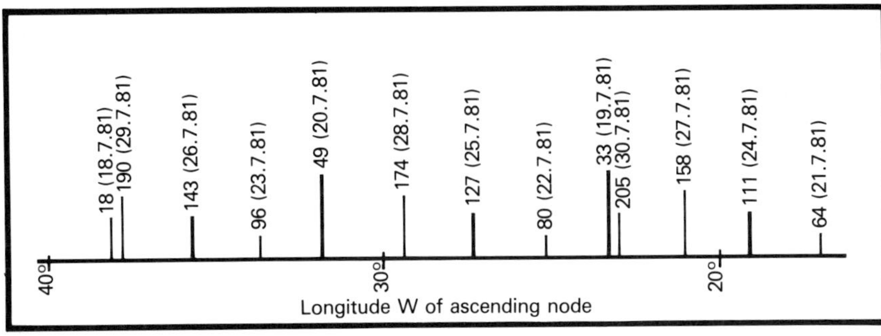


missions. They transmit on a frequency of 19.994 MHz and do not manoeuvre during the mission. Furthermore they transmit a TL recovery beacon which is unique to the sub-set, the usual beacon for the remaining third generation satellites being TF. The non-manoeuvring feature led to their initially being classified as successors to the first generation low resolution flights. As time passed and such flights appeared, in general, only two or three times a year at times close to the equinoxes and at different annual inclinations to the equator, some other role was sought and it is now thought that they perform a geodetic survey and mapping function. With the sun overhead close to the equator there will be average illumination in both northern and southern hemispheres. The Soviet Union has the need to be able to correlate the international mapping grids for precise targeting of its multiple independently targetable re-entry vehicles (MIRVs) and it has been suggested in *Aviation Week & Space Technology* that this is achieved by taking simultaneous photographs in opposite directions of ground targets and star backgrounds using laser technology [11].

### Fourth generation

These spacecraft, which do not transmit on any known shortwave frequency, appeared first in 1975 at a new inclination of 67°. The first and third of these fragmented in orbit but the system is now operational with eight such missions in the first nine months of 1981. Five of these have been at a new inclination of 64.9° from Tyuratam with the others at 67.2° from Plesetsk. Mission durations have ranged from 23 to 42 days. Apart from two 3-day flights in 1977 and 1979 at a 62.8° inclination from Plesetsk, all other flights in the series have flown at 67.2°. Initially it was thought that the new 67° inclination heralded the use of the F-vehicle at Plesetsk but all other evidence suggests continued use of the A-2, but from a new launch pad.

Perigees close to 175 km, and occasionally below 160 km, call for regular boosting to maintain orbit. Figure 2.3 shows the variations in apogee and perigee heights during the flight of Cosmos 1097.

**Figure 2.3. Variations in apogee and perigee heights during the flight of Cosmos 1097**

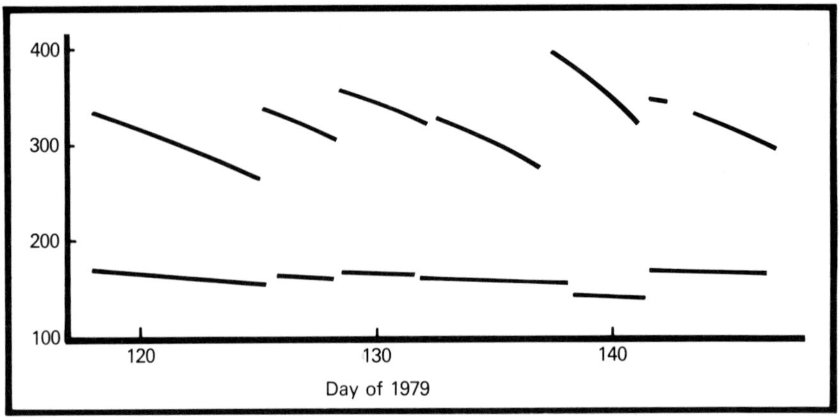


Imagery is probably returned by TV from orbit and film capsule ejection at regular intervals. Morse code transmissions of repeated TVs on 19.995 MHz are suspected as emanating from such capsules. They often persist for much longer than recovery beacons from the third generation satellites giving rise to speculation that they might be more difficult for the recovery teams to locate and retrieve. The very brief transmission observed between 0514 and 0516.15 UT on 29 September 1981 raises the possibility of a mid-air snatch-recovery. Since no recovery beacon has been identified as coinciding with the termination of one of these missions, there is some uncertainty as to whether the complete camera installation is recovered or whether the spacecraft is de-orbited over the Pacific Ocean in the manner of Progress spacecraft.

VHF/FM transmissions received from these spacecraft on a frequency of 240 MHz in Stockholm have exhibited both high and low data-rates [12].

The extended durations to a month or more require modifications to the flight patterns of earlier generations. Generally they are launched later in the day to ensure useful lighting conditions over the target areas at the end of the mission. Peter Wakelin has given a formula for the calculation of the 'longitude of local time', $L\,°$,

$$L = A - (T - 720)/4, \text{ corrected to lie between } \pm 180°$$

where $A\,°$ is the longitude of the ascending node of a given revolution measured in degrees West, and $T$ is the time of this equator-crossing expressed as minutes after midnight UT. Calculations of $L$ for these missions at times near to the launches and terminations reveal groupings by inclination and duration which can be used to decide whether or not a particular flight falls into this category. As expected, the value of $L$ increases in a westward direction throughout a mission and, for the flights with durations in excess of 40 days, has a value close to 180° at the end. For these flights manoeuvres are made after four weeks to relocate the perigee to the southbound pass in order to ensure suitable lighting conditions at the time of the perigee-pass over the target area. N. Johnson claims a very tight constraint on the initial lighting conditions for all recoverable photographic reconnaissance flights, pointing out that, independent of the inclination, these satellites reach their initial southern apex at almost identical local solar times [13].

**Summary**

Johnson produced an algorithm for classifying Soviet reconnaissance satellites in which he embodied techniques based on orbital parameters and manoeuvrability which had long been employed by analysts for that purpose [14]. To these he added a further criterion in which he related the perigee-velocity to the image-movement compensation of the photographic process. The high correlation between his classification and those of The Kettering Group based on criteria of transmission frequency, telemetry format and recovery beacon provides strong support for the contention that these missions fulfil a photographic reconnaissance role.

Estimates of the degree of resolution obtainable are difficult to provide in the absence of access to the actual imagery, none of which has come into the public domain from either the USA or the USSR. Grant Thomson suggests photoscales of: 1:3 000 000 that has been used from Salyut 6 for Earth-resources photography for the search-and-find type of mission; 1:250 000 for quick-look coverage; and 1:50 000 or 1:100 000 for close look missions [15]. In an earlier paper, working on a moderate resolution of 35 to 50 cycles per mm for a low-contrast target, he showed that one could expect the general outline of small vehicles, such as cars, or even rather smaller continuous features, such as well-used footpaths, to be discernible [16].

Commencing in 1979, certain flights were specially designated as Earth-resources payloads. There are no detectable differences in their two-tone tracking beacon transmission except for a tendency towards short cycle durations. Some of these payloads transmit the older TK recovery beacon rather than the standard TF. Some announcements refer specifically to operation with the Priroda (Nature) Centre but others omit such a reference. In 1979 flights were all at an 81.2° inclination but since then the inclination has been 82.3°, leading some to speculate that the change could be due to the introduction of an F-2 launch vehicle based on the SS-9 [17].

## *III. Electronic intelligence gathering*

The continued appearance of so many missions with the same orbital parameters, from which scientific results are never published, strongly suggests a

**Figure 2.4. Lifetimes of 74°, 95 min Cosmos satellites**
Note the dramatic effect of solar activity on atmospheric density during 1980 shortening the lifetimes.

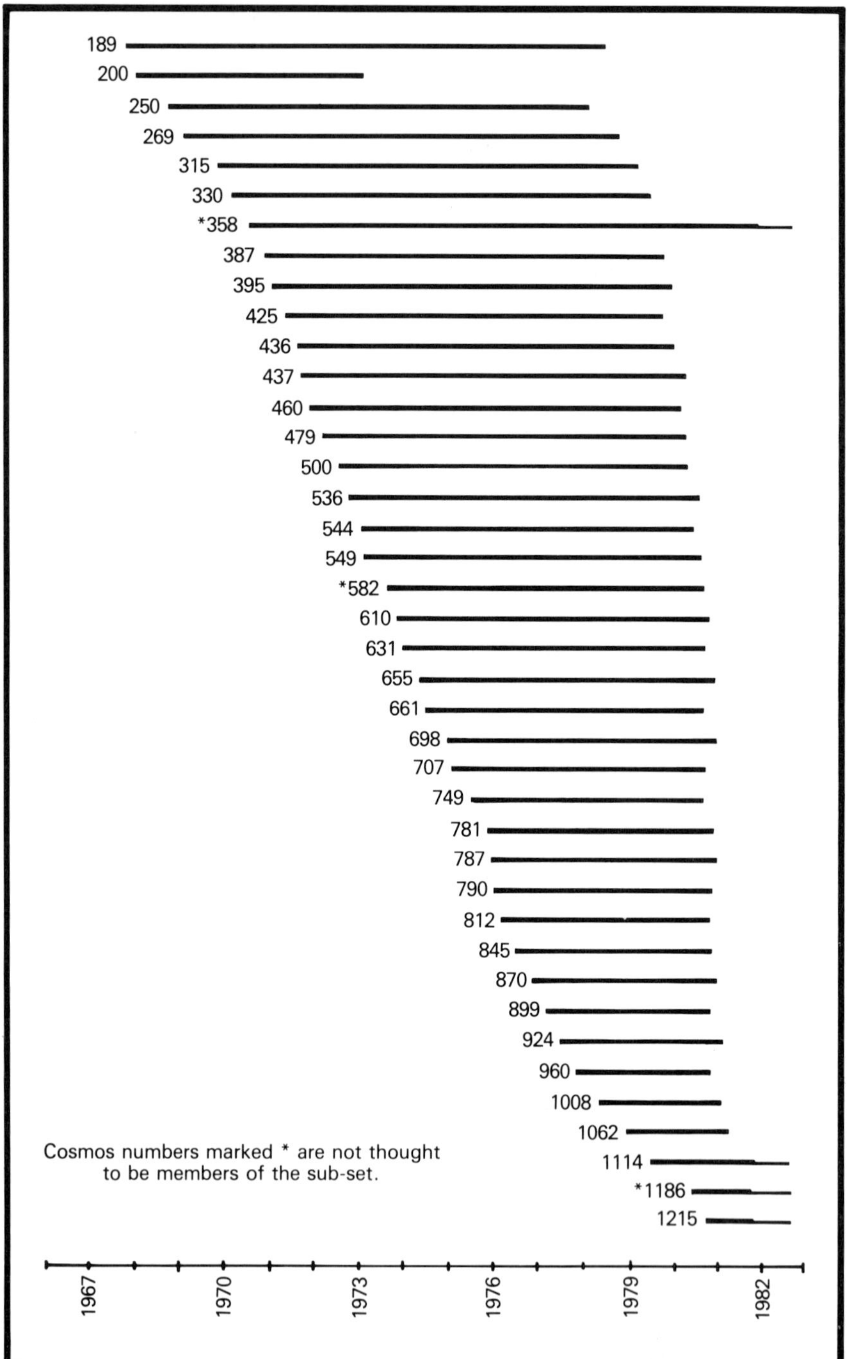

military role. Over the years it has become fashionable to classify such subsets as engaged in electronic intelligence gathering, in the manner of merchant ships which closely follow naval manoeuvres to intercept radio and radar transmissions, or aircraft which fly close to national borders for the same purposes.

As transmissions have not been intercepted from these satellites, it is only possible to demonstrate systems providing global coverage with constellations of satellites deployed in orbital planes spaced at regular intervals around the Earth.

In recent years such systems have included those at 74° inclination with periods of 95.2 min and those at 81.2° with periods of 97.4 min, reminiscent of the early meteors. Both systems were launched from Plesetsk, the first by the C-1 and the other by the A-1. It was shown that the smaller satellites flew in operational groups of four with their orbital planes spaced at 45° [18]. That system would appear to have been phased out since replacements for Cosmos 960, 1008, 1062 and 1114 have not been launched into similar orbits. The very recent launch of Cosmos 1311 into 82.9° inclination with a period of 94.5 min raises the possibility that it might be the forerunner of a replacement system. Satellites in the system, including those still operational, suffered in the period of maximum solar activity 1979–80. Figure 2.4 shows that few of them existed into 1981 due to the increased drag resulting from the greater atmospheric density. Possibly it was not felt worthwhile to launch replacements at a period when orbital decay made station-keeping impracticable.

The larger satellites fly in operational groups of six with their orbital planes spaced at 60°. Table 2.2, based on an analysis of orbital-plane spacing by Darren Conway of Kettering Boys School, shows the replacement

**Table 2.2. Replacement sequence of A-1 launched ELINT ferrets**

| | | | | | |
|---|---|---|---|---|---|
| | | | | 756 | |
| | | | 808 | | |
| | | 851 | | | |
| | | | | 895 | |
| | | | | | 925 |
| 955 | | | | | |
| | 975 | | | | |
| | | | 1005 | | |
| 1043 | | | | | |
| | | | | 1063 | |
| | | | | | 1077 |
| | | | 1093 | | |
| | | 1116 | | | |
| | 1143 | | | | |
| 1145 | | | | | |
| | | | 1154 | | |
| 1184 | | | | | |
| | | 1206 | | | |
| | 1222 | | | | |
| | | | | 1242 | |
| | | | | | 1271 |
| | | | 1315 | | |

sequence for these satellites since mid-1975 with each column separated from the next by 60° in right ascension of the ascending nodes.

Another sub-set of Cosmos satellites which constitutes a global system having a regular replacement policy is that which flies at 74° with periods of 100.8 min. Three of these, with orbital planes spaced at 60° intervals, are operational at any one time relying on both northbound and southbound passes to provide the global coverage. It is thought that these perform store/dump clandestine communications missions in what has become known as a Communications Intelligence (COMINT) role. In April 1977 it was alleged that an Iranian named Rabbani was arrested by the Shah's police. He was, apparently, receiving coded communications from the KGB via a satellite orbiting overhead and displaying the data on the LED output of a small calculator.

## *IV. Early warning satellites*

The Soviet Union was slow to exhibit the ability to place payloads into geosynchronous orbit stationed above the equator. In 1972 beginning with Cosmos 520, the Soviets implemented a series of satellites to provide early warning of rocket and missile launches from the USA, using semi-synchronous orbits with a long dwell-time in the northern hemisphere, based on Molniya technology. Cosmos satellites with Molniya-like parameters, such as Cosmos 41, 174 and 260, could be considered as tests or replacement failures related to the Molniya communications system, but others, including Cosmos 520, 606, 665 and 706 did not fall into the well-defined Molniya positions. Moreover, the argument of perigee for these satellites differed significantly from that of the Molniyas, having values close to 315° rather than 280° [19].

The relocation of perigee affects the shape of the ground-track considerably. The northern hemisphere loops near apogee are much broader than those of the Molniyas. However, it is the new position of the stabilized orbit which is more significant. The usual ascending node for Molniya satellites is chosen to be at 115° W. By moving this 25° to the east the satellites can easily perform an early warning role for a considerable part of their orbit, being able to see into the mid-western USA where the intercontinental ballistic missiles (ICBMs) are located and to communicate directly with the USSR simultaneously [20].

Orbital-plane spacing analysis by Mark Simmonds at Kettering Boys School established the history of the system. Following the initial development flights an orbital-plane spacing of 80° was adopted. At first the reason for this choice was obscure but it now appears that a system of nine satellites at 40° spacings to give complete coverage is gradually being established.

February 1981, coinciding with the launch of Cosmos 1247, saw the initiation of a complex set of manoeuvres designed to restabilize all ground-tracks of the operational satellites with their ascending nodes located close to

**Figure 2.5. Locations of ascending nodes of Cosmos 1191 and Cosmos 1247 during the first quarter of 1981**

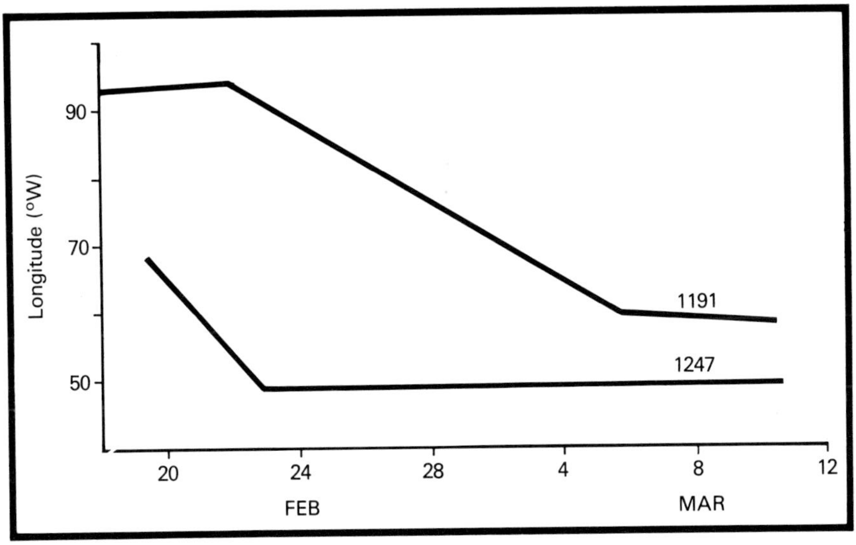


50° W [21]. Figure 2.5, originally produced for a meeting of the Gatwick Branch of the Royal Aeronautical Society, shows the manner of placing Cosmos 1247 at the desired location together with the relocation of Cosmos 1191. Plots for Cosmos 1172, 1217 and 1223 would be similar to that for Cosmos 1191 and are omitted to reduce confusion.

At present the system is again in a state of flux. Cosmos 1191 has drifted well off station since mid-year but the other four remain with ascending nodes close to 55° W. The three latest launches in the series have never been stabilized in that position. Cosmos 1261 has drifted steadily off station when failure to stabilize the ground-track correctly resulted in an orbital period of nearly 719 min with a consequent 0.5° per day drift. Cosmos 1278 reverted to the former practice of taking an initial orbital period in excess of 12 hours and was stabilized in the old position close to 90° W where it remains today. Cosmos 1285, launched to replace Cosmos 1261, was never stabilized at all and has a daily westward drift of nearly 5° per day. Cosmos 1317 became the eighth operational satellite in the constellation in October 1981.

## *V. Ocean surveillance*

Two sub-sets of Cosmos satellites are thought to fulfil ocean surveillance roles of differing types. The first to appear were those which flew in a low circular orbit at 65° inclination with an 89.5-min period but with a perigee close to 250 km rather than the 200 km of the photo-reconnaissance satellites. Initially the payload remains attached to the second stage of the launch vehicle, but after the operational period of the mission, which has gradually been increased with the passing of time, it separates from this stage and is manoeuvred to a circular orbit with a period of more than 130 min, leaving a platform and final rocket stage in the lower orbit from which they decay quickly. The end-of-mission manoeuvre is considered to be a safety precaution to prevent re-entry into the Earth's atmosphere of the radioactive power source which, following the uncontrolled re-entry of Cosmos 954 over northern Canada in January 1978 was confirmed to be a nuclear reactor fuelled with enriched uranium-235 [22]. The need for the great amount of electrical power provided by such a source pointed to the use of a side-looking radar for the detection of port and harbour installations and the location and movement of shipping.

Commencing with Cosmos 651 and 654, these satellites operated in pairs and station-keeping was rigidly enforced. Other pairs were Cosmos 723 and 724, Cosmos 860 and 861, and Cosmos 952 and 954. Following the Cosmos 954 accident there were no more flights of this type until Cosmos 1176 in 1980. This operated alone but there have been three such flights to date in 1981. Cosmos 1249 stayed in the low orbit for 106 days before being raised. Cosmos 1266 was launched into a co-planar orbit but stayed in the low orbit for only 8 days during which time Cosmos 1249 was still operational. Cosmos 1299 flew alone but once again stayed only a short while in the low orbit—13 days.

These satellites transmit on frequencies of 19.542 MHz and 166 MHz and shortwave signals were still received for a short period following the transfer to the high orbit from Cosmos 1266 and 1299.

They are launched from Tyuratam by the F-1-s vehicle where the 's' denotes a sustaining engine of the ion-thruster type which produces micro-thrusts at regular intervals to counteract orbital decay due to atmospheric drag. This type of engine was first tested on Cosmos 699 which operated in a near-circular orbit of 93.3 min [23]. This and its two successors fragmented in orbit at the end of their active phases when the small manoeuvring ceased. Although these satellites in the higher orbit appeared to be related in some way to the pairs in the lower orbits, it has not been possible to discover direct evidence of an ocean-surveillance role for them, but Johnson has shown that it is usual for there to be an initial plane separation of 132° to 150° between different types of satellite which are simultaneously operational [24]. His further conclusion—that this is to provide coverage of the band between latitudes 22° and 38° N by utilizing the intersection of the ground-tracks of the southbound passes of the side-looking radar satellites and the northbound passes of those in the higher orbit—seems less sound.

It is now generally agreed by Western analysts that the 93.3-min satellites have a passive role, collecting radio transmissions and radar signatures in an Electronic Intelligence (ELINT)/COMINT role. Following the Cosmos 954 accident, Cosmos 1094 and 1096 became the first operational pair of 93.3-min satellites [25]. Both of these decayed naturally from orbit after the sustaining microthrusts ceased. Their orbital planes were co-planar and a 23-min time difference was maintained during the operational phase. They were followed in 1980 by Cosmos 1167 and 1220 with launches separated by eight months. It was shown that these again operated as a pair rather than Cosmos 1220 being a replacement for Cosmos 1167 [26]. This time it was arranged that the ground-track of one satellite fell precisely mid-way between the consecutive ground-tracks of the other. Cosmos 1167 was replaced by Cosmos 1260 and then fragmented in orbit. Cosmos 1260 was, in turn, replaced by Cosmos 1286 and 1306 which are currently operating as a pair with ground-tracks spaced precisely mid-way between each other.

## *VI. Navigation*

The only Soviet satellite announced as specifically designed for navigation was Cosmos 1000, launched on 31 March 1978 [27]. The *Tass* announcement mentioned various non-military applications and a model of the 'Tsicada' satellite was displayed at the Paris Air Show in 1979.

It had long been established that there had been three separate systems of satellites providing the necessary global coverage for an operational system of navigation beacons [28]. Later, Wood had shown that the telemetry on frequencies close to 150 MHz from the third of these systems consisted of an f.s.k. binary code transmitted at a rate of 50 bits/s employing transitional encoding [29]. Within each minute there were eight blocks, each of seven lines. The first four of these blocks contained the geocentric co-ordinates and rates of change of those co-ordinates were updated at three-minute intervals. The remaining blocks contained the orbital parameters of the operational satellites in the system at an epoch close to midnight on Friday night, Moscow Time. By utilizing blocks in consecutive minutes it was possible to provide information on up to eight satellites at any one time. Immediately following the statement of Moscow Time in the third line of each parameter block was an identity number followed by the semi-major axis of the orbit in kilometers. These are not unique and originally related to a particular satellite only during its operational lifetime. They were then reallocated to replacement satellites but, until mid-1979, no replacement took the identity number of the satellite it was replacing. Complete reallocation of identity numbers occurred sometime between 5 August and 16 November 1979. This led to a logical sequence of numbering, from 1 to 6, of orbital-plane position round the equator. Since then, replacement satellites have assumed this logical number and the satellite being replaced has been renumbered 7 or 8 for the remainder of its operational life.

This third system consists of six operational satellites with planes spaced at 30° intervals. Cosmos 883 was not placed in any of these positions and had its ascending node in the opposite hemisphere. Although having an identical signal format, its identity number was 11. Numbers 9 and 10 have never been observed. It was followed by Cosmos 926 with an identity number 12 and ascending node 45° away from that of Cosmos 883 in the hemisphere opposite to that of the third operational system. When Cosmos 1000 was announced as having a navigation role and was found to be a further 45° away and numbered 13, it strongly suggested that these satellites with numbers greater than 10 were a civil system and, by implication, that those with numbers from 1 to 8 were mainly for military and naval use. Cosmos 1092, numbered 14, completed a four-satellite system with orbital planes spaced at 45° intervals giving complete global coverage.

It has been a strange fact that the 'civilian' navsats have had much longer operational lifetimes than their 'military' counterparts. Possibly this is because less precision is required for non-military purposes necessitating less frequent replacement due to some small degradation of performance.

## *VII. Communications*

As mentioned above when considering early warning satellites, the USSR adopted highly elliptical semi-synchronous orbits with periods close to 12 hours having a long dwell-time near apogee over the northern hemisphere. Three variants of the Molniya spacecraft have flown since 1965 and, although the Molniya-2 was phased out in 1977, Molniya-1 and Molniya-3 spacecraft are still currently employed.

It has been shown that these satellites were originally stationed in four positions with ground-tracks stabilized and orbital planes spaced at 90° intervals. Satellites followed the same track at six-hour intervals which produced continuous coverage for ground-stations within the Soviet Union with a generous overlapping at either end of each period [30]. The Molniya-1 spacecraft in the four groups A to D were joined in turn by Molniya-2 and Molniya-3 satellites so that, by the end of 1975, there was one operational satellite of each type in each group. This comfortable state of affairs was rudely disturbed early in 1976 by the launch of the 32nd Molniya-1 into an orbital plane mid-way between groups D and A [31]. Three more launches in March and July of that year completed a set of four satellites with planes spaced mid-way between each of the four original groups. It was suggested that the Molniya-1 had been transferred to a wholly military role with a subsequent relocation of positions [31]. No sooner had this been published than four more Molniya-1s were placed into the original groups.

S.J. Birkill has shown that current usage of the UHF channel is divided between two independently used sets of Molniya-1 satellites [32]. The four operational spacecraft in groups A to D employ a series of FM carriers for three hours each side of apogee on the Asian loop of the orbit, over the Soviet Union. They are also used, though with fewer carriers, in their North American loop, over Canada. The second set of four satellites, those with orbital planes intermediate between four standard groups, appears to operate only on the Asian loop radiating an estimated 12.5 kilobits/s digital f.s.k. signal.

These observations indicate that the Molniya-1 satellites in the main groups still carry domestic traffic but that the inter-group spacecraft operated in what may be presumed to be a military mode.

## *VIII. The anti-satellite (ASAT) activities test of 2 February 1981*

Joint operations between C-vehicle launched target spacecraft and F-1-m launched interceptor spacecraft have been shown to constitute a series of tests designed to perfect an inspection-destructive capability for use against hostile spacecraft in times of war [33]. Two such tests were conducted during the first quarter of 1981.

Cosmos 1241, the target satellite, was put into orbit from Plesetsk on 21 January with orbital parameters of 105 min, 1 012–978 km, and 65.8°. Basing his calculations on data from the 1980 interception of Cosmos 1171 by Cosmos 1174, Grahn deduced that the most probable time for an interception would occur early in the morning of 2 February. By 30 January he had discovered the transmission frequency of Cosmos 1241 and had determined that the signals were of a pulsed-radar type. On the morning of 2 February he picked up these signals again from Cosmos 1241 from 0345.49 to 0346.35 UT. He immediately began to search likely frequencies for signals from the expected interceptor and discovered those on a frequency well used by F-1 launched payloads at 0358.20. These were pulse-position-modulated AM signals and faded out at 0402.40. Cosmos 1243 had been launched from Tyuratam at approximately 0225 and these signals were received at the end of the first orbit. The initial orbital parameters were similar to those of the rocket: 88.2 min, 308–161 km, 65.1°. At the time of launch of Cosmos 1243, Cosmos 1241 was southbound near the Kamchatka peninsula. By virtue of the lower orbital period, Cosmos 1243 was catching up with Cosmos 1241 when, on reaching the equator at the end of the first orbit, the manoeuvring 'm' stage was fired to produce the necessary 0.7° plane change and initiate a transfer orbit with parameters of 97.9 min, 1 017–297 km, 65.8°. By the start of the third orbit the time difference had been reduced to almost zero and Cosmos 1243 began its climb to rendezvous with Cosmos 1241 near apogee. Grahn picked up signals from both spacecraft simultaneously shortly after 0530 and lost them both at 0544.56 when they passed beyond the Stockholm horizon [34].

Such interceptions are termed co-orbital and permit inspection as opposed to the earlier fast-fly-by type of interceptions. Cosmos 1243 was deorbited following the fly-by but Cosmos 1241 remains in orbit today, maintaining its orbit by use of microthrusts to overcome atmospheric drag. A second interception was made by Cosmos 1258 in 14 March following a similar flight plan. On that occasion reports were made that a Pentagon source had claimed that the target satellite had been successfully "knocked out", the interceptor exploding next to it. Following earlier tests it was indicated by Western sources that an approach to within a distance of 1 km could be considered as successful [35].

## *IX. Conclusion*

It has been demonstrated that certain aspects of the Soviet space programme are open to interpretation as serving military ends in the fields of photographic reconnaissance, electronic intelligence gathering, early warning, ocean surveillance, navigation, geodesy, communications, and anti-satellite inspection and destruction. The US Department of Defense has recently concluded that the Soviets have a vigorous and constantly expanding military space programme and that they appear to be interested in, and possibly developing, an improved ASAT [36]. They estimate that 70 per cent of Soviet space systems serve a purely military role and that another 15 per cent serve dual military/civil roles. They expect the predominantly military programme to continue to produce steady gains in reliability, sophistication and operational capability and the studies of The Kettering Group provide no grounds for disagreement with such an assessment.

**References**


1. SIPRI, *World Armaments and Disarmament, SIPRI Yearbooks 1973–1981* inclusive (Almqvist & Wiksell, Stockholm, and Taylor & Francis, London, 1973–1981).
2. *The RAE Table of Earth Satellites, 1957–1980*, (Macmillan Press, London, 1981).
3. Perry, G.E., 'Reconnaissance aspects of 8-day Cosmos satellites?' *Spaceflight*, Vol. 10, 1968, pp. 204–206.
4. Perry, G.E., Annex 2 in *Soviet Space Programs, 1971–75*, (Library of Congress, Committee on Aeronautical and Space Sciences, US Senate, Washington, D.C., 1976).
   (a) —, p. 469.
   (b) —, p. 461.
5. SIPRI, *Outer Space—Battlefield of the Future?* (Taylor & Francis, London, 1978), p. 18.

6. Perry, G.E. and Flagg, R.S., 'Telemetry from Russian spacecraft', *Journal of British Interplanetary Society*, Vol. 23, 1970, pp. 451–464.
7. Perry, G.E., 'Two recent Cosmos recoveries', *Flight International*, Vol. 95, 22 May 1969, pp. 844–845.
8. Perry, G.E. and Grahn, S., 'Recovery beacon signals from Cosmos satellites', *Spaceflight*, Vol. 10, 1968, pp. 142–143.
9. Perry, G.E., *Spaceflight*, Vol. 16, 1974, p. 69.
10. Perry, G.E. and Grahn, S., 'In-flight frequency changing of Cosmos satellites', *Spaceflight*, Vol. 10, 1968, p. 431.
11. *Aviation Week & Space Technology*, Vol. 113, No. 12, 22 September 1980, pp. 14–15.
12. Grahn, S., Private communication.
13. Johnson, N.L., 'Soviet satellite reconnaissance activities and trends', *Air Force Magazine*, March 1981, p. 92.
14. Johnson, N.L., 'An algorithm for classifying photographic reconnaissance satellites', *Journal of British Interplanetary Society*, Vol. 34, 1981, pp. 197–199.
15. Thomson, G.H., 'Soviet photo-reconnaissance satellites—what are they used for?' *British Journal of Photography*, Vol. 127, 21 November 1980, pp. 1164–1165.
16. Thomson, G.H., 'Soviet spy satellites—what can they see?' *British Journal of Photography*, Vol. 126, 10 August 1979, pp. 774–775.
17. *TRW Space Log*, (TRW Defense and Space Systems Group, Redondo Beach, Calif. 90278, USA, 1980), pp. 85, 87–88 and 91–94.
18. Perry, G.E. and Mobbs, S.M., '74°, 95 min Cosmos satellites', *Spaceflight*, Vol. 22, 1980, pp. 38–39.
19. Perry, G.E., 'Soviet early warning satellites', *Journal of British Interplanetary Society*, (in press).
20. Perry, G.E., 'Russian geostationary and early-warning satellites', *Royal Air Force Quarterly*, Vol. 17, Autumn 1977, pp. 273–279.
21. Perry, G.E. and Parker, S.J., Unpublished data.
22. Perry, G.E., 'Russian ocean surveillance satellites', *Royal Air Force Quarterly*, Vol. 18, Spring 1978, pp. 60–67.
23. Perry, G.E. and Attwood, R.J., 'Cosmos 699 explodes', *Flight International*, Vol. 107, 1 May 1975, p. 718.
24. Johnson, N.L., 'Orbital phasings of Soviet ocean surveillance satellites' (to be published).
25. Perry, G.E., Sharpling, G.P. and Sharpling, T.A., 'The joint mission of Cosmos 1094 and 1096, 1979–33A & 36A', *Observers' Forum*, No. 62, S.O.N. 738/79, Appleton Laboratory, 9 August 1979.
26. Perry, G.E., Sharpling, G.P., Sharpling T.A. and Grahn, S., 'Ocean ELINT Cosmos satellites', *Flight International*, Vol. 118, 20 December 1980, p. 2278.
27. *Tass* (in English), 0935 G.M.T., 1 April 1978.
28. Perry, G.E., 'Soviet navigation satellites', *Royal Air Force Quarterly*, Vol. 18, Autumn 1978, pp. 276–284.
29. Wood, C.D. and Perry, G.E., 'The Russian satellite navigation system', *Philosophical Transactions of the Royal Society of London*, Vol. A 294, 1980, pp. 307–315.
30. Perkins, P.J. and Perry, G.E., 'A pattern of Molniyas', *Flight International*, Vol. 107, 16 June 1975, p. 79.

31. Perry, G.E., 'The Molniya communication satellites', *Royal Air Force Quarterly*, Vol. 17, Summer 1977, pp. 154–162.
32. Birkill, S.J., 'Channel frequencies and utilization in the Soviet communication satellite systems, with particular reference to the television traffic', Annex in *Soviet Space Programs, 1976–80*, (Library of Congress, Washington, D.C., to be published in 1982).
33. Perry, G.E., 'Russian hunter-killer satellite experiments', *Royal Air Force Quarterly*, Vol. 17, Winter 1977, pp. 328–335.
34. Grahn, S., Kettering Group, Internal communication.
35. Covault, C., 'New Soviet antisatellite mission boosts backing for US tests', *Aviation Week & Space Technology*, Vol. 112, No. 17, 28 April 1980, p. 20.
36. *Soviet Military Power*, (U.S. Dept. of Defense, Washington, D.C., 1981), pp. 79–80.

# Paper 3. Navigation satellite systems: their characteristics, potential and military applications

**K.D. McDonald***

*Federal Aviation Administration, 800 Independence Avenue, S.W., Washington, D.C. 20591, USA*

## *I. Introduction*

The science of navigation and position determination has progressed from the use of natural satellites, such as the Moon and the planets, for primitive navigation of ships, to the recent use of man-made or artificial satellites for navigation, position location, geodesy, survey, mapping and position surveillance. Observations of the Doppler signal from the first satellite, the Soviet Sputnik 1, launched in 1957, provided the basic concept for the first navigation satellite system, the US Navy's TRANSIT. So we see that the origins of man-made navigation satellite systems go back to the beginnings of the space age. Further, we find that progress in navigation satellite system concepts and technology has evolved hand in hand with the rapid developments which have characterized the satellite field in general, and with which we are all familiar.

This paper primarily considers and describes those navigation satellite systems, both planned and deployed, that have military implications. The principal systems discussed are the operational TRANSIT system of the US Navy, and the NAVSTAR Global Positioning System (GPS) being developed by the US Department of Defense (DoD). Emphasis is placed upon the description of the NAVSTAR GPS because of its advanced stage of development, its strong potential for fulfilling a wide variety of military requirements and missions and the size and sophistication of its space segment. This also appears appropriate in view of the length of time the TRANSIT system has been deployed and its comparatively restricted availability and military applications.

Descriptions of the systems are provided including their principles of operation, performance characteristics, and applications, as well as a few of

---

* The opinions expressed in this paper are those of the author and do not necessarily reflect the official policy of his professional affiliation.

the more important factors involved in their development. Special emphasis has been given to the spacecraft characteristics, the orbital configurations and the military system applications consistent with the stated objectives of the SIPRI Symposium. The ground control and user segments of the space-based navigation systems are also addressed in moderate detail to provide an overall perspective of the systems covered.

## *II. System requirements and performance objectives*

Navigation system limitations or deficiencies frequently constrain their effectiveness and use in military and other operations. It seems appropriate to briefly address the system characteristics which are important to the accomplishment of defence-related missions. These navigation system characteristics or requirements may be categorized as either essential or desirable for defence navigation systems.

The *essential* defence navigation system characteristics include: (*a*) world-wide coverage, (*b*) high accuracy, (*c*) common grid capability, (*d*) continuous availability in real time, (*e*) passive (non-radiating) user, (*f*) unlimited capacity, (*g*) acceptable survivability, security and anti-jamming features, (*h*) operation with dynamic users, (*i*) satisfactory portability, size and weight, (*j*) minimal frequency allocation problems and (*k*) freedom from ambiguities.

The characteristics, or requirements, which are not absolutely essential, but are normally found to be *desirable*, for a military system include: (*a*) all ground stations on own territory, (*b*) minimum propagation limitations, (*c*) acceptable operation under water, (*d*) provision for world-wide time reference, (*e*) compatibility and integrability with other military systems, and (*f*) evolutionary growth from research and development to operational capability.

A principal attraction of satellite-based systems for position location is their line-of-sight operation over very large coverage areas which allows considerable flexibility in the selection of frequency and signal structure. The propagation considerations of Earth-based systems do not hold for satellite systems. It is not necessary to select very low frequencies of transmission, that is, those that diffract around the Earth at the cost of poor accuracy, in order to obtain large areas of coverage. The factors influencing the frequency selection for a satellite position-determination system relate more to the minimization of propagation and related errors (primarily from trans-ionospheric group path delay for ranging systems), reasonable antenna sizes and gains for users of the system, the bandwidth required for the selected signal structure, the spacecraft transmitter power requirements and, of course, the availability of an internationally acceptable frequency allocation [1]. Satellite-based systems, such as TRANSIT and the NAVSTAR GPS, can satisfy a large number of the identified essential and desirable military navigation characteristics.

## *III. The US Navy Navigation Satellite System, the TRANSIT System*

The US Navy Navigation Satellite System (TRANSIT) provides all-weather, passive-user navigation capabilities on a world-wide basis to both military and civil users. The TRANSIT system is composed of three subsystems, as shown in figure 3.1 [2, 3]: a constellation of satellites which continuously radiate navigation signals for reception by system users; a network of tracking and control stations that monitor and update the satellite data; and the user receivers, with their associated antennae, computers and displays.

**Figure 3.1. The Navy Navigation Satellite System**

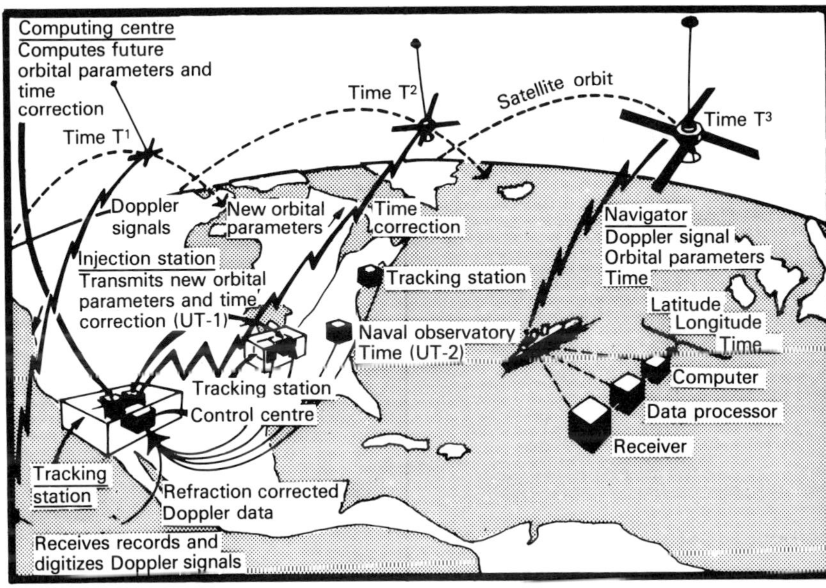


The TRANSIT system began its operational phase in January 1964 and it has been in continuous operation since then. Both the characteristics of the TRANSIT satellite signal and the details of the shipboard receiver were released by the US government for civil use in July 1967. There are now about 4 000 military receiver systems in operation and, as shown in figure 3.2, more than 30 000 commercial systems using the navigation data from TRANSIT [4]. This represents a substantial and impressive increase in the user base, indicating a user growth rate of about 50 per cent per year for the past 10 years or so.

The basic idea behind the TRANSIT system was developed by two scientists at the Johns Hopkins Applied Physics Laboratory (APL), William Guier and George Wieffenbach [5, 6]. In attempting to determine the orbit of Sputnik 1, launched in October of 1957, they were struck by the very

**Figure 3.2. TRANSIT (NNSS): Current and projected user sets**

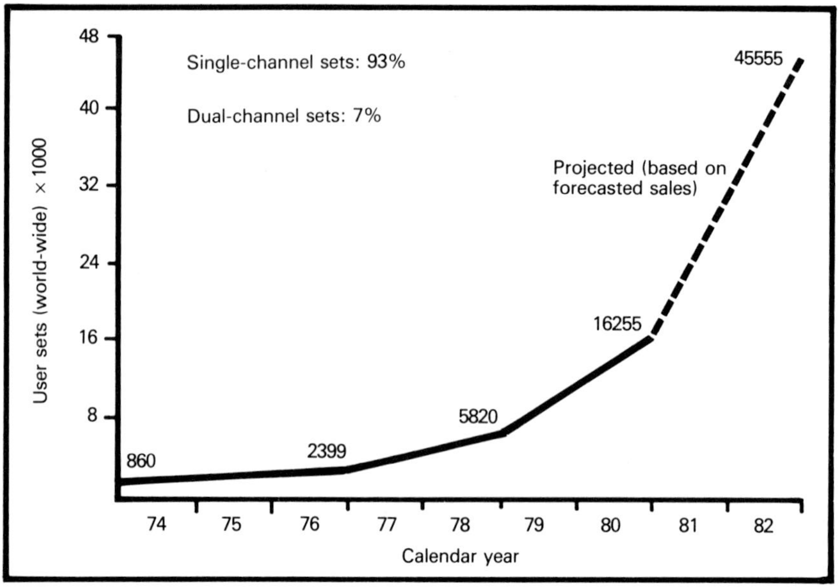


marked and predictable Doppler frequency shift of the signals received from the satellite. It was found that the entire satellite orbit could be established by means of the Doppler measurements made during a single satellite pass. Using principles from dynamic geodesy, Guier and Wieffenbach demonstrated that a satellite trapped in the Earth's gravitational field moves in an orbit which both satisfies the normal constraints placed on orbiting bodies and the restraints imposed on the satellite by its operation in the complex gravitational field near the Earth. They found that only one specific and well-defined orbit could both satisfy these considerations and correspond to the measured Doppler curve.

The TRANSIT system concept was formulated when Dr Frank McClure, Director of the research centre at APL, realized that if Guier and Weiffenbach could determine a complex satellite orbit with Doppler measurements from a single receiving station, then a receiver's unknown position could be determined accurately by Doppler measurements of a known satellite orbit. This reversal of the Doppler orbit determination process is the basic principle of the TRANSIT system.

At about this time the Polaris submarine fleet needed an accurate position-fix capability to update their inertial navigation systems, and McClure suggested to Dr Richard Kershner of APL that a Doppler navigation satellite system could provide this capability. Dr Kershner, who had been working with the Navy on the Polaris programme, assembled a team of APL scientists to study the satellite concept. A navigation satellite development group was established and the first phase of the TRANSIT programme was initiated in December 1958. The three major tasks initially facing the developers of the TRANSIT system were: (*a*) development, fabrication, launch and successful space operation of appropriate spacecraft; (*b*) precise modelling of the Earth's gravitational field so as to permit accurate determination of low altitude satellite orbits; and (*c*) development of user equipment to determine and display the navigation results.

After the initial TRANSIT development funding was provided by the Navy in December 1958, the first spacecraft, TRANSIT 1A, was designed and built in seven months. It was unsuccessfully launched on 17 September 1959, with a Thor-Able vehicle at Cape Kennedy, but useful data was obtained. TRANSIT 1B was successfully launched on 13 April 1960 into a somewhat lower orbit than planned but one which permitted satisfactory experimental operation of the system [7]. Subsequent satellites have been built by APL and the Radio Corporation of America (RCA).

The first operational TRANSIT satellite was launched in December 1963, and since that time the TRANSIT navigation system has always been available to the US submarine fleet and other users.

## System configuration

The TRANSIT satellite constellation is normally made up of four to six satellites in circular polar orbit at an orbital altitude of 1 075 km (figure 3.3) [8]. The satellites orbit the Earth every 107 minutes, continuously transmitting signals at approximately 150 and 400 MHz. The users determine their positions by measuring and examining the Doppler shift of the signals over the duration of a TRANSIT pass. There are four ground stations which track and update each satellite's ephemeris and time data. The control centre is located at Point Mugu, California, which provides the system tracking and data injection facilities, the central computer centre and the operations centre.

The satellite constellation provides world-wide coverage; a single satellite in view allows a user the opportunity for a position fix. Precession of one orbit with respect to another that changes the constellation configuration is minimized by the use of polar orbits. An orbit at 1 075 km was selected as a compromise between the lower orbital altitudes, where the air drag becomes significant, and higher altitudes, which reduce the magnitude of the observed Doppler shift.

The physical configuration of the TRANSIT satellite is shown in figure 3.4 [9]. Four solar cell panels charge the satellite's internal batteries, providing extended eclipse condition operation. The TRANSIT navigation signals are transmitted to Earth by the satellite's antenna which is kept pointing towards Earth by a gravity gradient stabilization boom. The fact that elongated objects in orbit tend to align with the gravity gradient provides a simple and reliable means for obtaining single axis attitude control for the spacecraft. Magnetic hysteresis rods placed along the solar panels interact

**Figure 3.3. TRANSIT orbital constellation**

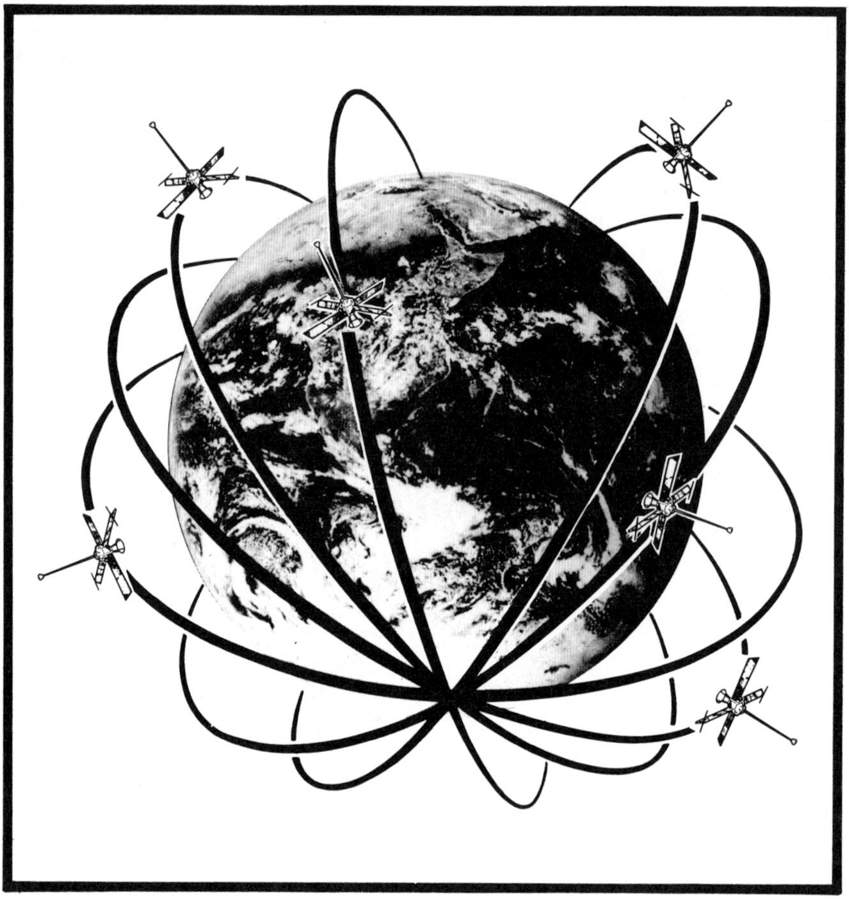

with the Earth's magnetic field and effectively damp out the gravity gradient boom oscillations. The satellites transmit their orbital parameters from an onboard memory system which is updated about every 12 hours by transmissions from ground injection stations. The satellites weigh about 61 kg and are normally placed in orbit with a Scout launch vehicle.

The correction of ionospheric refraction error, and in particular group path delay, is accomplished by transmitting two phase-coherent signals with a substantial frequency separation. The two frequencies of about 150 and 400 MHz were selected to provide signals for which reasonable transmitter power could be achieved at the satellites and omni-directional receiving antennae could be employed by the users.

The average time interval between TRANSIT fixes for a constellation of five satellites varies from about 30 to 100 minutes, as shown in figure 3.5 [8]. A user is able to compute his position based on the data received during a single satellite pass as shown in figure 3.6 [8]. The basic data employed are:

Figure 3.4. Physical configuration of the TRANSIT satellites

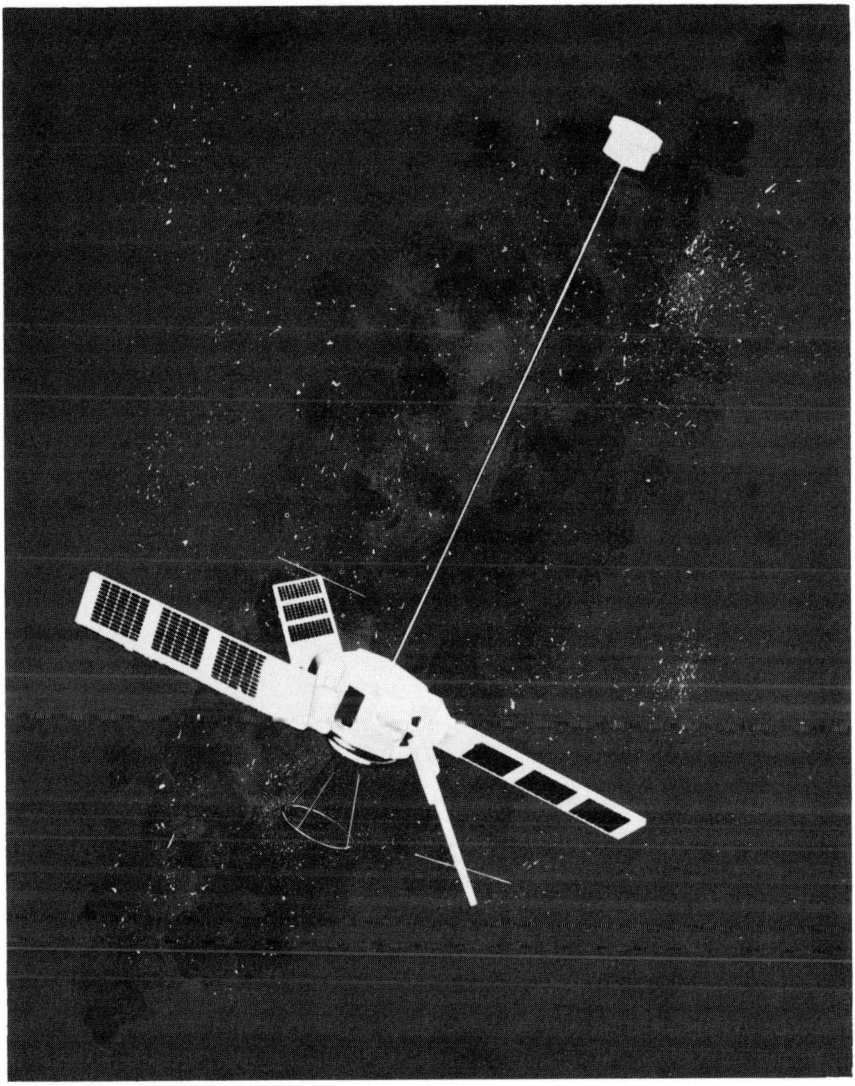

(*a*) the measured Doppler shift over the pass duration of 10–16 minutes—during this period the satellite travels 4 000–7 000 km—which provides an excellent measurement baseline; (*b*) the two-frequency ionospheric refraction correction; (*c*) the satellite orbital parameters transmitted as data modulation on the signal carriers; (*d*) the accurate time marks, also transmitted by the satellite; and (*e*) the vehicle motion for the period of the satellite pass.

Operation of the ground support network is the responsibility of the Navy Astronautics Group, headquartered at Point Mugu, California. It continually tracks the TRANSIT satellites, computes and predicts their orbits

**Figure 3.5. Mean time between position fixes**

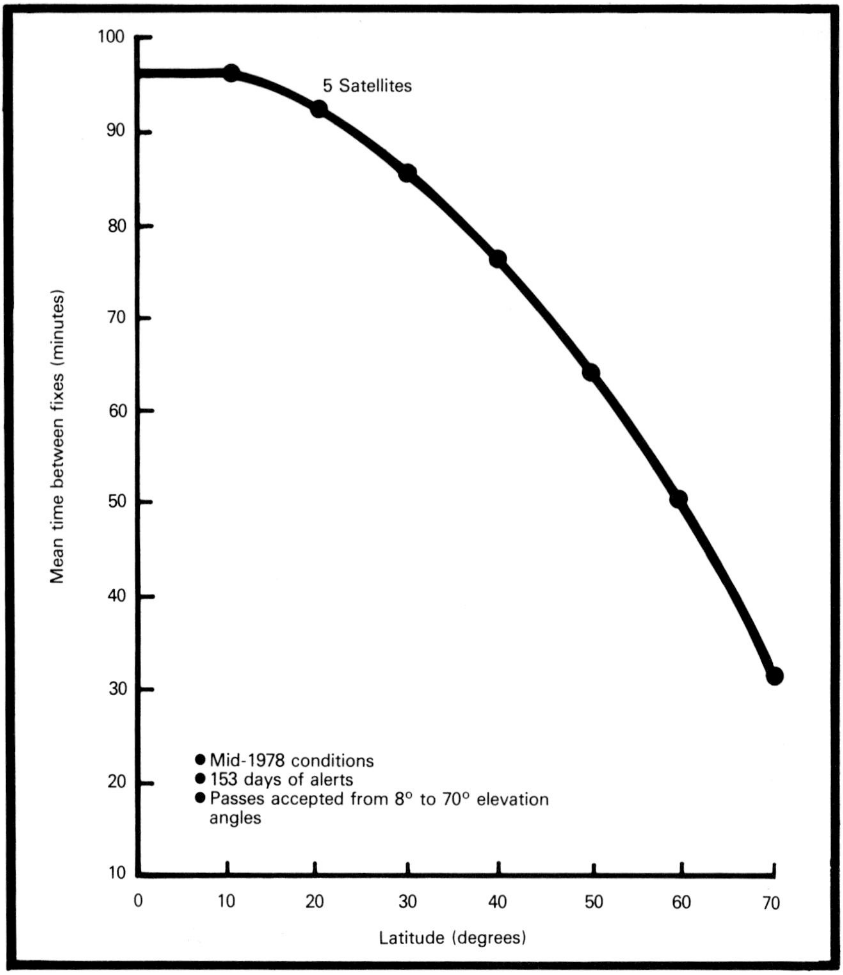


and maintains the orbital parameters in the satellite memory system. The ground support network currently consists of four tracking stations located in Maine, Minnesota, California, and Hawaii, a central computing centre at Point Mugu and injection stations at Point Mugu and in Minnesota. The Naval Weapons Laboratory at Dahlgren, Virginia shares with APL the task of evaluating and updating the model of the Earth's gravitational field. In addition they normally participate in the reduction of Doppler field data which is used to establish world-wide geodetic control points.

A number of improvements have been incorporated into the TRANSIT

**Figure 3.6. Satellite pass geometry and slant range change Doppler count measurement**

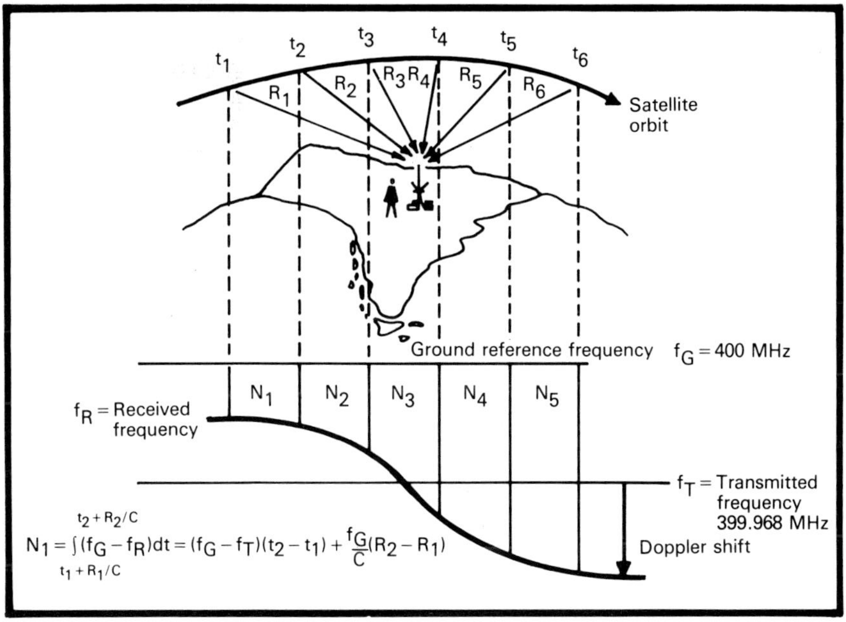


system since its inception. A new TRANSIT spacecraft called NOVA (shown in figure 3.7[8, 9]) has incorporated a satellite-based ranging signal which employs a pseudo-random noise (PRN) phase code modulation for the transmission of precise time and for direct precision range determination. The PRN code allows the TRANSIT satellites to be tracked individually even though their Doppler frequencies may cross or interfere with each other. The PRN modulation is designed to be compatible with and transparent to current TRANSIT navigation user equipment. The existing equipment can continue to operate with both the old and new signals with no change in performance.

The NOVA satellites have also been hardened to radiation effects and the accuracy of the satellite constellation ephemerides has been improved considerably. The technique for enhancing the reliability and accuracy of the satellite orbital predictions uses a system called DISCOS, which stands for Disturbance Compensation System. DISCOS is a drag sensing device which consists of a small proof mass contained within a hollow enclosure within the satellite. Since the enclosure is an internal part of the satellite it is not influenced directly by the drag forces acting upon the spacecraft. The ball is in 'free fall' in a drag-free environment and by sensing its motion inside the satellite it is possible to compensate for the drag (and other) disturbances on the satellite. This is done by applying small thrusts to the satellite which counter the effects of the drag forces and recentre the ball. The overall effect

**Figure 3.7. New generation NOVA satellite**

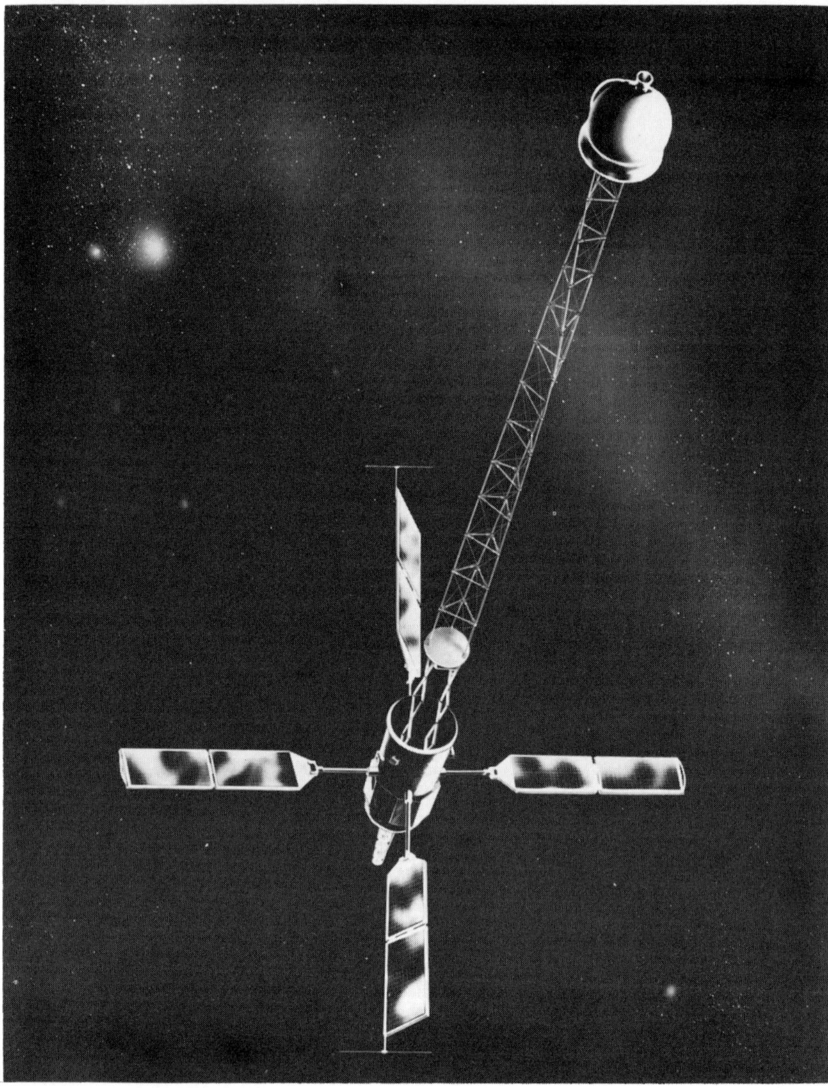

is to correct the orbit to one which is essentially free from drag. This process significantly enhances the stability and predictability of the satellite orbit.

A simplified block diagram of the TRANSIT satellite electronics is shown in figure 3.8 [8]. The transmitted signals are obtained by direct multiplication of a reference oscillator which has a stability of about one part in $10^{11}$ during the period of a satellite pass. This results in coherently related transmissions which have negligible error during a user position determination. The reference oscillator also drives the satellite memory causing the stored navigation message to be read out precisely every two minutes and phase modulation encoded onto both signals at a carefully controlled rate. In this way the satellite transmissions provide users with signals required for Doppler measurement, two frequencies for ionospheric error correction and a data stream modulated onto both carriers which contains satellite ephemeris, precise UTC time and other information.

**Figure 3.8. Block diagram of the TRANSIT satellite electronics**

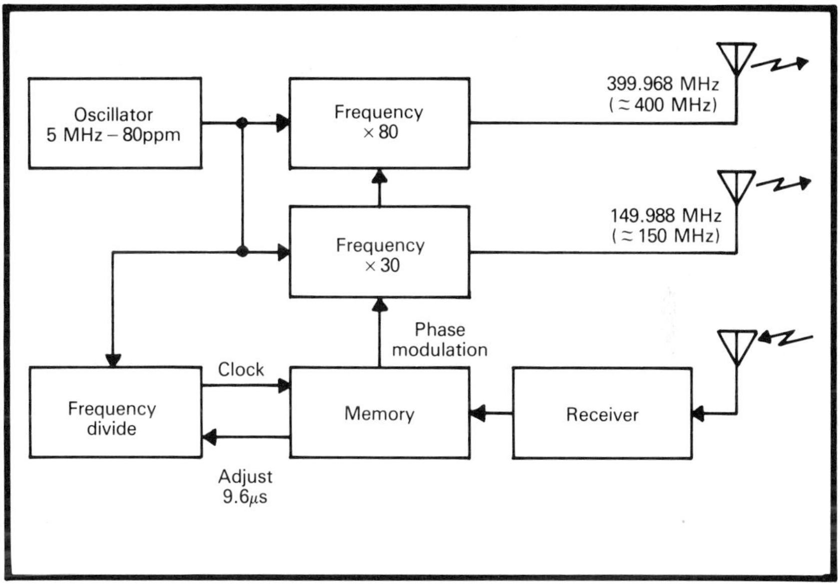


**TRANSIT performance**

A TRANSIT navigation fix contains two principal components of error, as shown in figure 3.9. The first of these is the system error which includes contributions due to uncertainties, or error, in: (*a*) the satellite ephemeris and the system time, (*b*) the propagation correction, (*c*) knowledge of the user's altitude, (*d*) the geopotential (gravity) model used to generate the satellite orbit, (*e*) the modelling of satellite drag and radiation pressure effects and (*f*) the instrumentation and measurement noise of the user receiver. The inherent system error can be measured by operating a TRANSIT set at a fixed location and observing the scatter of the resulting position data. Dual-channel receivers typically obtain navigation results in the range of 25–40 metres rms. Single channel (single frequency) receivers do not remove ionospheric errors and typically have results in the range of 80–100 metres rms, as shown. The second source of navigational position error arises from the uncertainty in the movement of the user vehicle during a satellite pass. This error is a complex function of the geometry of the satellite pass and the direction of the user motion, but it can be approximated as about 370 metres (0.2 nautical miles) of position error for each knot of user velocity error. This degradation in navigation accuracy is shown for both receiver types in figure 3.9 [8].

**Figure 3.9. Approximate satellite position fix error as a function of user velocity uncertainty**

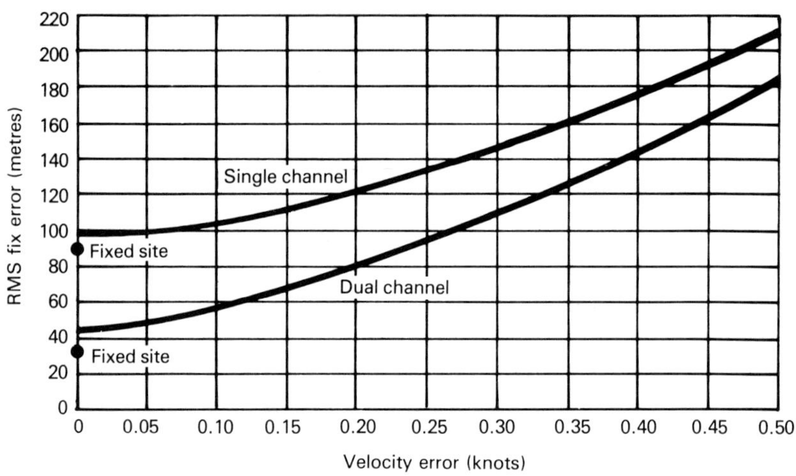


Notes: 1. Maximum single-channel fix error can reach 200 to 500 metres due to ionospheric refraction versus only 90 metres of maximum fix error for dual-channel results.
2. Solving for velocity north can limit fix error to the range of 100-200 metres when velocity is poorly known.

## Military applications

As mentioned earlier, the TRANSIT system was initially developed to provide precise position updates for the Polaris submarine fleet. The system continues to be operated for this specific Navy application. To accomplish this, a submarine exposes its TRANSIT antenna at the appropriate time for a satellite pass and receives the position fix. This information is then used to update and maintain the accuracy of its inertial navigation systems. The US Navy attack submarines and many surface ships are also provided navigation data by TRANSIT. Submarine use of TRANSIT requires that the satellite navigator set provide satellite alert information which advises the user of the appropriate times to expose the antenna. Receiver sets are available which also automatically tune to the appropriate satellite frequency.

Surface ships normally have user sets (such as the MX-1112 navigator) that track every available satellite pass. Applications range from navigating major combat ships to US Coast Guard patrol vessels. TRANSIT is also used for military land survey, geodesy and mapping. The US Defense

Mapping Agency and several NATO nations have co-operated in satellite-based survey operations across Europe.

TRANSIT is an extremely reliable system which provides world-wide coverage with periodic fix availability. It is an all-weather system that achieves excellent accuracies which are dependent on the receiving equipment employed and precise knowledge of user velocity and direction. The system performance is generally competitive with short-range radio-location systems, and for many applications is clearly superior.

## *IV. NAVSTAR Global Positioning System*

The NAVSTAR Global Positioning System is a space-based radio navigation system which provides highly accurate three-dimensional position, velocity, and time information to an unlimited number of suitably equipped users [10, 11]. The DoD programme is in full-scale development and is planned for full operational status by late 1987. At that time the GPS will provide military users with continuous horizontal accuracies of 10 metres circular error probable (CEP) corresponding to an accuracy of 25 metres or better for 95 per cent of the time. A three-dimensional accuracy of about 16 metres spherical error probable (SEP) corresponding to 55 metres or better with a 95 per cent probability should be obtainable world-wide. Similarly, velocity accuracies of 0.1 metres per second or better should be available.

The GPS programme is a joint undertaking of the US military services with the Air Force serving as executive agent within the Department of Defense. A Joint Program Office has been established for the programme at the Space Division of the Air Force Systems Command, with participation by the Army, Navy, Marine Corps, Air Force, the Defense Mapping Agency and the Department of Transportation. Although the GPS has been designated as a navigation satellite system—and indeed it does provide navigation capabilities—it has been primarily justified within the defence community for its weapon-related features. The GPS capabilities provide very substantial improvements in a wide variety of military missions.

The NAVSTAR GPS has been developed to satisfy several long-standing concerns of the DoD. Its implementation should: (*a*) provide multiple users with accurate, continuous, world-wide, all-weather, common-grid, three-dimensional position and velocity data for navigation and other functions; (*b*) reverse the present trend towards proliferation of radio navigation systems used for various missions; (*c*) increase military force effectiveness by enhancing accuracy and other capabilities for weapon delivery; and (*d*) provide a cost-effective satellite host for complementary payloads.

The capabilities provided by the GPS and its potential for widespread use by the United States has resulted in considerable interest in the system by NATO. The USA has signed a formal Memorandum of Understanding with nine NATO countries who each have one or more representatives participating in the system development and in residence at the Joint Program Office. The NATO activities to date have been primarily in understanding how the system operates and in defining standardization and inter-operability specifications. With the progress of the GPS development cycle, additional emphasis will be placed on procurement and acquisition options which may afford the NATO representatives with a greater role in the programme.

### System description

The NAVSTAR GPS implementation [12, 13] consists of three basic parts or segments: space, ground and user equipment. The operational space segment consists of 18 satellites in 12-hour circular orbits at an orbital altitude of 17 540 km. The satellites are to be equally spaced in six orbital planes, each located at a nominal inclination of about 63°. The orbital constellation of the operational system is to provide a minimum of four satellites continuously in view of ground-based users anywhere on Earth. Each satellite transmits the same radio frequencies with a pseudo-random noise (PRN) signal structure and with precise-timing information. The satellite transmissions contain a navigation message which includes ephemeris data and satellite clock bias information. The constellation of satellites is managed and directed by the control segment which consists of several geographically dispersed monitor stations for tracking the satellites and a master control and satellite upload station. This part of the system transmits and receives information to and from the satellites, tracks the satellites and determines their ephemerides, controls the timing of the satellite transmissions, and establishes the navigation message which is transmitted from the satellite to the users. The user segment consists of a family of user sets, or receivers, located in aircraft, ships, land-based vehicles, manpacks, and fixed sites for which a GPS receiver can establish position, velocity and time with great accuracy.

### Space segment

The 18 satellites in the GPS constellation provide coverage of the Earth as indicated in figure 3.10 [13, 14]. The satellite is a solar-panel-powered three-axis stabilized spacecraft as shown in figure 3.11 [15], which transmits two PRN spread spectrum signals at frequencies of 1575.42 MHz (L1) and 1227.6 MHz (L2). The use of two frequencies permits the user to determine ionospheric propagation group path delay. Figure 3.12 [14] shows the general spectrum characteristics of these two signals and demonstrates the inherent jamming immunity of the spread spectrum signals. The use of the L-band frequency range allows good all-weather operation with aircraft which may use hemispherical coverage antennae. Two signal structures are

**Figure 3.10. A GPS space segment constellation**

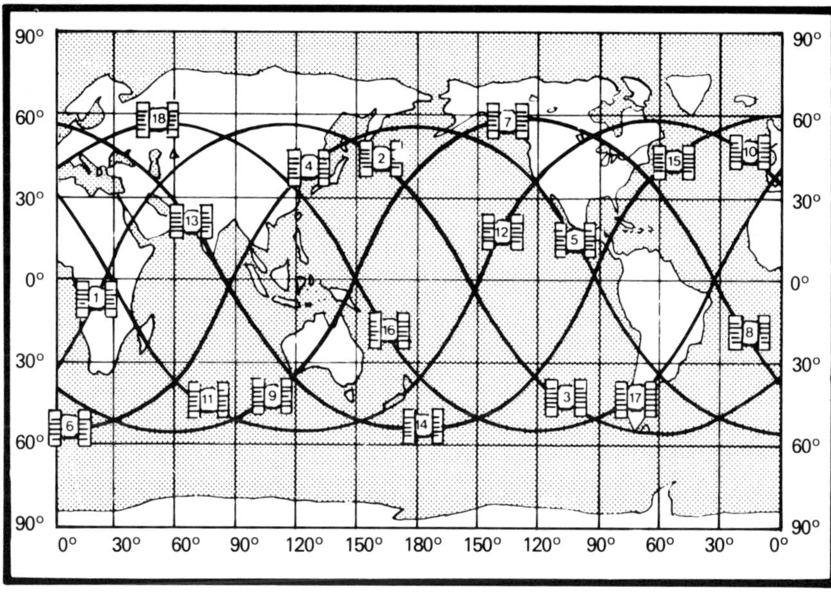


**Figure 3.11. The GPS Block II Satellite**

**Figure 3.12. The spread spectrum signals with jamming immunity**

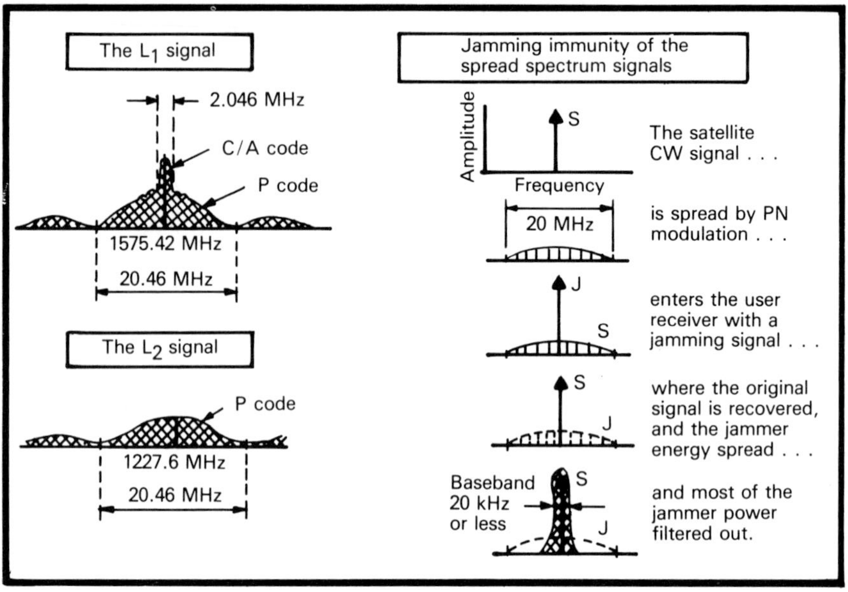


**Figure 3.13. The GPS signal modulation techniques**[a]

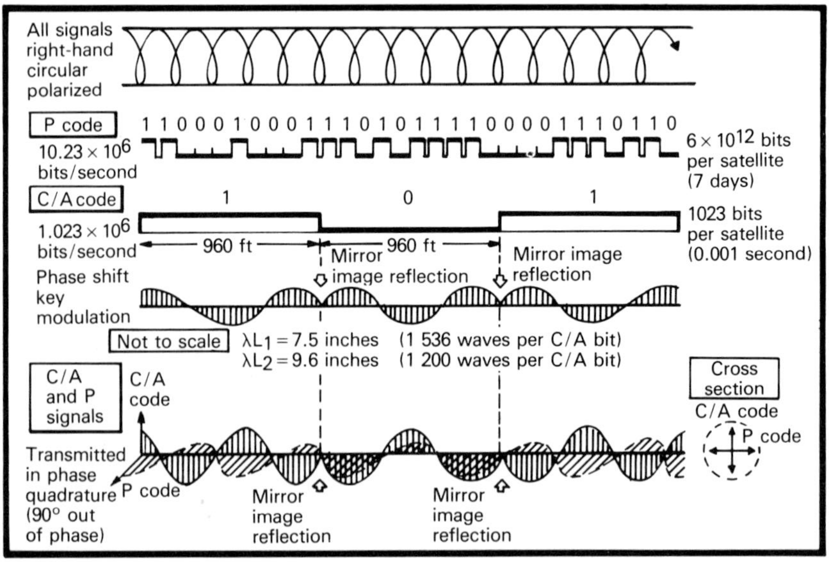


[a] 1 ft = 30.480 cm (12 inches = 1 ft).

**Figure 3.14. The GPS pseudo-random noise codes and data characteristics**[a]

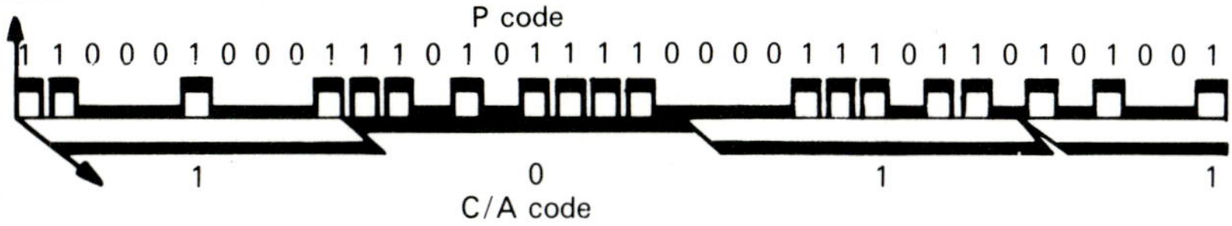


| Parameter | C/A code | P code | Data |
|---|---|---|---|
| Chipping rate | $1.023 \times 10^6$ bits/second | $10.23 \times 10^6$ bits/second | 50 bits/second |
| Spatial length (per bit) | 960 ft | 96 ft | 3 720 miles |
| Repetition interval | 0.001 second | 7 days | Not applicable |
| Code type | Gold code | 280-day pseudo random code | Not applicable |
| Total no. of codes | 37 unique gold codes | 37 seven-day sections | Not applicable |
| Special properties | Easy to acquire | Slightly more accurate navigation resistant to jamming and spoofing. Rejection of multipath. | Provides handover from C/A to P code. Ephemeris data and clock correction. Rejection of multipath. |

[a] 1 ft = 30.480 cm

provided from the satellite: a precise or P-code signal, and a coarse/acquisition or C/A signal, as shown in figure 3.13 [14, 16, 17]. Each of these signals consists of a pseudo-random sequence of binary digits which modulate the carrier frequencies at rates of 10.23 Mbps (megabits per second) for the P-Code and 1.023 Mbps for the C/A Code. The P-Code provides the capability for highly precise navigation and is resistant to electronic countermeasures and multipath interference. The C/A signal provides a rapid acquisition capability for P-Code high precision users and a coarse ranging signal for users who require less precise navigational accuracy. Fifty bps data modulation is transmitted by the satellites on both the P and C/A Code sequences. The characteristics of the GPS pseudo-random noise codes and the data modulation are shown in figure 3.14 [8]. The low data rate navigation message is used to provide the users with the necessary satellite ephemerides and clock parameters so that the receivers can perform their respective position, velocity and time calculations. As shown, the P-Code and C/A Code signals are quadraphase modulated on the L1 and L2 signals.

**Figure 3.15. The control segment**

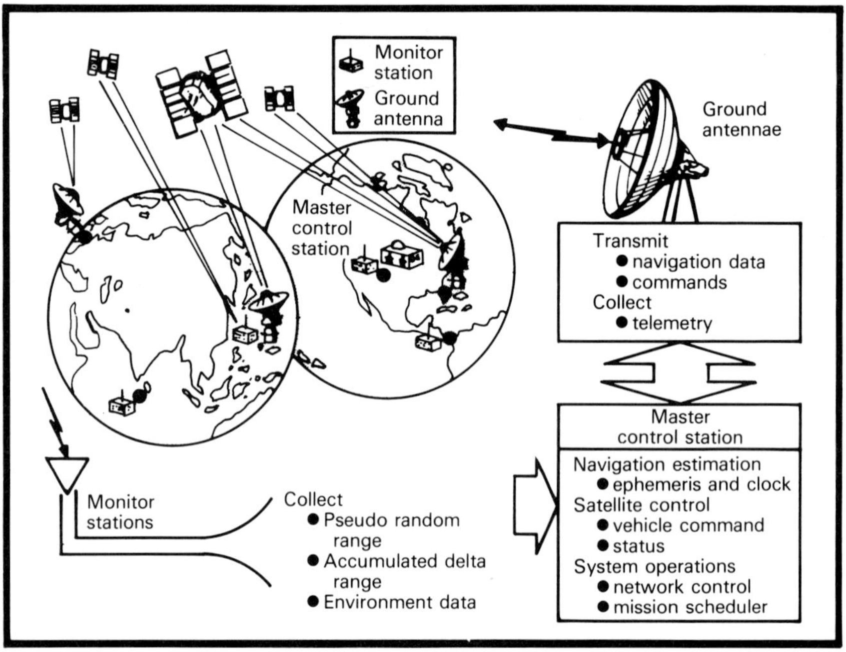


**Figure 3.16. The control segment signal linkages**

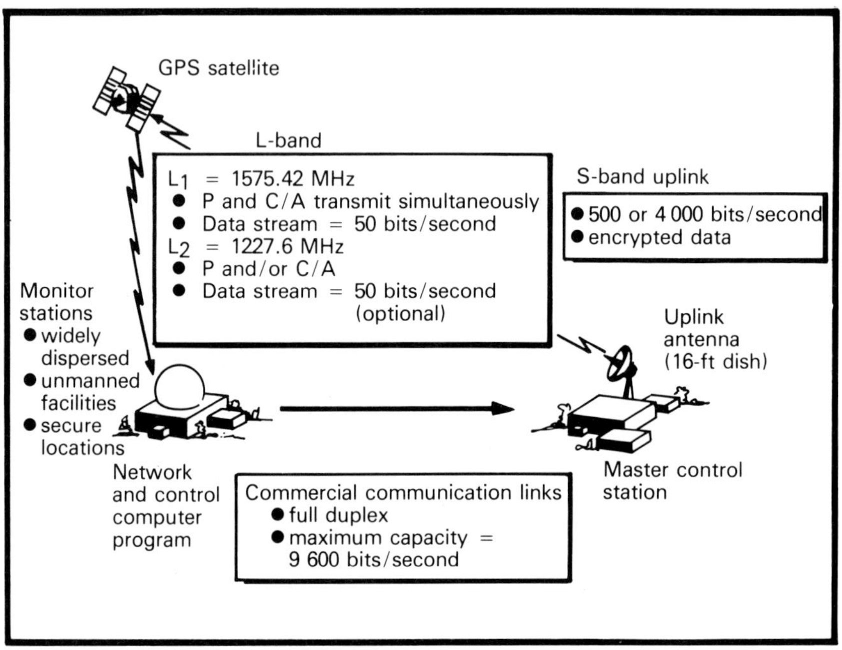

## Control segment

The control segment's primary function is to track each of the satellites in the constellation and to update the satellites daily. Passive monitor stations are utilized with a master control station to collect satellite ranging data, compute the errors, generate a new navigation message and upload each satellite. This general operation is shown in block diagram form in figure 3.15, and the control segment signal linkages are indicated in figure 3.16 [14]. The navigation message is intended to keep the satellite ephemeris and the clock or timing errors within precise limits. The satellite clocks are rubidium or caesium beam atomic standards and have sufficient stability so that even without a daily upload the system stability provides highly accurate navigational data for several days.

## User segment

The GPS user segment is made up of a family of receiver systems. A user set typically consists of a receiver, a navigation processor and some form of control/display unit or other input/output device [18]. The application and sophistication of a user's requirements will determine which of the various GPS user sets should be employed. These requirements will vary depending upon whether the user is a dynamic, high-velocity aircraft or a slow-moving

**Figure 3.17. Navigation signal processing procedures for the high accuracy users**

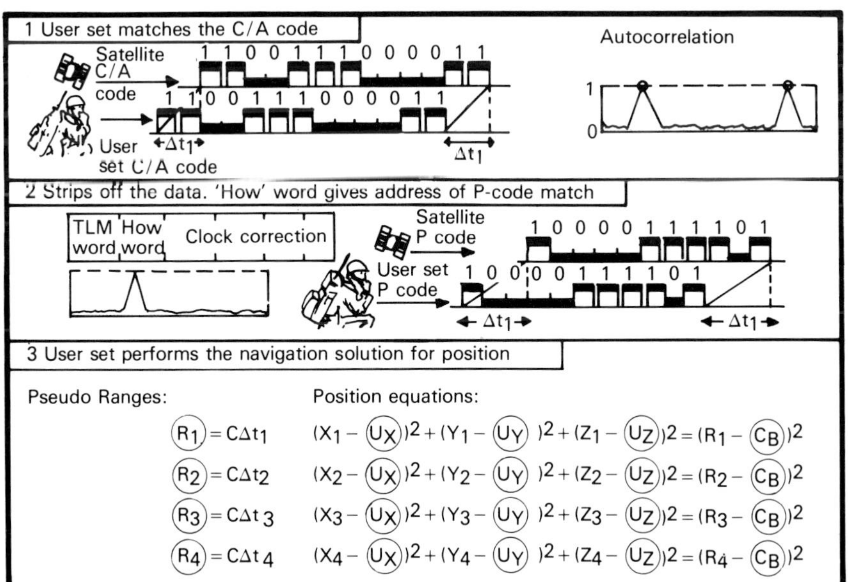

**Figure 3.18. A GPS receiver block diagram**

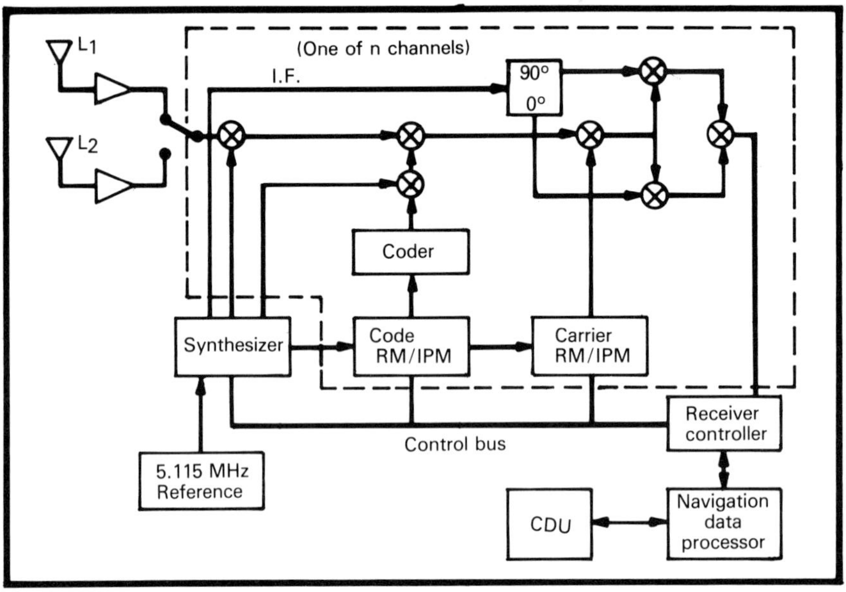


vehicle, and whether highly accurate or only moderately accurate position data are required.

The user set will normally require signals from four satellites simultaneously to determine the time and the position of the user in three dimensions. The receiver equipment does this by determining the distances between the observed satellites from measurements of the propagation time from the satellite to the user. By making four separate ranging measurements from the four satellites, the receiver can solve four simultaneous equations for the four unknowns of user position (in three dimensions) and user time. This process is described in figure 3.17, in which the user set initially acquires the C/A code and strips off the data which provide the information for him to acquire the P-Code. The position solution equations and the pseudo-ranges measured are shown in the figure. The time is normally determined as the user clock bias error. Measurements of the user velocity and the user clock drift can be obtained by determining the carrier frequency Doppler shift for each satellite. Figure 3.18 shows a GPS receiver block diagram for a two-frequency system. A GPS avionics system with cockpit display provides the same general type of display as that of inertial navigation systems.

A common grid co-ordinate system is formed by the GPS measurements. However, the geodetic frame utilized by the system is the 1972 World Geodetic Survey (WGS-72) co-ordinates and the receiver equipment converts the measurements to these co-ordinates or to any other reference system desired [19].

## GPS programme status and organization

The development and implementation schedule for the GPS programme is shown in figure 3.19 [20]. The programme has three distinct phases: the concept validation phase which has been completed; the full-scale development phase currently in process; and the operational phase which follows the current activity if the programme remains successful.

**Figure 3.19. NAVSTAR GPS programme evolution**

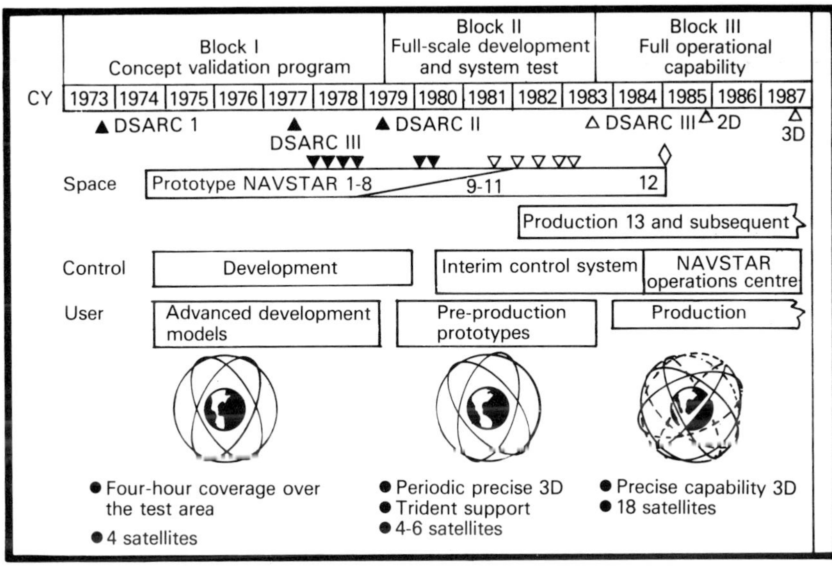


*Phase 1: Concept validation*

The concept validation phase has effectively proven the viability of the system concept. This was done by developing the subsystems and by testing equipment at the US Air Force Yuma Proving Ground Test Range. Satellites were placed in medium altitude orbit such that maximum visibility was obtained at the Yuma site. A control segment consisting of a master control station and four monitor stations was constructed and operated with the satellites and user equipment. The general configuration of the GPS Phase 1 system is shown in figure 3.20 [21]. Advanced development models of a number of the user equipment configurations were designed, built, integrated into user vehicles and tested. The results of this initial phase of the programme validated the system concept and provided sufficient data to justify full-scale development, or Phase 2, of the programme. Field test results were obtained for a large number of mission areas, and several of those results are shown in figure 3.21 [14].

**Figure 3.20. The GPS Phase 1 system**

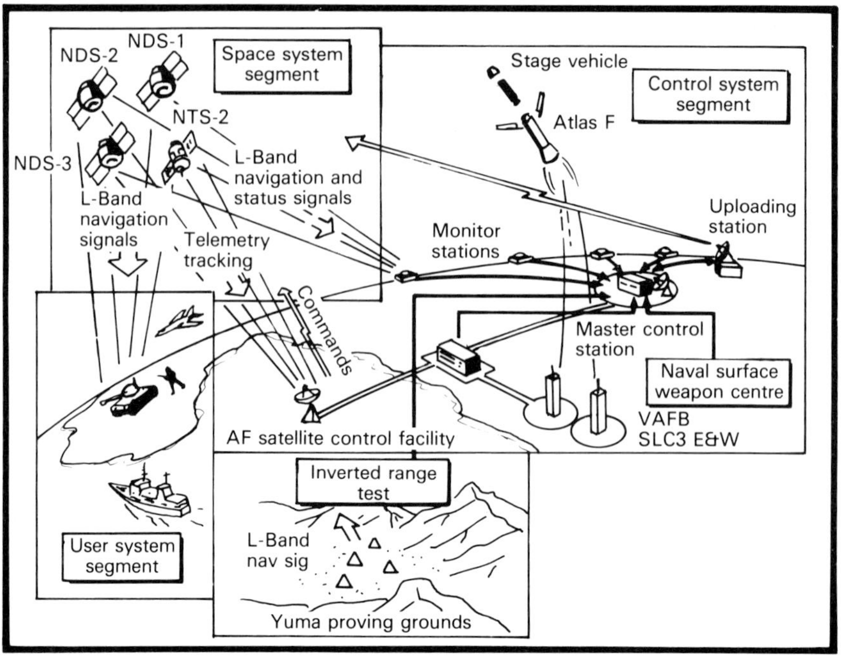


**Figure 3.21. Field test results**[a]

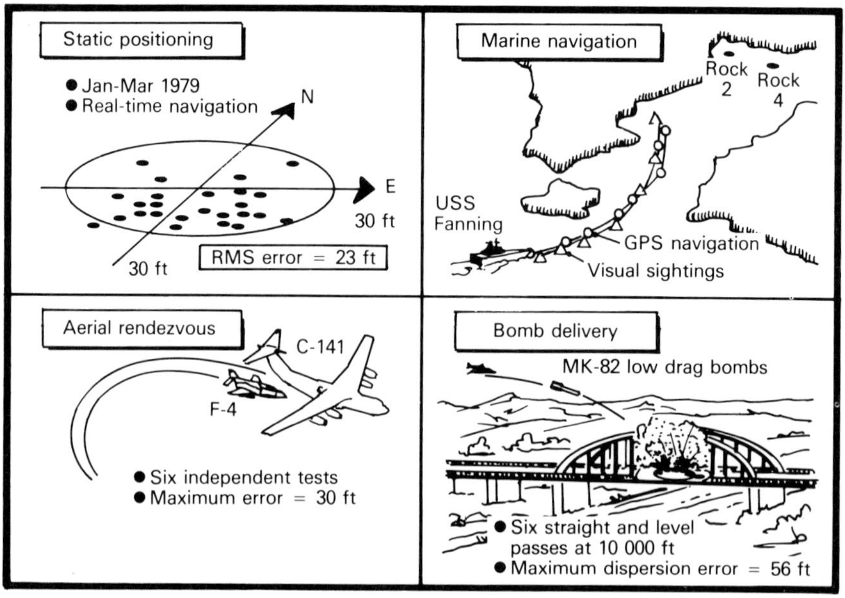


[a] 1 ft = 30.480 cm

*Field test results.* For the GPS to reach its current full-scale development phase, it has completed a test and evaluation phase which incorporated over 600 different test missions. Data from these have provided impressive evidence that the GPS both provides performance improvements and meets performance requirements consistent with the substantial enhancement of operational missions. Additional tests have been conducted since the initiation of full-scale development, and some of these will be discussed with the more significant test results obtained from the test and evaluation phase.

As shown in figure 3.22 [21], the three-dimensional positioning accuracy from all user sets tested was better than 13 metres for 50 per cent of the time, corresponding to about 25 metres for 90 per cent of the time. Overall velocity accuracy for the tests ranged between 0.12 and 0.8 metres per second. The results of the use of GPS for precise weapon delivery of low-drag bombs delivered from a Navy F-4J aircraft demonstrated an accuracy improvement of an order of magnitude over conventional weapon delivery systems. Precision rendezvous between an Air Force C-141 cargo aircraft and a Navy F-4J, between an Army helicopter and a manpack, and between a Navy P-3B and a landing craft, were accomplished utilizing GPS. The overall results of this test activity demonstrated a rendezvous capability of about 30 metres. The US Army performed a number of tests using four manpacks applied to infantry tactics in the field with excellent results; in a dramatic demonstration of GPS capabilities, all four manpacks homed in to within 3 metres of a surveyed rendezvous point on the Yuma Test Range.

Testing of the GPS system on board ship included the comparison

**Figure 3.22. Phase 1 system accuracy test results**

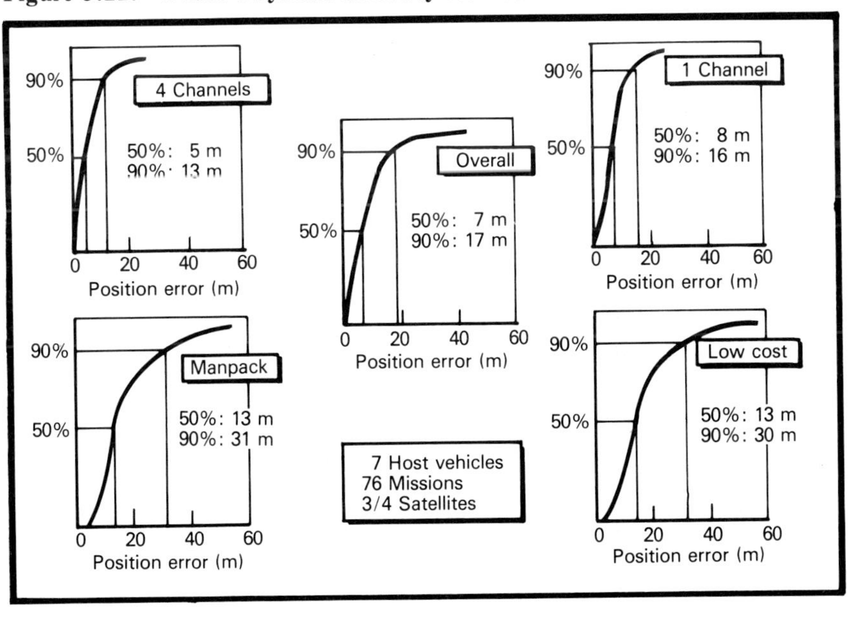

between GPS fixes and visual sitings on board a Navy frigate sailing out of Acapulco Bay. The ship also sailed through the San Diego Channel and out to sea at low visibility with NAVSTAR providing fixes to within about 30 metres of the channel buoys. A variety of other tests have been àccomplished with the GPS, including the use of the system to steer a C-141 and F-4J aircraft on landing approaches independent of any ground landing aids, and *en route* use in flying a P-3B aircraft from Hawaii to California, as well as navigating a C-141 aircraft along Vortac airways on cross-country flights.

The influence of a variety of environmental and other effects were investigated relating to the performance of the GPS. These included the effects of propeller modulation, signal multipath situations, foliage attenuation and abnormal atmospheric conditions. It was found that none of these tests proved to have any significant adverse effect upon GPS performance. The system also demonstrated moderate resistance to jamming and its signal structure provides an inherent capability for the denial of accurate navigation data to unauthorized users.

The orbital constellation currently consists of six satellites in two orbital planes inclined at 63° to the equator. Four of those satellites are fully usable in the tests. This configuration allows about four hours of testing a day in the region of the Yuma test site, and less time at other locations. The test constellation will be used for additional extensive test and development of GPS equipment in Phase 2; this activity will proceed at least until mid-1982. The tests will also support the US Navy's Improved Ballistic Missile Accuracy Program and will evaluate GPS performance in applications to an expanded

**Figure 3.23. GPS projected battlefield benefits**

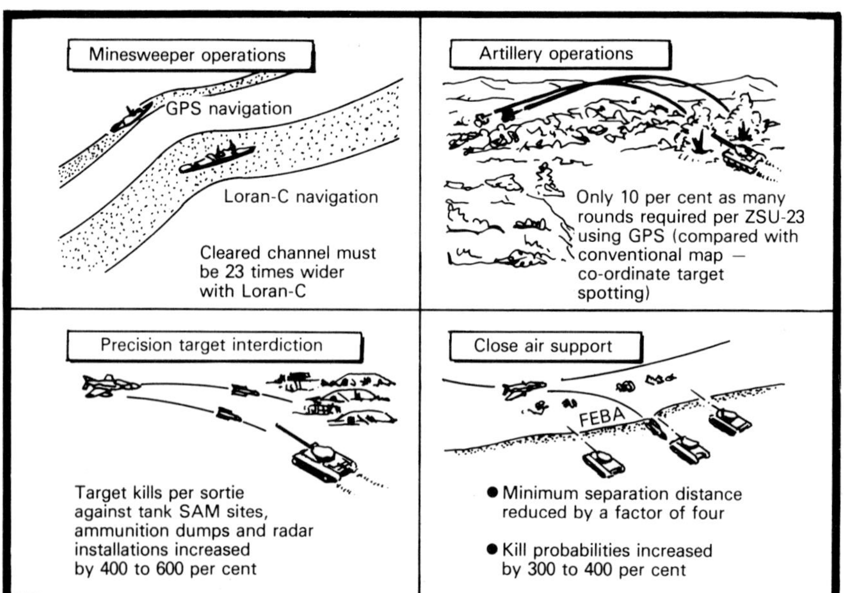

group of military missions. Studies have been performed to determine the projected battlefield benefits obtainable through the use of the GPS and some of these estimates are shown in figure 3.23 [14].

*Phase 2: Full-scale development*

The programme is currently in the full-scale development phase in which all three segments are under development and deployment. With the successful completion of the concept validation phase, and approval by senior US Defense Department officials through the Defense System Acquisition Review Council (DSARC) process, full-scale development commenced in June 1979. Six developmental spacecraft are successfully in orbit. Operational prototype spacecraft are under development, and satellite production has begun. The operational ground control segment is also under development. The control and management of the orbital constellation will be accomplished in Phase 2 by the Phase 1 master control station at Vandenberg Air Force Base until an operational system is fully deployed. The contractors are developing about 50 user equipment sets for testing and evaluating during Phase 2. This equipment will be tested on a number of vehicles with the participation of the various military services.

The additional satellites planned for launch during the Phase 2 programme have minor modifications to those satellites developed during Phase 1. Some changes are planned to improve their nuclear hardening and to provide compatibility with the space shuttle. Also planned is the incorporation of an operational nuclear detonation detection system (IONDS), a system carried as a secondary payload on the GPS spacecraft. The IONDS concept for global nuclear burst localization is shown in figure 3.24 [14]. These space components will be deployed on all satellites in the operational

**Figure 3.24. IONDS concept for global nuclear burst localization**

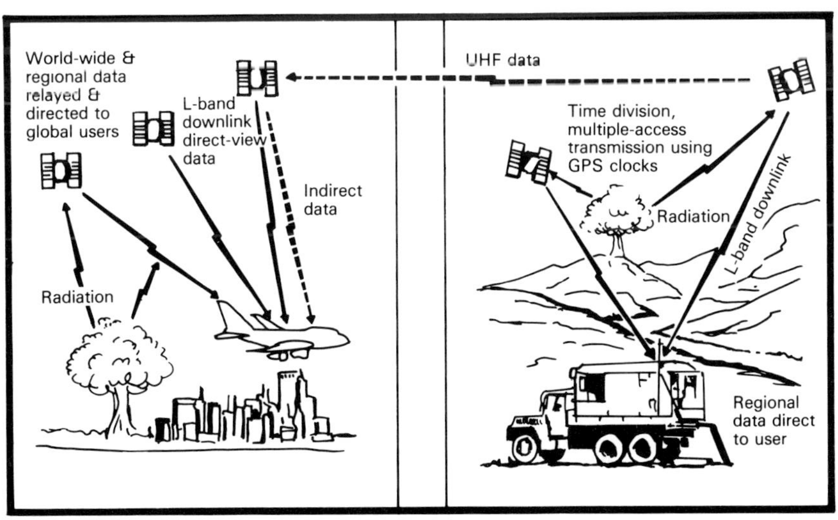

GPS constellation. IONDS operates with the same PRN modulation as the GPS navigation signals but it radiates only part of the time. The IONDS terminals are to be similar to standard GPS user sets and will be capable of receiving simultaneous IONDS data from all GPS satellites in view as well as having a satellite cross-link data capability.

The initial shuttle launch of the GPS spacecraft is planned to take place in mid-1985, and subsequent launches of production spacecraft will be with the shuttle. Current plans call for achieving a two-dimensional position determination capability for the GPS system with 12 satellites in orbit by the end of 1986. For the full three-dimensional world-wide capability, 18 satellites are required, and this is to be achieved about one year later.

Full-scale development of the control segment is proceeding through a contract awarded in late 1980. This basically upgrades the Phase 1 control segment at Vandenberg AFB and involves continued operation of the Phase 1 constellation. Also, development of hardware and software is proceeding for the operational control segment which is to assume responsibility for the functions performed at Vandenberg and become operational in 1986.

The user equipment for Phase 2 involves the competitive parallel development of a family of user equipment to be tested and selected on host vehicles. This equipment is being developed primarily by two contractors who are designing, manufacturing and testing about 50 production prototype user sets with associated instrumentation and support equipment. This equipment emphasizes common modules for the different user sets as a means of reducing production and support costs. The equipment will be subjected to a variety of demanding tests to demonstrate thoroughly its ability and readiness to enter into production.

*Phase 3: Operational capability*

Another thorough review of the programme status of GPS will be accomplished by the DSARC process late in 1983 or early 1984. If this milestone is completed successfully, then operational user set quantity production will be initiated and the full constellation of satellites will be deployed. The objective is to achieve an operational system with world-wide three-dimensional capability by about 1987.

## Military applications

The excellent accuracy and world-wide availability of GPS provides a system which is attractive for a wide variety of military applications.

*Air operations*

The GPS can provide *en route* navigational capabilities for tactical and strategic airborne operations either as the prime navigation system, or more likely, operating with an inertial measurement unit (IMU) or other navigational aid. Integration with the GPS provides significant performance improvements in inertial positioning, in-flight IMU alignment and auxiliary sensor calibration. The precise information available continuously from the GPS provides an aircraft with both its co-ordinates of position and its velocity vector, thereby improving its weapon delivery accuracy and related capabilities. The GPS provides a means for 'blind' or visual co-ordinate bombing, and can assist substantially in close air support missions, airborne refuelling, rendezvous, cargo drops, ferry missions and a variety of other tactical applications [14]. The position data available from the GPS are also useful in reconnaissance, photo-mapping, targeting and damage assessment, as well as for passive electronic intelligence and emitter location determination. Use of the GPS navigational information with appropriate terminal guidance techniques provides remotely piloted vehicles with a world-wide deployment capability for automated missions.

*Sea operations*
GPS applications to sea operations include naval patrolling, task force rendezvous and harbour or beach-head pilotage. Additionally, many of the benefits outlined for air operations are also appropriate for the naval area. By using information from maps and charts and incorporating obstacle and reference data into the GPS co-ordinate frame, the GPS can provide an additional capability for hazard avoidance, including reef, shoal and buoy location. Substantial improvements in harbour and sea mine placement operations and in sensor emplacement may also be obtained. Other benefits would include the co-ordination of ship and aircraft rescue operations and carrier aircraft recovery. Possibly very significant improvements in sea-based missile launching accuracy may be attained as well as improvements in shore bombardment effectiveness when target information is available. Anti-submarine operations can be co-ordinated more effectively and may benefit appreciably from the highly accurate co-ordinate grid system provided by the GPS.

*Land operations*
GPS can provide for the precise positioning of troops, armour and mobile or fixed artillery. The system has the capability for improving the geodetic accuracy and mapping of global locations of interest. The all-weather, full-time, passive navigation capability provides a useful system for mobile air cavalry troop deployments and clandestine assault operations. Improved and more effective fire support co-ordination can be provided by the common positioning and time reference data available from GPS and associated real-time communications links. These capabilities appear to have application even to the most fluid tactical mission plans.

*Space operations*
The potential use of the GPS by space vehicles for accurate positioning and velocity determination is particularly significant. The space shuttle, for example, requires accurate, instantaneous position and velocity data to

provide operational efficiency during launch and in-flight operations and during the re-entry phases. Precise GPS positioning data can facilitate orbit transfer and the placement in appropriate trajectories of the free-flyer payloads. Low orbiting military surveillance and other satellites could be equipped with GPS as well as NASA's LANDSAT and similar spacecraft, to alleviate the need for ground-tracking and the requirement for extensive onboard computational capabilities. The use of GPS for the updating of missile guidance systems, including possible mid-course correction, for other manoeuvrable re-entry vehicles and for ballistic missile instrumentation are applications of obvious benefit to the DoD.

**Civil applications**

The GPS has been justified and is proceeding through development as a DoD programme. However, it is clear that there are many potential civil uses for GPS systems [22]. The programme has been structured to encourage participation by the civil sector and there has been a substantial amount of interest shown. The experience gained from the Navy's TRANSIT system has demonstrated the willingness of civil users internationally to exploit the potential of satellite navigation.

The civil potential of GPS is also beginning to be exploited. There are an increasing number of investigations being carried out into the various applications of GPS, in *inter alia* the fields of sea navigation, helicopter and air navigation, harbour control, mapping, surveying and space applications, including the space shuttle. Although a good deal more work needs to be done to investigate the GPS applications in the civil arena, it is conceivable that GPS could become the principal general purpose position location system for air, land and sea applications.

**Comparison of system capabilities**

Table 3.1 [23] provides a comparison of selected performance characteristics with a number of military systems, including 11 operational systems and 3 developmental systems (PLRS, JTIDS and GPS). The characteristics of coverage, accuracy, common grid, jamming immunity, etc., are plotted against the 10 ground-based systems, the 2 self-contained systems (inertial and Doppler) and the 2 satellite navigation systems. The highest scoring characteristics are circled to show clearly the merits of the systems. Matrix notes indicate specific system limitations and features. The TRANSIT, GPS and inertial systems all provide extremely desirable characteristics, as noted by the circled characteristics.

The range and accuracy comparison results are graphically shown in figure 3.25 [23]. The two sigma (rms) ranging error is plotted against the ranging capability of the systems considered. The platform mounted inertial

**Table 3.1. Comparison of military navigation systems**

| System | Coverage | Accuracy | Common grid | Passive user | User no. limitation | Jamming immunity | Ambiguity | Selective denial |
|---|---|---|---|---|---|---|---|---|
| Inertial | **World-wide** | Time variable | Singular | **Yes** | **None** | **Infinite** | **None** | NA |
| Doppler | **World-wide** | Time variable | Singular | No | **None** | None | **None** | NA |
| Loran-C | Limited ~ 14% | Moderate – 150 m | None | **Yes** | **None** | None | Slight | No |
| Loran-D | Theatre only | Moderate | None | **Yes** | **None** | Minimal | Slight | No |
| Decca | Limited | Moderate | None | **Yes** | **None** | None | Slight | No |
| Omega | **World-wide** | 2000 m | None | **Yes** | **None** | None | Severe | No |
| TRANSIT | **World-wide** | 50 m | **WGS-72** | **Yes** | **None** | None | Velocity aid | No |
| Radio Beacon | Regional | Low | Singular | **Yes** | **None** | None | Slight | No |
| VOR/ DME | Regional | Low | Singular | No | Limited | None | **None** | No |
| TACAN | Regional | Moderate | Singular | No | Limited | None | **None** | No |
| MLS/ ILS | Terminal area | Range-dependent | Singular | **Yes** | **None** | None | **None** | No |
| PLRS | Sector net | 30 m | Singular | No | Limited | 25 + dB | Slight | **Code key** |
| JTIDS RELNAV | Sector net | 30 m | Singular | No | Limited | 30 + dB | Slight | **Code key** |
| NAVSTAR GPS | **World-wide** | 16 m | **WGS-72** | **Yes** | **None** | **55 dB** | **None** | Yes |

Important for combined forces operations (Coverage to Ambiguity)

system shown, which has about a one nautical mile per hour drift rate, corresponds to a typical airborne inertial system. The inertial error degradation shown assumes an aircraft velocity of about 500 knots. The TRANSIT accuracy shown is somewhat pessimistic but may be considered valid for a moving platform with about one knot of velocity uncertainty. The TRANSIT and GPS performances are indicated on the maximum operating range line and in the high accuracy region. These charts provide a useful perspective of the overall capabilities and performance comparison of the military navigation systems in use or planned.

## *V. Status, issues and future prospects*

### Status

The TRANSIT system has been expanding rapidly during the past decade. There has been a steady and dramatic increase in the types of equipment available for military and civilian users. There should be over 30 000 user sets world-wide by the end of 1981. A new class of TRANSIT spacecraft, the NOVA satellite, was introduced operationally by a successful launch in May 1981. In general, the satellites have been extremely reliable with lifetimes frequently in excess of 10 years.

**Figure 3.25. Navigation systems range and accuracy comparison**[a]

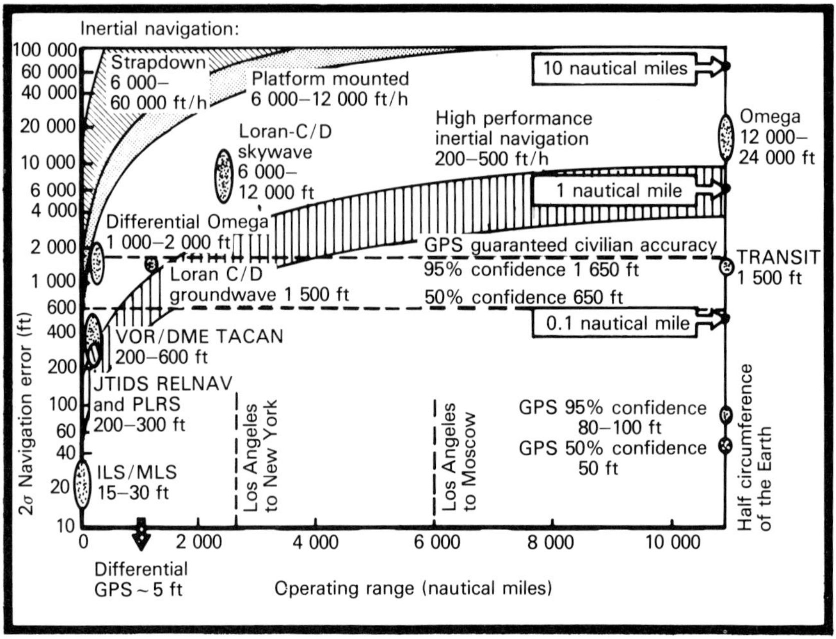


[a] 1 ft = 30.480 cm

The NAVSTAR GPS recently completed a very successful test and demonstration phase which has carried it into full-scale development. Six developmental satellites are in medium altitude orbit providing four-satellite visibility for continued test operations at the Air Force's Yuma test range. A variety of user equipment has been built, achieving excellent performance in the test series. Several early spacecraft developed problems or failures in their caesium beam or rubidium atomic clocks, but these difficulties have been remedied. The system operation is excellent and the Phase 2 equipment fabrication and test preparations appear to be progressing on schedule.

## Issues

It is apparent that the features which make the GPS attractive to the USA and allied military forces, that is, its world-wide coverage, timeliness and extreme accuracy for rapidly moving vehicles, also make it attractive to potential adversaries. Also, if the system is in fact of great military benefit, it is likely that attempts would be made to prevent the use of the system in time of conflict. The DoD has recognized the need to limit the potential of the GPS being used militarily against the USA and its allies.

## Selective availability

The principal concern has been to provide some form of access denial or degradation capability to limit the accuracy of the 'precise' signals that are transmitted by the satellites. However, it has been found that the C/A signals are also of sufficient accuracy to be militarily useful and therefore may require denial of access provisions.

The DoD has examined a number of techniques not only to selectively deny the use of the precise signal but also to degrade the accuracy available from the C/A signal. The Office of the Joint Chiefs of Staff has studied the national security implications of the GPS capabilities and has recommended a course of action. In developing this, the DoD has worked closely with the Department of Transportation, the Federal Aviation Administration, the US Coast Guard, the National Aeronautics and Space Administration and other agencies within the federal government to establish a range of capabilities consistent with specific requirements for civil navigation services. An attempt has been made to strike a reasonable balance between the national security concerns and the national/international needs and potential benefits of the GPS system.

The determination that has been made, primarily by the DoD, is that the GPS system should be made available for civil use at an accuracy of approximately 200 metres for a 50 per cent confidence level. It is believed that this accuracy would not seriously jeopardize national security in the mid-1980s. An annual review of the level of performance is to be made by the DoD and modifications may be incorporated to accommodate any changes commensurate with national security needs. The expectation is that the accuracy available will improve with time. It is recognized that this available accuracy may have a good deal to do with, and have critical impact on, the civil acceptance and future widespread civil use of the GPS capability.

## Future prospects

There is a large group of military missions that await the capabilities of the GPS, for example, the use of time transfer techniques from GPS for the synchronization or control of time division multiple access (TDMA) communications or other command, control and communications ($C^3$) techniques. The continuing advances in solid-state devices and large-scale integration (LSI) indicate that the sophisticated signal processing requirements of GPS-user equipment may be met at considerably less cost. Concurrently, we may expect to see higher speed micro-processors accomplish with software what is now done by analogue systems, further reducing cost and complexity. Advances in display technology and micro-miniaturization may take GPS-user equipment to very small size and widespread use.

## *VI. Conclusions*

It is clear that the potential and demonstrated capabilities of navigation satellite systems have many important military and civil uses. The TRANSIT system was initiated for a special purpose, that of accurately updating the position of the inertial navigators on board the US nuclear submarine fleet. TRANSIT has done this job well for nearly 20 years and has become in many respects a general purpose system serving over 4 000 military vessels and over 30 000 civil users. The TRANSIT satellite constellation has recently been upgraded with the initial NOVA satellite, which has a number of performance improvements. The TRANSIT system can be expected to remain viable as long as the US Navy continues to provide the space constellation. With the planned introduction of the NAVSTAR GPS operational system in the late 1980s, the TRANSIT system is planned for a 10-year phase-out period ending in about 1995–2000.

The NAVSTAR GPS has been intended from its inception as a general purpose military system which has the potential for fulfilling essentially all of the DoD's position, velocity and time determination functions. It recently completed an extensive test and demonstration programme which validated the GPS concept and established the operational value of the system to a variety of military missions. Some enhancements to user force effectiveness include: precise weapon targeting and delivery with greater single pass impact; missions less affected by adverse weather; availability of a common co-ordinate grid for operations with combined forces; weapon stand-off ranges are increased; improved reconnaissance and surveillance position annotation; more efficient rendezvous and refuelling operations; world-wide time availability for $C^3$; and forcing opponents into a large expenditure commitment to counter the navigation satellite capability. The NAVSTAR GPS is now in the full-scale development phase, which is planned to provide additional validation of equipment designs and operational tests from a limited space constellation. The system is planned to be fully operational with an 18-satellite constellation by 1987–1989.

**Acknowledgements**

The information presented is predominantly based on published data for the system concerned. However, to some extent, unpublished notes and personal discussions with a number of the principals involved in the development of satellite-based systems have been used in this paper. In particular, the author would like to express his appreciation to Mr Thomas Stansell, Jr of Magnavox Corporation, and to Dr Richard Kershner of the Applied Physics Laboratory of Johns Hopkins University, for their assistance in the background and current status, including graphics, of the TRANSIT system; also to Col John Martel, USAF in the Office of the Under-Secretary of Defense, Research and Engineering, for his interest, assistance and current data on the GPS. Acknowledgement is also given to Messrs Tom O'Connor and Charles Helms of Rockwell Corporation, and to Mr Robert Losey of Radio Corporation of America for their assistance and graphics support. Responsibility for errors in fact or interpretation is solely that of the author.

## References

1. McDonald, K.D., 'A survey of satellite-based systems for navigation, position surveillance, traffic control, and collision avoidance', *Navigation*, (Institute of Navigation), Vol. 20, No. 4, Winter 1973–74.
2. Kershner, R.B., 'The U.S. Navy Navigation Satellite System', Paper prepared for United Nations Conference on the Exploration and Peaceful Uses of Outer Space, Vienna, 14–27 August 1968.
3. Kershner, R.B. and Newton, R.R., 'The TRANSIT system', *Navigation*, (Institute of Navigation), Vol. 9, No. 2, Summer 1962.
4. 'NAVSAT survey update (by the Navy Astronautics Group)', *Navigation*, (Institute of Navigation), Vol. 28, No. 1, Spring 1981, pp. 74–75.
5. Guier, W.H. and Weiffenbach, G.C., 'A satellite Doppler navigation system', *Proceedings, Institute of Radio Engineers*, Vol. 48, No. 4, April 1960.
6. Stansell, T.A., Jr, 'The Navy Navigation Satellite System: description and status', *Navigation*, (Institute of Navigation), Vol. 15, No. 3, Fall 1968.
7. Freitag, R.F., 'Project transit: a navigation satellite,' *Navigation*, (Institute of Navigation), Vol. 7, No. 2–3, Summer-Fall 1960.
8. Stansell, T.A., Jr, *The TRANSIT Navigation Satellite System*, Magnavox Government and Industrial Electronics Company, Report R5933, October 1978.
9. RCA Astro Electronics, Personal communications relating to TRANSIT and NOVA, spacecraft illustrations provided by RCA, 1981.
10. 'Radionavigation plans and policy', *Federal Radionavigation Plan*, Vol. 1, (US Department of Defense and the US Department of Transportation, Washington, D.C., July 1980), DoD Report No. 4650.4 p, 1; DOT-TSC-RSPA-80-16, 1.
11. *Department of Transportation National Plan for Navigation*, (US Department of Transportation, Washington, D.C., November 1977).
12. McDonald, K.D., 'Navigation satellite systems: status and potential', Paper prepared for the IEEE National Telecommunications Conference, Dallas, Texas, 1 December 1976.
13. *Global Positioning System*, (Institute of Navigation, Washington, D.C., 1980).
14. *Information Briefing on GPS*, (Office of the Under-Secretary of Defense for Research and Engineering, The Pentagon, Pentagon City, Va., 1981).
15. Rockwell International Corporation, Personal communication relating to GPS, Block II spacecraft illustration provided by Rockwell, 1981.
16. Spilker, J.J., 'GPS signal structure and performance characteristics', *Global Positioning System*, (Institute of Navigation, Washington, D.C., 1980).
17. Cahn, C.R. and Martin, E.H., 'Design considerations for a spread spectrum navigation receiver (NAVSTAR)', Paper prepared for the International Conference on Electronic Systems and Navigation Aids, Paris, France, 14–18 November 1977.

18. Jacobson, L., 'User equipment for the NAVSTAR global positioning system', Paper prepared for the IEEE National Telecommunications Conference, Dallas, Texas, 1 December 1976.
19. Senus, J. and Hill, R.W., 'GPS application to mapping, charting and geodesy', *Navigation*, (Institute of Navigation), Vol. 28, No. 2, Summer 1981.
20. Martel, J.H., 'Navstar global positioning system', Paper prepared for the AIAA/NASA Symposium on Space Tracking and Data Systems, Pentagon City, Va., 16–18 June 1981.
21. Henderson, D.W. and Coriat, H., 'Status report—global positioning system', *Navigation*, (Institute of Navigation), Vol. 27, No. 1, Spring 1980.
22. McDonald, K.D., 'The satellite as an aid to air traffic control', Chapter 31, *A Survey of Modern Air Traffic Control*, Vol. 11, (NATO Advisory Group for Aerospace Research and Development, AGARDograph No. 209, July 1975).
23. *Comparisons Between the NAVSTAR GPS and Competing Navigation Systems*, (Office of the Under-Secretary of Defense for Research and Engineering, The Pentagon, Pentagon City, Va., 1981).

# Paper 4. System aspects in military satellite communications

**E. Biglieri***

*Istituto di Elettronica e Telecomunicazioni, Politecnico di Torino, Corso Duca degli Abruzzi 24, 10129 Turin, Italy*

## *I. Introduction*

Satellite communication systems have a number of advantages which are being extensively exploited for commercial applications. They include the following:

1. A single satellite can provide communications coverage over a large geographical area.
2. Satellite transponders have large transmission capacities that enable them to support high data rates.
3. The inherent flexibility of satellite systems allows communication between a large number of widely dispersed users.

Other characteristics of satellite systems are highly desirable for military applications. A feature peculiar to these systems is that satellites are physically remote and hence less vulnerable to direct attack. They can house easily transportable tactical communications terminals which remain operational when moving from one location to another. Generally, these small tactical terminals are characterized by: (*a*) high mobility, (*b*) low radiated power, (*c*) poor receiving sensitivity, (*d*) limited capability for precise pointing of an antenna, (*e*) limited data rate, and (*f*) inability to meet highly sophisticated equipment requirements [1].

Satellite-borne tactical terminals have the following advantages for military communications:

1. They can provide communications in areas where coverage by terrestrial links is inadequate.
2. Mobile terminals allow communications to be rapidly established in areas where facilities are absent or destroyed.

---

* The opinions expressed in this paper are those of the author and do not necessarily reflect the official policy of his professional affiliation.

3. They provide a service to mobile, shipboard and airborne terminals which is more reliable than HF terrestrial communications.

4. They surpass the conventional systems by simplifying logistics for critical command and control networks serving forces deployed in a combat area [2].

A military satellite communications system must meet the following demands:

1. It should be difficult to locate an Earth station transmitter by means of its radiated signal. The need to conceal necessary communications is especially important for those transmitters that are difficult to detect and locate by other means. For example, highly mobile military terminals are open to physical attack if their location can be determined, whereas larger, less mobile terminals have more physical protection and their detection is therefore less critical.

2. The system must be operable in the presence of hostile jamming intended to disrupt communications.

3. Data security is required for both communications and systems control. The system must incorporate design features that reduce its vulnerability to unauthorized interception.

4. The increasing flow of digital data places demands on compatibility between a variety of services and users with different terminal sizes, data rates, types of modulation and connectivity needs.

5. They must be flexible and capable of rapid reconfiguration, independent of political or diplomatic negotiations.

6. The nuclear-capable users require maximal physical and electronic survivability under extreme conditions for command and control of the forces.

This paper is intended to focus primarily on military satellite communications system aspects that derive from the first two of these requirements, that is, low probability of interception (LPI) and anti-jamming (AJ) capabilities [3].

## *II. Anti-jamming techniques*

A satellite communications link can be jammed on both the uplink and the downlink (see figure 4.1.). In a typical scenario it is assumed that the jammer has a full knowledge of the system to be jammed and will use the best possible strategy to disrupt communications, and that the interfering source has much more power than the communicator.

Users needing protection against jamming will use spread-spectrum modulation, possibly combined with adaptive techniques intended to direct a null of the antenna gain pattern towards the jammer ('antenna nulling techniques').

Figure 4.1. Jamming and intercepting satellite communications

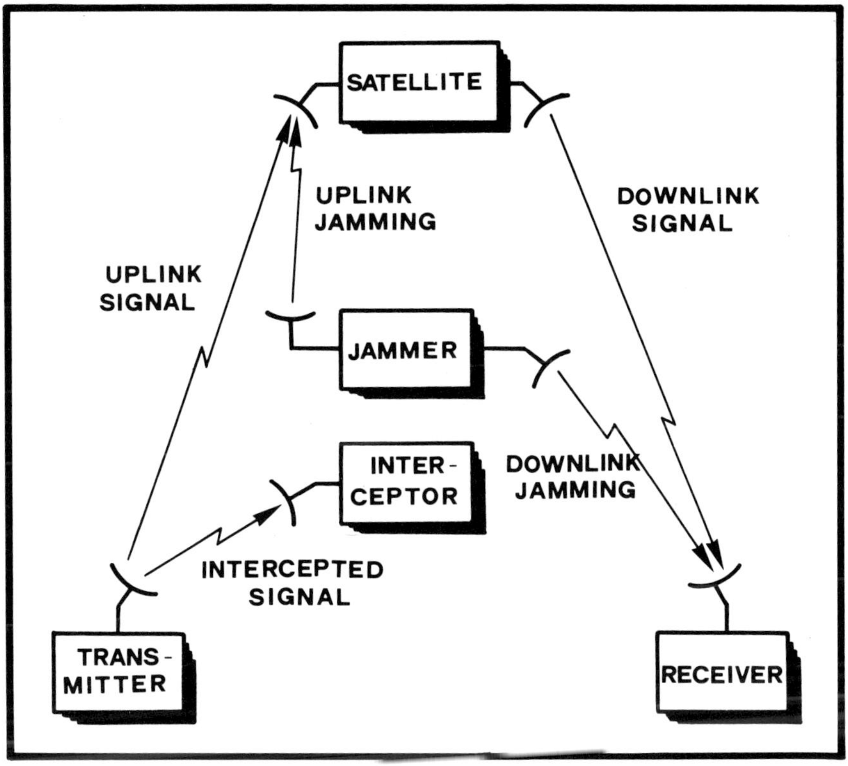


**Spread-spectrum modulation**

Essentially two spread-spectrum modulation techniques are suitable for military satellite applications (although a combination of both is also possible): (*a*) direct-sequence spread-spectrum, and (*b*) frequency-hopped spread-spectrum.

These techniques share the common feature of deliberately employing bandwidths wider than would be needed to transmit the same information using 'conventional' modulation techniques such as amplitude modulation (AM) or frequency modulation (FM) [4]. Moreover, the transmitted bandwidth results from a function which is generated locally and is different from that carrying the useful information. This function—a long pseudo-noise (PN) sequence—is assumed to be available to the transmitter and receiver, but not to the jammer.

To understand how an increase in bandwidth can provide protection against jamming, let us assume that the jamming signal has a bandwidth at least equal to $W$, the total modulated signal bandwidth, and a power $J$ watts. Let the useful component of the received signal have power $S$ watts, and the data rate be $R$ bits/second. The received energy ($E$) per information bit is:

$$E = S/R$$

Now, the relevant parameter in determining how well a communications system can perform is the bit error rate, which in turn is a direct function of the ratio $E/N_0$, where $N_0$ is the average density of the noise. Assuming $N_0 = J/W$ watts/Hertz, for modulated signals the value of $E/N_0$ operationally required can be written as:

$$E/N_0 = (S/J) . (W/R)$$

or, in dB

$$(E/N_0)_{\text{dB}} = (W/R)_{\text{dB}} - (J/S)_{\text{dB}}$$

The ratio $J/S$ between the jamming power and the signal power is known as the 'jamming margin', and the ratio $W/R$ as the 'processing gain'. The jamming margin is the amount of interference that a receiver can withstand while operating at an acceptable level. For example, with a system which has a 40-dB processing gain, if the minimum acceptable $E/N_0$ is 10 dB and implementation losses are 2 dB, then the jamming margin is $40 - 12 = 28$ dB. A 28-dB jamming margin would permit a receiver to operate in an environment in which the desired signal power is about 600 times smaller than the interference at its input. Expressed another way, an interfering transmitter can have 600 times more power output than the desired signal's transmitter (if their distances are equal) before affecting the receiver's operation.

**Direct-sequence spread-spectrum**

Although there are many techniques which can be used for spreading the transmitted signals over large bandwidths—and hence achieving a large processing gain—spread-spectrum modulation is the most commonly used. With this type of modulation, a long PN sequence phase-shift keys (PSK) a radio-frequency sinusoidal carrier. Then, the binary information phase-shifts successive groups of $K$ PN symbols (see figure 4.2).

**Frequency-hopping spread-spectrum**

In these systems, the PN sequence provides a pattern used to shift the carrier frequency in discrete increments instead of directly modulating the carrier. Here the bandwidth $W$ over which the energy is spread is essentially independent of the chip duration $T$, and depends on the number and the size of the frequency hops. Ideally, the power spectrum of the transmitted signal extends over the spreading bandwidth $W = nD$, where $n$ is the

**Figure 4.2. Direct-sequence spread-spectrum modulation**

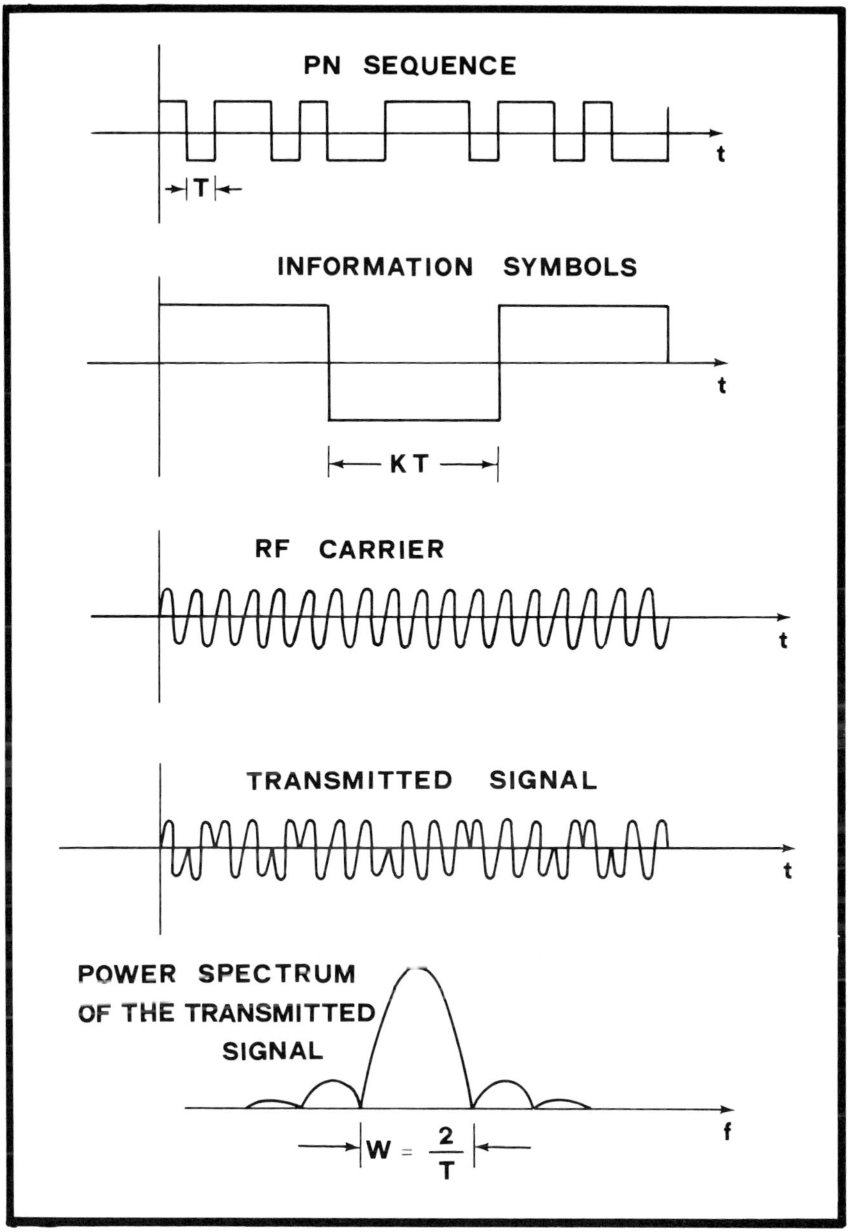


number of frequencies used in the hopping and $D$ is the spacing among them.

Frequency-hopping systems can afford a bandwidth larger than direct-sequence systems, and are more robust (e.g., they require no phase estimate and are more tolerant to Doppler shift).

### Advantages of spread-spectrum modulation

Spread-spectrum modulation offers the following advantages:

1. Spreading the signal power over a large bandwidth is advantageous for providing a low probability of interception, that is, for minimizing the risk that the communications signal might be used to detect and locate the signal-emitting platform. In conventional systems, since minimum transmitted energy helps reduce the risk of interception, covert signals must be minimum data rate signals, or relatively short digital messages.
2. Message privacy is inherent in spread-spectrum signals because of the presence of the PN sequence in their transmission format. In this case the PN sequences must be secure in the cryptographic sense, in accordance with the application and the assumed threat.
3. Spread-spectrum systems can provide an interference rejection capability that can hardly be matched in any other way.

### Antenna nulling

It has recently been realized that conventional AJ and LPI techniques via bandwidth spreading can be supplemented by inclusion in the system of antenna nulling capabilities. In fact, shaping the antenna radiation pattern can further improve resistance to jamming and the LPI performance of the mobile users.

With various techniques [5, 6] it is possible to produce high gain beams in certain directions and nulls in other desired directions. Thus, a receiving antenna can minimize the power received from the jammer and a transmitting antenna can minimize the power radiated in directions other than towards the intended recipient.

Adaptive antenna processing based upon correlation with expected received signals can be implemented to control the location of pattern nulls. Figure 4.3 illustrates such a result.

## *III. EHF spectrum allocation*

In commercial satellite communications, severe crowding of the geostationary orbital arc and the radio-frequency spectrum, coupled with requirements for increased bandwidth (due to the need for accommodating higher data rates and a larger number of circuits) motivates the need for moving to operating frequencies above 10 GHz.

Current military satellite communications in the USA operate in the UHF (225–400 MHz) and SHF (7.2–8.4 GHz) frequency bands, and share these bands with terrestrial systems and with satellite services of other

**Figure 4.3. Adaptive antenna nulling for protection against a jammer**

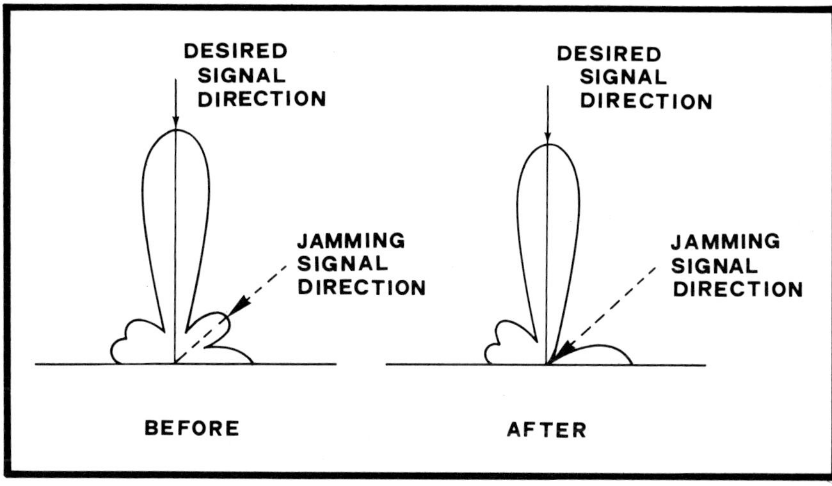


nations. The availability of higher frequency bands (in particular, EHF) for military satellite communications, made accessible by the development of a technology base, besides allowing traffic to be enhanced, permits increased protection from jamming and lower probability of intercepting transmissions from a covert terminal. The frequency bands which are considered for military satellite communications systems above 10 GHz are: 14/11 GHz band; 30/20 GHz band; 43–48 GHz band; and 50/40 GHz band [7, 8].

The following comments can be made [7, 8]:

1. At EHF, it is possible to achieve very large antenna gain with antennae restricted to a manageable size. The possibility of achieving high resolution antennae with physically small structures permits a significant improvement in anti-jamming capabilities with adaptive nulling techniques.

2. Operation at EHF reduces significantly the probability of intercepting covert terminals. Since larger bandwidths must be monitored by the interceptor, receiver noise is more likely to overcome the signal power received from high resolution antennae.

3. Increased bandwidths allow increased processing gain in spread-spectrum modulation, and hence enhanced anti-jamming capabilities.

## References

1. Sullivan, D.P., 'Future trends in military communication satellite repeaters', *IEEE Transactions on Aerospace and Electronic Systems*, Vol. AES-6, No. 2, March 1970, pp. 129–36.
2. Bond, F.E., 'Overview: military satellite communication systems', in *Communications Satellite Developments: Systems*, eds. G.E. La Vean and W.G. Schmidt, Progress in Astronautics and Aeronautics Series, Vol. 41 (MIT Press, Cambridge, Mass., USA, 1974).

3. Dayton, A.D. and Jain, P.C., 'MILSATCOM architecture', *IEEE Transactions on Communications*, Vol. COM-28, No. 9, September 1980, pp. 1456–59.
4. Dixon, R.C., *Spread Spectrum Systems* (J. Wiley and Sons, New York, 1975).
5. Mayhan, J.T., 'Adaptive nulling with satellite antennas', *Proceedings of 1980 IEEE International Conference on Communications (ICC'80)*, pp. 59.1.1–59.1.8.
6. Compton, R.T., 'An adaptive array in a spread-spectrum communication system', *IEEE Proceedings*, Vol. 65, No. 3, March 1978, pp. 289–98.
7. Cummings, W.C., Jain, P.C. and Richardi, L.J., 'Fundamental performance characteristics that influence EHF MILSATCOM systems', *IEEE Transactions on Communications*, Vol. COM-27, No. 10, October 1979, pp. 1423–35.
8. Jain, P.C., 'Use of EHF frequency bands in future military satellite applications', *Proceedings of 1979 IEEE International Conference on Communications (ICC'79)*, pp. 33.4.1–33.4.4

# Paper 5. Image analysis and sensor technology for satellite monitoring

**T. Sakata* and H. Shimoda***

*Tokai University, Research and Information Center, Tomigaya 2–28, Shibuya, Tokyo 151, Japan*

## *I. Introduction*

Since the launching of the first Landsat satellite by the United States in 1972, the Landsat series has proven the most effective system for Earth observation. Further, research and development efforts to improve the operational use of remote sensor data obtained from satellites and aircraft have been made by many nations. Considerable advances in the development of devices for obtaining improved viewing and recording images from space and high altitudes have been made over the past 10 years. The applications of such data have also increased rapidly.

The application of data obtained from remote sensors is expected to provide the most economic means of management and decision-making in the fields of agriculture, environmental surveying, city planning, geological studies, land use, and so on. Thus an increasing variety of research projects have been undertaken at national and local government levels over the past 10 years.

Remote sensor technology is also expected to provide practical and effective approaches to studying the distribution of natural resources and monitoring changes in an effort to avert crises. Recent developments in image information technology have provided many data processing systems and related equipment for handling remote sensor data obtained from satellites and aircraft. These systems required much research and development to reach such high levels of performance.

* The opinions expressed in this paper are those of the authors and do not necessarily reflect the official policy of their professional affiliations.

## II. *Photo-interpretation by the human eye*

Image information data may be analysed either by visual processing ('photo-interpretation') or by computer processing. Pattern recognition for remote sensing can be divided into two categories: spatial and spectral.

In carrying out photo-interpretation of Earth-resources data, the image analyst must make a series of complex judgements. The difficulty in making these judgements and the frequency and importance of the associated errors are a function of many factors. The human visual system is superior to any man–machine equipment. It has a high sensitivity and can maintain high resolution on an absolute scale with relatively small variations in sensitivity over an enormous range of scene brightness. An initial problem is to distinguish objects from their surroundings. This problem is best understood by looking at a picture in which objects are difficult to distinguish (e.g., a spotted pattern against a dappled background). It is difficult to see the original pattern; although the contours, texture and colour all help, the boundaries of objects are often not sharp, and colour differences can be misleading. The image of an object can be considered as a two-dimensional distribution of light intensity which has been converted from the three-dimensional distribution of the natural scene by a sensor on a spacecraft or aircraft.

'Image processing' refers to operations performed upon an image in order to modify it in some useful way or to extract information on image properties for further analysis. The qualitative and quantitative description may take either of two forms: analogue or digital.

## III. *Analogue image processing*

Spatial pattern information in the form of pictures taken by a multispectral scanner (MSS) or conventional airborne camera can be easily analysed by the human eye. Analysis of analogue information is done mainly by interpretation using cathode ray tube (CRT) displays, additive colour viewers and other optical instruments; it may also be combined with digital image processing.

Spectral information, on the other hand, is analysed automatically. Consequently, image information processing is usually carried out by a hybrid method combining digital and analogue, together with various kinds of input equipment. This results in a reduction of total access time and operator cost. Analogue information processing uses optical, opto-electronic and photographic systems. The methods available have several variations.

The film images obtained using large-format cameras, such as the photogrammetric camera or multi-band camera, have high image quality. A suitable configuration combining optical and electronic methods will give very effective results. Processing can be expanded to cover the following items: adding, subtraction, grey level control, colour coding, colour combination, superposing, enhancement, threshold and enlargement.

The overall performance of a photographic system depends on several factors: contrast, shadow effect, platform altitude, atmosphere, characteristics of the camera and image input.

## *IV. Digital image processing*

### Data acquisition and handling

At present most image information data from spacecraft are received at ground stations and then passed on to image processing facilities or research institutions.

Since remote sensing satellites are still at the development stage, the processed data, which include formatting, geometric and radiometric corrections, are supplied in such forms as films, printed papers and computer-compatible tape (CCT).

Digital data requirements have been increasing for several years because of the growing use of computer-aided information processing techniques.

### Preprocessing of image data

Image processing can be broadly classified into two categories: image correction and information extraction.

Image correction includes correction of radiometric errors, geometric distortion resulting from the sensor itself, and internal and external effects, such as those occurring during space flight.

Internal distortions are due to the optical and mechanical system; external effects are due to rolling, pitching and yawing. These corrections are based upon sensor data before launching, the spacecraft attitude data and ground control data. Other preprocessing includes removal of noise and shading, and restoration.

### Image processing

Digital image processing systems contain the following: source data, image data, processor and image data output.

Generally, image processing computer systems require various data-input devices and system consoles for man–machine conversation. When the source data involve, for example, optical scanner or radar data, they must be suitably converted onto tape or film. Also, film and other materials can be converted to computer-compatible tapes after processing with an analogue-digital converter.

The type of processor mainly used for image data is the conventional mini-computer with various kinds of input and output devices and large-scale memory discs. However, the computer capabilities depend on both the programmes and the requirements. Table 5.1 shows common programmes for image processing applications.

Image output from the computer requires several kinds of output devices,

**Table 5.1. Common image processing application programmes**

| | |
|---|---|
| *Support programmes* | |
| Histogram | Displays frequency of occurrence of each intensity value in image |
| Grid | Superimposes graduated grid on image data |
| Grey scale | Adds grey scale wedge to image |
| Annotate | Adds alphanumeric annotation to image |
| Statistic | Computes image data statistics |
| Report | Provides processing time and costs |
| Error analysis | Computes accuracy of geometric and radiometric correction |
| *Geometric modification* | |
| Magnify | Expands image size |
| Reduce | Reduces image size |
| Geometric | Implements two-dimensional image geometry correction |
| Rotate | Rotates image through specified angle |
| Mosaic | Geometrically combines two or more images and removes overlap |
| *Radiometric modification* | |
| Intensity | Implements uniform intensity modification to image |
| Intensity (X, Y) | Implements spatially variable intensity modification to image |
| Normalize | Normalizes data so that average equals one |
| *Image manipulation* | |
| Combine | Adds (or subtracts) two or more images |
| Output | Outputs image on tape, display, recorder, plotter or printer |
| Scroll | Advances displayed image at selected rate |
| Register | Translational adjustment to cause two images to be in registration |
| *Measurement programmes* | |
| Area | Determines area (in appropriate units) of polygon defined by 3 or more points |
| Length | Computes length (in appropriate units) between two points |
| Contour | Generates isodensity contour line plot |
| *Enhancement programmes* | |
| Filter | Performs Fourier transformation and filtering of input image |
| Noise removal | Eliminates noise from image data |
| Derivative | Computes directional derivative or Laplacian of image |
| Ratio | Divides each element of one image by corresponding element of second registered image |
| *Information extraction* | |
| Classify | Multispectral classification/feature extraction |

*Source:* Reference [1]

**Figure 5.1. A block diagram of software system—TIPE—for processing remote sensing data**

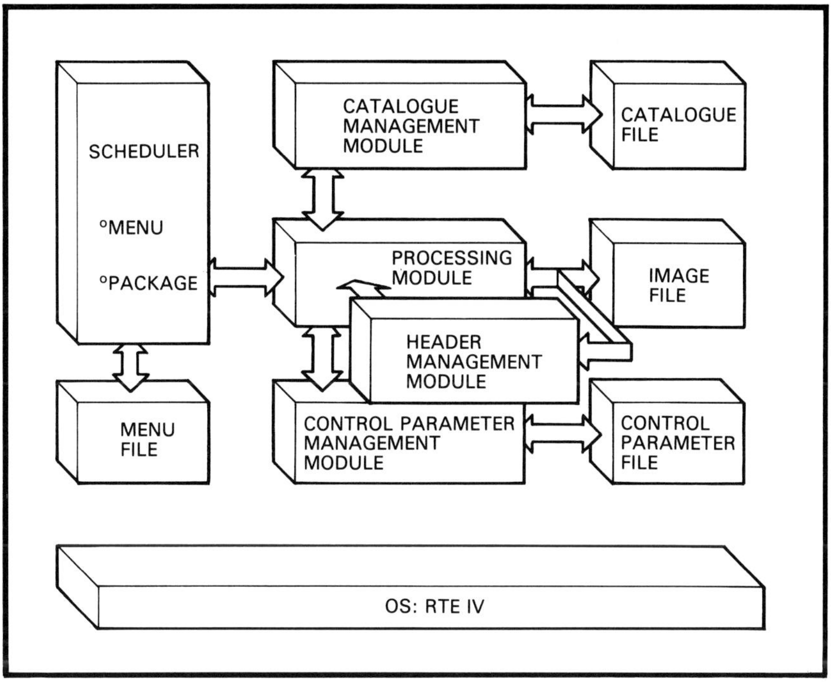


*Source:* Reference [6]

such as, CRT, teletypewriters, inkjet printers, film recorders, high-speed non-impact printers and electron beam recorders.

## *V. Software and hardware*

Software systems for remote sensing must have flexibility of operation. Improvement of the available systems is expected to decrease both the access time and errors, hence giving good performance.

An ideal type of software system for remote sensing use can be illustrated (figure 5.1) using the example of the *T*IAS *I*nteractive *P*rocessing *E*xecutive (TIPE) System. In this system the scheduler has been developed with a 'menu' which has a flexible structure for research and development, and a package designed to contain standard programmes for remotely sensed data processing.

This system can also be operated for parallel processing, and a *For*mula *Tran*slation computer system (FORTRAN) program can be used for easier maintenance.

**Figure 5.2. Sensors used for earth observation from space**

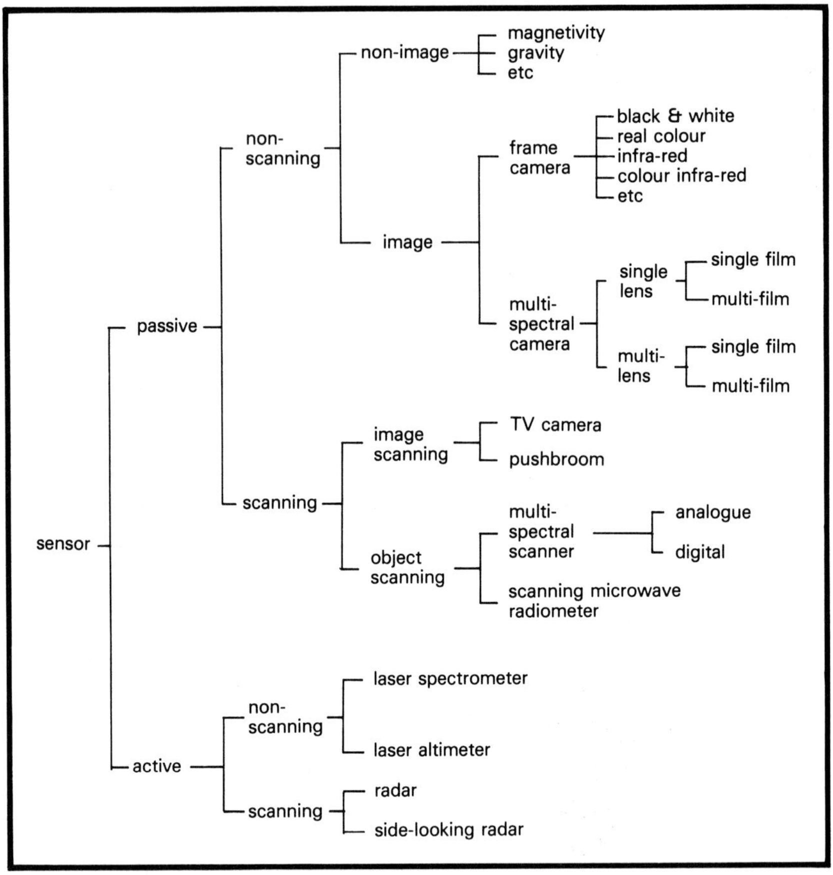


*Source:* Reference [6]

The hardware that makes up the computer system, such as the tape drives, discs, drums, central computer units and main memory, are discussed below. The image processor facilities consist of a flexible system with several parallel processors which operate on different items. At present, however, computers used for remotely sensed data processing are too costly for routine use; therefore, simpler low-cost computer systems must be developed for remote sensing.

## *VI. Sensors*

### Sensor classifications

Figure 5.2 shows sensors which are used for observation of the Earth's surface from satellites. Passive types collect electromagnetic (EM) waves emitted

**Figure 5.3. Sensitivity of various sensors to different regions of the electromagnetic spectrum**

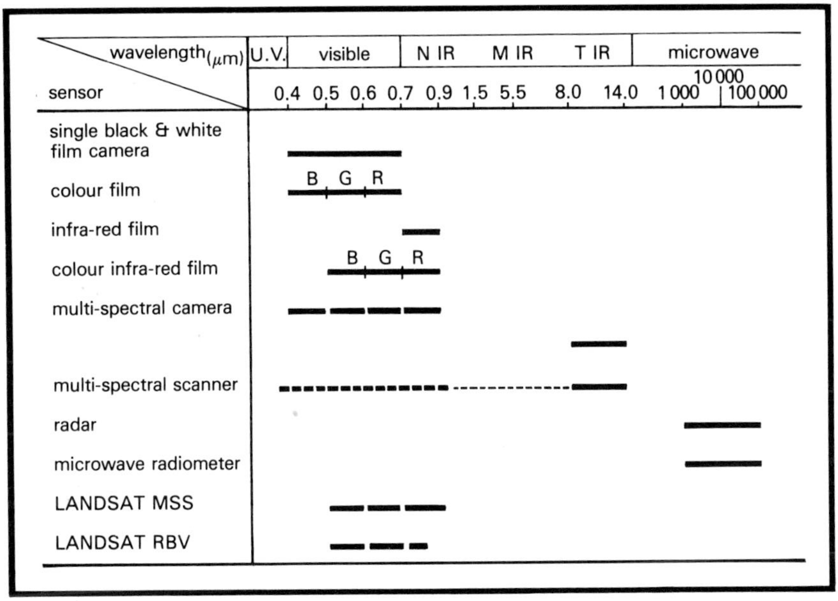


*Source:* Reference [6]

**Figure 5.4. Types of photographic camera used for earth observation from space**

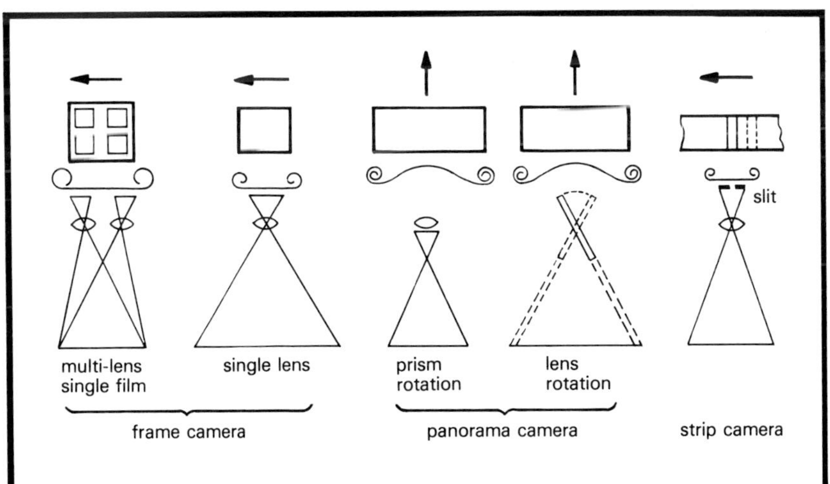


*Source:* Reference [6]

## Table 5.2. Various types of mapping—frame aerial cameras

| Camera | F-639 | KC-1A | KC-1B | KC-4 | KC-6A | T-11 & T-11B |
|---|---|---|---|---|---|---|
| Manufacturer | Fairchild | Fairchild | Fairchild | Fairchild | Fairchild | Fairchild |
| Primary use | med/hi alt mapping | med/hi alt mapping | med/hi alt mapping | med/hi alt mapping | med/hi alt mapping | med/hi alt mapping |
| Format (cm) | 23 × 23 | 23 × 23 | 23 × 23 | 23 × 23 | 23 × 23 | 23 × 23 |
| Film length (m) | 215 | 120 | 120 | 120 | 120/240 | 120 |
| Base thickness (μm) | thin | standard | standard | standard | standard/thin | standard |
| Frames/roll | 770 | 450 | 450 | 450 | 470/940 | 450 |
| Magazine | integral | integral | integral | integral | integral | integral |
| Cycle time (s) | 2.0 max | 2.5 | 2.5 | 2.5 | 2.0 max | 3.0 |
| FMC type | moving platen | none | none | none | moving platen | none |
| FMC rate (mm/s) | 1.3–30.5 | NA | NA | NA | 2.5–20.3 | NA |
| FMC error | NA | NA | NA | NA | NA | NA |
| AWAR, 1000:1 (lp/mm) | 55 | 20 | 20 | 38 | 45* | 20 |
| Film | Pan-X | 8401 | 8401 | 8401 | 8401 | 8401 |
| Processing | NA | NA | NA | NA | NA | NA |
| Type of measurement | laboratory dynamic | dynamic, inflight | dynamic, inflight | dynamic, inflight | laboratory dynamic | dynamic, inflight |
| Compatible lenses | Geocon I | Planigon II | Planigon II | Geocon I | Geocon IV B** | Metrogon II |
| Manufacturer(s) | Fairchild | B&L, Curtis, Goerz | Goerz | Fairchild, J.G. Baker | Kollsman | B&L |
| Focal length (mm) | 150 | 150 | 150 | 150 | 150 | 150 |
| f/no. range | 5.6 | 6.3–11 | 6.3–11 | 5.6–11 | 5.0–11.2 | 6.3–11 |
| Wavelength range (μm) | 0.4–0.7 | 0.4–0.7 | 0.4–0.7 | 0.4–0.7 | 0.4–0.7 | 0.4–0.7 |
| Filters | W12, W25, anti-vignetting | clear or yellow anti-vignetting | clear or yellow anti-vignetting | yellow anti-vignetting | yellow anti-vignetting | yellow anti-vignetting |
| Distortion (μm) (R-radial T-tangential) | <10 | NA | <10 | <10 | ±8R, <8T | NA |
| Can be used with normal colour film? | yes | yes | yes | yes | yes | yes |
| Can be used with IR colour film? | NA | NA | NA | NA | yes | NA |
| Change of focus required for IR colour film? | NA | NA | NA | NA | no | NA |
| Angular coverage | 74° × 74° | 74° × 74° | 74° × 74° | 74° × 74° | 74° × 74° | 74° × 74° |
| Shutter speed (s) | 1/25–1/900 | 1/10–1/500 | 1/10–1/500 | 1/25–1/500 | 1/50–1/800 | 1/10–1/500 |
| Shutter type | intralens Rapidyne | intralens Rapidyne | intralens Rapidyne | intralens Rapidyne | intralens Rapidyne | intralens Rapidyne |
| Weight (kg), without film and mount | 45 w/film | 43 w/film | 43 w/film | 43 w/film | 66 w/film | 39 w/film |
| Mount(s) | hard w/vibration isolators | A-8, -11, -25, -28, -32, ART-21, -25 | A-8, -11, -25, -28, -32, ART-21, -25 | A-28, ART-25, Model 6 | LC-7A | A-8, -11, -25, -28, -32, ART-21, -25 |
| Data annotation | counter, clock, card, etc. | counter, clock, card, etc. | counter, clock, card, etc. | counter, clock card, etc. | CRT, counter, fixed data | counter, clock, card, etc. |
| Intervalometer | yes | yes | yes | yes | yes | yes |
| Exposure control | auto (AFI 6, 20, 64) | manual | manual | manual | auto (AFI 20–2 000) | manual |
| Remarks | platen reseau | no reseau | no reseau | platen & peripheral reseau | platen reseau | no reseau |

* Minimum required by specification
** High contrast AWAR, lens only, is 60 lp/mm on Pan-X process in D-19 w/$\gamma$ = 1.4
† Measured to NBS specifications, lens only

*Source:* Reference [6]

| Model VI | MRB 15/2323 | MRB 9/2323 | MRB 11 5/1818 | | RC-10 | |
|---|---|---|---|---|---|---|
| Galileo Corp. of Amer. | Jena | Jena | Jena | | Wild Heerbrugg | |
| mapping | med/hi alt mapping | med/hi alt mapping | med/hi alt mapping | | general mapping | |
| 23 × 23 | 23 × 23 | 23 × 23 | 18 × 18 | | 23 × 23 | |
| 110 | 120–150 | 120–150 | 120–150 | | 150 | |
| NA | NA | NA | NA | | 100 | |
| NA | 450 | 450 | 540 max | | 580 | |
| interchangeable | interchangeable w/MRB 9 | interchangeable w/MRB 15 | NA | | cassettes | |
| 5.0 | 2.0 | 2.0 | NA | | 1.5 | |
| NA | NA | NA | NA | | none | |
| NA | NA | NA | NA | | NA | |
| NA | NA | NA | NA | | NA | |
| NA | NA | NA | NA | 55† | NA | NA |
| NA | NA | NA | NA | NA | NA | NA |
| NA | NA | NA | NA | NA | NA | NA |
| NA | NA | NA | NA | static | NA | NA |
| Orthogon Officine Galileo S.p.A. | Lamegon PI Jenoptic Jena GmbH | Super Lamegon Jenoptic Jena GmbH | Lamegon Jenoptic Jena GmbH | Universal Aviogon Wild Heerbrugg | Universal Aviogon II Wild Heerbrugg | Super Aviogon II Wild Heerbrugg |
| 150 | 150 | 90 | 115 | 150 | 150 | 90 |
| 5.6–11 | 4.5–8 | 5.6–11 | 4.0–8.0 | 5.6–22 | 4 | 5.6–22 |
| NA | NA | NA | NA | 0.45–0.85 | 0.4–0.8 | 0.45–0.85 |
| NA | anti-vignetting | anti-vignetting | yellow, orange | yellow & red (1.4 A.V.) | NA | yellow & red (2.2 A.V.) |
| ±20 | ±6R, <5T | ±7 | ±8R, <5T | ±10R, <5T | <4 | ±10R, <5T |
| yes | yes | yes | NA | yes | yes | yes |
| yes | yes | yes | NA | yes | NA | yes |
| no | no | no | NA | no | no | no |
| 74° × 74° | 74° × 74° | 104° × 104° | 90° × 90° | 74° × 74° | 74° × 74° | 104° × 104° |
| 1/125–1/300 | 1/100–1/1 000 | 1/50–1/500 | 1/100–1/1 000 | 1/100–1/1 000 | 1/100–1/1 000 | 1/100–1/1 000 |
| NA | rotating disc | rotating disc | rotating disc | rotary | rotary | rotary |
| 36 | 63 | 52 | 37 | 78 | NA | 75 |
| special | Universal MRB-A | Universal MRB-A | Universal MRB-A | PAV-10 | NA | NA |
| NA | clock, altitude, grey scale, etc. | clock, altitude, grey scale, etc. | clock, altitude, grey scale, etc. | clock, altitude, counter, etc. | clock, altitude, counter, etc. | clock, altitude, counter, etc. |
| yes | yes | yes | yes | yes | yes | yes |
| NA | optional | optional | optional | manual, remote | manual, remote | manual, remote |
| reseau optional | peripheral reseau | peripheral reseau | peripheral reseau | no reseau | no reseau | no reseau |

or reflected by the Earth's surface, while active types emit EM waves and receive back scattered portions of emitted energy.

Image-scanning types scan the image generated by the optical systems of sensors, while object-scanning types scan the Earth's surface.

Figure 5.3 shows EM regions sensed by the typical remote sensing sensors.

The sensors which are used for satellite monitoring are photographic cameras, image tubes, multispectral scanners (MSS) and synthetic aperture radar (SAR).

### Photographic cameras

Photographic cameras can be classified into several categories, as shown in figure 5.4.

### Frame cameras

Frame cameras have a conventional camera structure. The advantages of this type are (*a*) the very small distortions, and (*b*) a stereo measurement capability obtained by overlapped images.

### Panorama cameras

This type of camera can cover very wide areas perpendicular to the flight direction by super-wide angle lenses with the aid of lens rotation. Images taken by this kind of camera contain large distortions and are mostly used for aircraft reconnaissance.

### Strip cameras

Strip cameras have no shutters. Instead, images which have gone through a small slit are recorded on the films, which move synchronously with the movement of the Earth. The construction of these cameras is simpler than that of a frame camera, but they cannot be used for stereo measurements.

Films which are used for these cameras are fairly large compared to those used for conventional cameras. The most frequently used size is 9½ inches (23¼ cm) wide, with a very thin base for greater film capacity.

Film types used for satellite monitoring are (*a*) panchromatic films, (*b*) infra-red films, (*c*) real-colour films and (*d*) colour infra-red films. Colour infra-red films contain the largest amount of spectral information, but panchromatic films have the highest spatial resolution.

Typical cameras are listed in table 5.2, and films are listed in table 5.3.

## Multispectral scanners

Multispectral scanners scan the Earth's surface perpendicular to the flight direction by the combination of a telescope and a rotating mirror. Luminous flux passed through this assembly is then divided by spectrometers, such as dichroic mirrors or diffraction gratings, and passes into photo-detector arrays, such as photo-multipliers or photo-diodes. This type of multispectral scanner is called the optical mechanical scanner. Its structure is shown in figure 5.5.

**Figure 5.5. Components of a multispectral scanner**

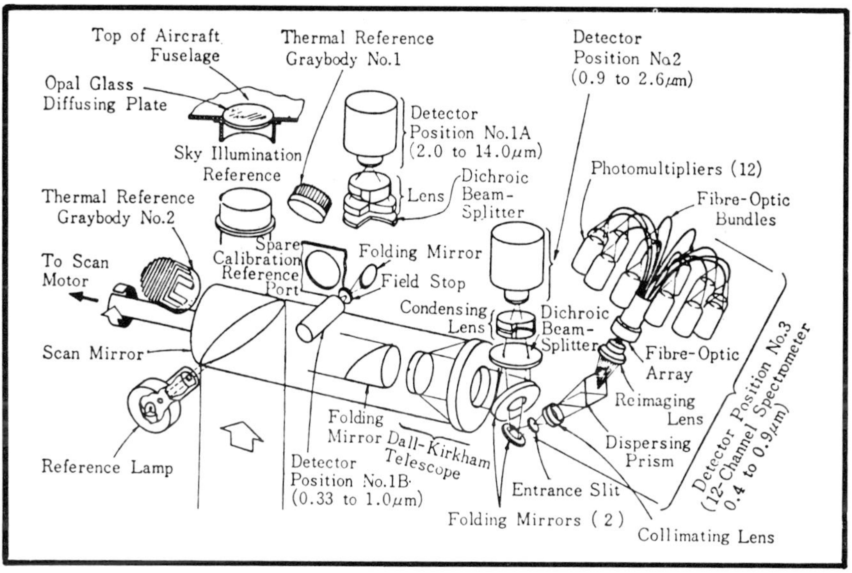


*Source:* Reference [6]

The advantages of this type of MSS are: (*a*) it can detect a wide range of the EM spectrum; and (*b*) it facilitates data transmission from spacecraft to the Earth. On the other hand, it has some disadvantages, namely: (*a*) geometric distortions are larger than with cameras; and (*b*) its spatial resolution is poorer than that of cameras.

Another type of MSS is a pushbroom-type scanner. It utilizes linear detector arrays, such as charge coupled devices (CCD) rather than the rotating mirrors of an optical mechanical scanner. Its principle is shown in figure 5.6. Its advantages over an optical mechanical scanner are: (*a*) its structure is simpler and contains no moving parts; and (*b*) geometric distortions are fewer. However, because of the limitations of array detectors, the EM wavelengths are limited to visible and near infra-red regions.

Typical optical mechanical scanners are listed in table 5.4.

**Table 5.3. Various types of aerial photographic films**

| Film name | Film number | Nominal base thickness (μm) | Aerial film speed (AFS) | Aerial exposure index (AEI) | Developer or processes | Resolving power, lines/mm target-object contrast 1000:1 | 1.6:1 | Diffuse RMS granularity[f] |
|---|---|---|---|---|---|---|---|---|
| Plus-X Aerographic | 2402 | 100 | 200 | 80 | D-19 | 100 | 50 | 19 |
| Tri-X Aerographic | 2403 | 100 | 650[c] | 250 | D-19 | 80 | 20 | 33 |
| Double-X Aerographic | 2405 | 100 | 320[c] | 125 | DK-50 | 80 | 40 | 26 |
| Panatomic-X Aerial | 3400 | 60 | 64 | 20 | D-19 | 160 | 63 | 16 |
| Plus-X Aerial | 3401 | 60 | 200 | 64 | D-19 | 100 | 40 | 30 |
| High-Definition | 3414 | 60 | 8 | 2.5 | D-19 | 630 | 250 | 9 |
| Aerial | 1414 | 40 | 8 | 2.5 | D-19 | 630 | 250 | 9 |
| Infrared Aerographic[a] | 2424 | 100 | 200 | 100 | D-19 | 80 | 32 | 33 |
| Multispectral Infra-red Aerial[a] | SO-289 | 100 | 100 | | | 200 | 80 | 12 |
| Aerochrome | 2443 | 100 | 40 | 10 | EA-5 | 63 | 32 | 17 |
| Infrared[b] | 3443 | 60 | 40 | 10 | EA-5 | 63 | 32 | 17 |
| High definition | SO-127 | 100 | 6[c] | | EA-5 | 160 | 50 | 9 |
| Aerochrome | SO-131 | 60 | 6[c] | | EA-5 | 160 | 50 | 9 |
| Infrared[b] | SO-130 | 40 | 6[c] | | EA-5 | 160 | 50 | 9 |
| Aerocolor Negative | 2445 | 100 | 100[c] | 32[e] | Aero-Neg. colour proc. | 80 | 40 | 13 |
| Ektachrome MS Aerographic | 2448 | 100 | 32[c] | 6[e] | EA-4 | 80 | 40 | 12 |
| Aerial colour | SO-242 | 60 | 6 | 2 | ME-4 (modified) | 200 | 100 | 11 |
| | SO-255 | 40 | 6 | 2 | ME-4 (modified) | 200 | 100 | 11 |
| Ektachrome EF | SO-397 | 100 | 64[c] | 12 | EA-4 | 63 | 32 | 13 |
| Aerographic | SO-154 | 60 | 64[c] | 12 | EA-4 | 63 | 32 | 13 |
| Water Penetration Aerial Colour | SO-224 | 100 | 40[c] | | EA-5 | 125 | 50 | 24 |
| Anscochrome | D/200 | 100 | 90[d] | 30 | AR-1C | 100 | 50 | 25 |
| Anscochrome | D/500 | 100 | 230[d] | 75 | AR-1C | 80 | 40 | 45 |

[a] Speeds are for no filter
[b] Speeds are for a Wratten number 12 filter
[c] Effective film speeds are relative speeds, found by comparison to films whose speeds are determined according to ANSI Standard PH2.34, 1969
[d] At time of compilation the manufacturer had not released an AFS for this film. The number shown is an estimate determined by multiplying the AEI by a factor of about 3
[e] Haze filters such as HF-3, HF-4, and HF-5 may be used without changing exposure
[f] The RMS-Diffuse granularity value represents 1000 times the standard deviation in diffuse visual density produced by scanning a uniformly exposed and developed sample with an f/2.0 system of 48-μm scanning spot size

*Source:* Reference [1]

## Image tubes

Image tubes scan electronic images generated on a photo-detector plane by electron beams. Most TV cameras have similar devices, but the image tube used in space monitoring has a special structure to improve the image quality. This tube is called a return beam vidicon (RBV). Its advantages over an MSS are: (*a*) it contains no moving parts; and (*b*) it is lighter than an MSS although it contains fairly large geometric and radiometric distortions.

**Figure 5.6. An illustration of the basic principle of a pushbroom-type scanner**

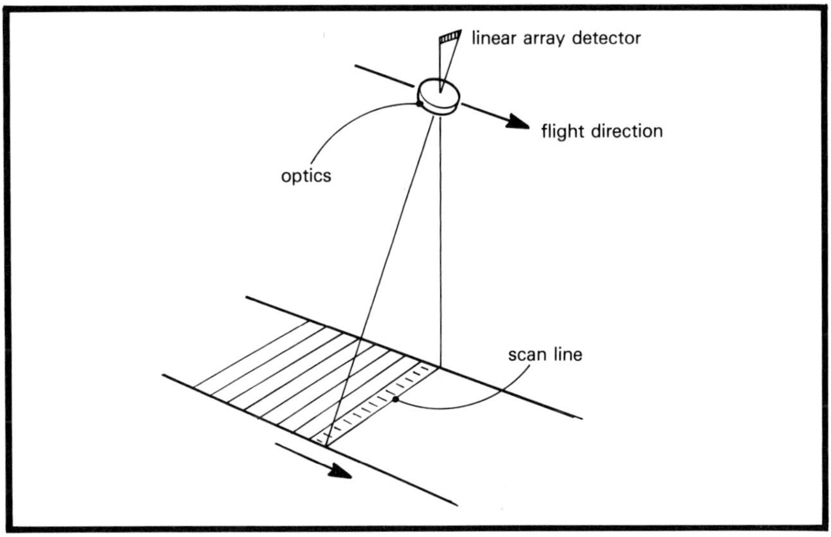


*Source:* Reference [6]

**Figure 5.7. An illustration of the principle of side-looking radar**

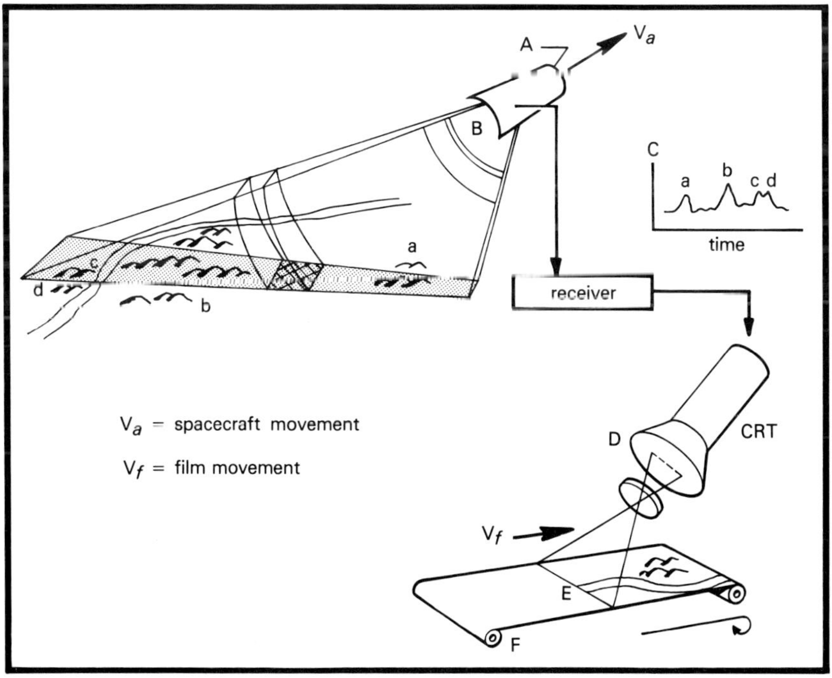


*Source:* Reference [6]

## Table 5.4. Various types of optical mechanical scanners

| Manufacturer | ITT Aerospace Optical Div. | | | ITT Aerospace Optical Div. | Hughes/Santa Barbara Research Center | | Santa Barbara Research Center | |
|---|---|---|---|---|---|---|---|---|
| Scanner model | High resolution IR radiometer | | | Surface composition mapping radiometer | Medium resolution infrared radiometer | | Spin scan cloud camera | |
| Satellite | Nimbus<br>I | II | III | Nimbus E | Nimbus II, Nimbus III<br>III | | ATS I | ATS II |
| Method of scanning | Single 45° mirror | | | Single 45° mirror | Single 45° mirror | | Spinning spacecraft + tilting mirror | |
| Scan lines/s | 0.8 | 0.8 | 0.8 | 10 | 0.133 | | 1.66 | 1.66 |
| Swathwidth degrees | 180 | 180 | 180 | 90 | 180 | | 15 | 17.3 |
| Orbit | Sun synchronous | | | Sun synchronous | Sun synchronous | | Earth synchronous | |
| Resolution IFOV (mrad) | 7.9 | 7.9 | 7.2 | 0.6 | 50 | | 0.1 | 0.1 |
| Type of collecting optics | Modified Cassegrain | | | Afocal Dall-Kirkham | Cassegrain for each band | | Paraboloidal Newtonian | Wynn-Rosin |
| Focal length and no. | 102 mm f/1 | | | 166 mm f/0.8 | 107 mm f/2.44 | | 254 mm f/2 | 380 mm f/3 |
| Diameter of aperture | 102 mm | | | 203 mm | 43.7 mm | | 127 mm | 127 mm |
| Effective aperture area | 5 300 mm$^2$ | | | 30 795 mm$^2$ | 1 160 mm$^2$ | | — | — |
| Calibration in flight | Blackbody, outerspace | | | Blackbody, outerspace | Blackbody, space, sun | | — | — |
| Method of recording ground | Photofacsimile film, digital or analog | | | Photofacsimile film, digital or analog | Analog, output chart digital tape photo display | | Photofacsimile | |
| Method of spectral separation | Filters | | | Filters, dichroic | Filters | | Filters | |
| Size | — | | | — | — | | 10 × 11 × 17 in | 12 × 11 × 17 in |
| Weight (kg) | 8.45 | 8.45 | 8.67 | 22.7 | 6.9 | 6.9 | 9.0 | 10.4 |
| Power (W) | 4 | 4 | 4 | 15.0 | 7.5 | | 24 max<br>3.08 avg | 22.5 max<br>10.8 avg |
| Source of specifications | IRIA | | | IRIA | IRIA | | IRIA | |
| Detector | PbSe | | | Hg:Cd:Te | Bolometer | | Photomultipliers | |
| Detector cooling | Radiation cooled (200°K) | | | Radiation cooled (≃ 100°K) | no | | no | |
| Spectral bands (μm) | 3.4–4.2 | 3.4–4.2 | 3.4–4.2 | 8.3–9.0<br>10.2–11.2<br>0.8–1.1 | 0.2–4.0<br>5–30<br>6.4–6.9<br>10–11<br>14–16 | 0.2–4.0<br>6.5–7.0<br>10–11<br>15.4–15.5<br>20–23 | 0.47–0.63 | 0.38–0.48<br>0.48–0.58<br>0.55–0.63 |

*Source:* Reference [1]

## Synthetic aperture radar (SAR)

Side-looking radars (SLR) generate Earth-surface images by emitting narrow fan beams perpendicular to the flight direction. The principle of SLR is shown in figure 5.7. Signals are resolved by the returning time duration of the radar pulse for the range direction and by the beam width for the azimuth

| Hughes/<br>Santa Barbara<br>Research Center | Santa Barbara<br>Research Center | Santa Barbara<br>Research Center | Santa Barbara<br>Research Center | RCA | Honeywell |
|---|---|---|---|---|---|
| Multispectral<br>scanner | Temperature-<br>humidity IR<br>radiometer | ITOS High resolu-<br>tion scanning<br>radiometer | Visible IR spin<br>scan radiometer | Very high res. rad. | Multispectral<br>scanner |
| ERTS B | Nimbus IV, E, F | NOAA 1,<br>ITOS B-E | SMS | ITOS E, ITOS D<br>TIROS M | SKYLAB A<br>(S-192) |
| Pivoting mirror | Single 45° mirror | Single 45° mirror | Spinning spacecraft<br>+ tilting mirror | 45° mirror | Image plane<br>conical |
| 13.6 scans/s<br>6 lines/scan<br>(refl. bands)<br>2 lines/scan<br>(thermal bands) | 0.8 | 0.8 | 1.6 scan/s<br>8 lines/scan (visible)<br>2 lines/scan (IR) | 6.67 | 100 |
| 11.5 | 180 | 180 | 18 | – | 10 |
| Sun synchronous<br>(925 km) | Sun synchronous<br>(1 110 km) | Sun synchronous<br>(1 460 km) | Geostationary | 1 460 km | 435 km |
| 0.086 (reflective | 7 (10.5–12.5 μm) | 2.7 (0.55–0.73μm) | 0.21 (0.55–0.75 μm) | 0.6 | 0.182 |
| 0.26 (thermal band) | 21 (6.7 μm) | 5.6 (10.5–12.5 μm) | 0.25 (10.5–12.6 μm) | | |
| Spherical +<br>corrector mirrors | Cassegrain | Cassegrain | Ritchey Chrétien | Dall-Kirkham | – |
| 825 mm f/3.6 | 430 mm f/3.38 | 430 mm f/3.38 | 2 913 mm f/7.2 | – | 635 mm |
| 229 mm | 127 mm | 127 mm | 406 mm | 127 mm | – |
| 32 900 mm² | 10 200 mm² | 10 200 mm² | 109 000 mm² | – | 14 500 mm² |
| Lamp, sun | Blackbody, space | Blackbody, space,<br>lamps | Blackbody, space,<br>sun | Blackbody, space,<br>sun | – |
| computer-<br>compatible<br>magnetic tape | Magnetic tape | Magnetic tape<br>and facsimile | Magnetic tape | – | computer-<br>compatible<br>magnetic tape |
| Filters | Dichroic | Dichroic | Filters | Dichroic | Spectrometer |
| 15 × 24 × 35 in | 7 × 7.5 × 15.6 in | 6.4 × 8.4 × 15.9 in | 26 × 26 × 60 in | 19 × 12 × 10 in | – |
| 31 | 9.03 | 0.00 | 64.3 | 9.0 | 136 |
| 42 max | 7.5 | 6.5 | 22 avg–52 max. | 5.1 | 266 |
| Mfg lit. | Mfg lit. | Mfg lit. | Mfg lit. | Mfg lit. | NASA lit. |
| Photomultiplier,<br>Si photodiode | Bolometer | Bolometer,<br>Si photodiode | Si photodiode (8),<br>Hg:Cd:Te (2) | Hg:Cd:Te/Si<br>photodiode | Hg:Cd:Te |
| yes (Hg:Cd:Te) | no | no | yes | no | Closed cycle<br>cryostat (77°K) |
| 0.5–0.6<br>0.6–0.7<br>0.7–0.8<br>0.8–1.1<br>10.5 12.6 | 6.7<br>10.5–12.5 | 0.55–0.73<br>10.5–12.5 | 0.55–0.75<br>10.5–12.6 | 0.6–0.7<br>10.5–12.5 | 0.410–0.460<br>0.460–0.510<br>0.520–0.556<br>0.565–0.609<br>0.620–0.670<br>0.680–0.762<br>0.783–0.880<br>0.980–1.080<br>1.090–1.190<br>1.200–1.300<br>1.550–1.750<br>2.100–2.350 |

direction. However, in order to get a high spatial resolution to azimuth direction from spacecraft, the antenna aperture size becomes impossibly large.

Synthetic aperture SLR or SAR can avoid this problem. It synthesizes the signals in along-track directions by matching the signal coherence. The original image can be reconstructed by optical or digital processing. Figure 5.8 shows its principle.

**Figure 5.8. An illustration of the principle of synthetic aperture radar**

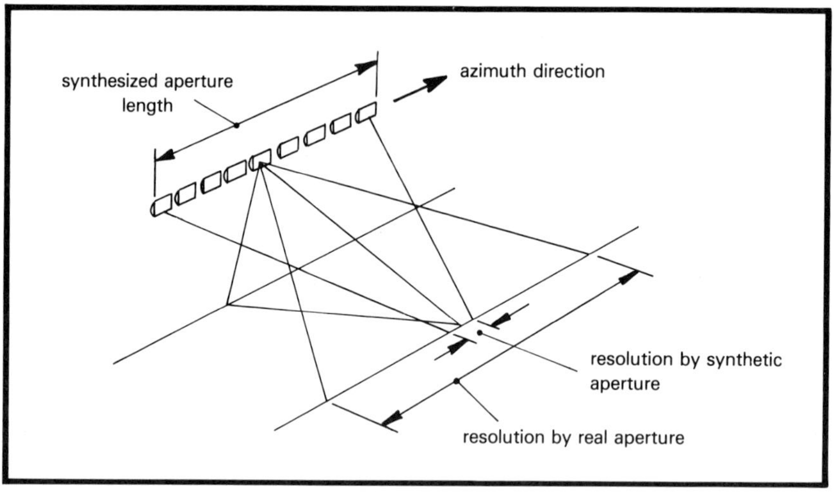


*Source:* Reference [6]

The advantages of SAR are: (*a*) it can work under cloudy or rainy conditions; and (*b*) it can detect small water turbulences. The disadvantages of SAR are: (*a*) it has large geometric distortions; (*b*) the reconstructed image contains speckle noises; (*c*) reconstruction of the image needs much processing; and (*d*) the interpretation of the images is more difficult than optical EM regions.

## Resolutions of sensor systems

Many factors are used to evaluate the sensor system performance, for instance: spatial resolution, spatial coverage, spectral resolution, radiometric dynamic range, spectral region, signal-to-noise ratio, geometric accuracy, data access speed, and so on. The most important factor for reconnaissance will be the spatial resolution. Therefore, in this paper, only the spatial resolution wil be discussed.

The resolution of the different sensors is determined by different factors.

### *Camera*

In the camera system, its spatial resolution is determined by the resolving power of optical systems and films. The resolving power of optical systems of space-borne cameras will be in the range of 30–200 line pairs per millimetre (lp/mm). Film resolving powers are listed in table 5.2, and they are in the range of 30–300 lp/mm. Hence, the total resolving power will be in the range 15–100 lp/mm.

With the size of the films used (i.e., about 200 mm wide), the total number of resolved points will be 6 000–40 000 for each line.

*RBV*

The resolution of RBV can be determined by the spot size of the electron beam, its diffraction and the size of the photo-cathode. The resolving power will range from 45 to 70 lp/mm, while the size of the photo-cathode will be in the range 25 × 25mm to 50 × 50 mm. Hence, the total number of resolved points will be 1 125–3 500 for each line.

*MSS*

The resolution of optical mechanical scanners is determined by the instantaneous field of view (IFOV) and the field of view (FOV). However, IFOV and FOV have a strong correlation. In reality FOV/IFOV will be determined by the engineering requirements. From this viewpoint, the number of resolved points will range from 2 000 to 6 000.

For pushbroom scanners the situation is different. In this type of sensor the numbers of detector elements of linear array determine the resolution. Existing maximum array elements are 2 048 and the highest value achievable by recent technology will be about 8 000.

*SAR*

In the synthetic aperture radar the resolution is determined in a completely different way compared to other sensors. In the SAR system the azimuthal resolution is only defined on the target surface.

The theoretical limitation is half of the real aperture. For spacecraft SAR it will be about 2–5 m. However, the actual ground resolution will be 10–25 m.

### Ground resolution and trade-off

Ground resolution is a function of the resolving power of the sensor systems and field of view, except for synthetic aperture radar. Consider a camera system which has 40 000 resolved points. If the spacecraft altitude is 120 km, and the focal length of the lens is 150 mm, then the ground resolution is 2 m. The resulting coverage is about 80 × 80 km. If we change the focal length to 1 500 mm, the ground resolution will be 20 cm and the coverage will be 8 × 8 km.

A similar thing will happen when the altitude of the spacecraft is changed. In other words, if one wishes to increase the ground resolution, a trade-off must be made with the coverage area. A decrease in coverage induces an increase in processing times.

Other kinds of trade-off exist. For instance, if there is a margin in the spacecraft payload, one can double the ground resolution by simply adding another sensor of the same type. Ground resolution is also influenced by spectrum resolution. Panchromatic films have higher resolving power than colour films.

Therefore, it is not meaningful to discuss the ground resolution on its own. The optimum design of the sensor system is determined not only by the physical performance of the sensor systems but also by the specific purpose of the observation, the spacecraft's performance, processing capacities, and so on.

## References


1. Reeves, R. (editor-in-chief), *Manual of Remote Sensing*, (American Society of Photogrammetry, Falls Church, Virginia, 1975), in two volumes.
2. Rosenfeld, A., 'Digital and picture analysis', *Topics in Applied Physics* (Berlin). Vol. 11, 1976.
3. 'Data processing research', Remote Sensing Technical Center, (Contract Research Report of the Science Technological Agency of Japan, Tokyo, 1978–80).
4. 'Digital image analysis—for application technology', (Research Report of the Science Technological Agency of Japan, Tokyo, 1976).
5. Sakata, T., 'Application to land use planning', The United Nations Seminar of Remote Sensing, Tokyo, Japan, September 1980.
6. Japan Association of Remote Sensing, *Remote Sensing Note* (Giho-do, Tokyo, Japan, 1975).

# Paper 6. Information extraction from images

**T. Orhaug* and G. Forssell***

*National Defence Research Institute, Box 1165, S-581 11 Linköping, Sweden*

## *I. Introduction*

Information collection for the purpose of verification for arms control and crisis monitoring can be accomplished in different ways. The use of remote sensors has become more important for several reasons. Among the various remote sensors, imaging sensors are the most versatile.

In this paper some general remarks on remote observations are made and distinctions between various sensor modes are presented. The relationship between the image content and the scene represented by the image is briefly discussed. Since images are always ambiguous with respect to interpretation, the need for realistic models or *a priori* knowledge to solve ambiguities is stressed. It is often assumed that the detecting medium of an imaging sensor is a photographic plate. More and more often, however, imagery data are acquired and stored in digital form. Therefore the special features associated with finite element data should be taken into account. Such features are, for example, spatial sampling and radiometric quantization. Some of the more important effects, like spectral aliasing and quantization associated with digital imagery, are discussed.

For all practical uses of imagery, the (ground) resolution is an important characteristic. The concept of resolving power for sensor characterization is presented, and the importance of defining resolution capability in statistical terms is emphasized. Some relevant figures with respect to detection performances are also given.

Some of the problems related to imagery of limited resolution have been studied by means of producing computer-simulated images (starting from high resolution photographic imagery). Some of the results of this

---

* The opinions expressed in this paper are those of the authors and do not necessarily reflect the official policy of their professional affiliation.

experiment are presented and discussed. They indicate that the use of computer-simulated imagery is a convenient method for studying detection identification performances.

## *II. Information collection by remote sensing*

Information can be collected from the physical environment by either *in situ* or remote observations. In the latter case, the more important methods are the ones that utilize travelling waves. Two main types of remote sensor can be identified: sensors producing non-imagery data and sensors producing images. Also, for each type of sensor, we can make a further distinction between passive and active ones.

The total sensor system with its output data can be considered as a kind of generalized filter connected in front of the physical environment. The information content of the data, including of course the processing and analysis that must be applied to extract information from the data, depends upon the mode of operation.

In some cases the kind of information needed is simple, and 'alarm' sensors can then be designed. Examples of such sensors can be found in all fields. In the military field, traffic recorders of various kinds and satellite infra-red (IR) sensors to detect intercontinental ballistic missile (ICBM) launchings are examples of such sensors.

Alarm sensors can be considered as sharply tuned filters. Such sensors are very efficient for the task for which they were designed; on the other hand, they are also rigid in the sense that they are inefficient for slightly different tasks.

Active sensors are technically more complicated than passive sensors since illumination of the scene has to be provided. On the other hand, such sensors often have a larger degree of freedom with respect to what kind of information is collected. This also makes it possible to specify and to narrow the sensor-filter. One classical example is the Doppler radar giving a binary response with respect to the presence of targets having line-of-sight velocity vectors in a predefined range. It should be noted, however, that the Doppler radar may be used in an imaging mode displaying a thematic map with information of targets and their velocities. Such thematic maps are not images in the sense that the term will be used here; instead, they are the two-dimensional representations of symbolic information.

Images, on the other hand, are the two-dimensional representations of raw, environmental data (radiance, reflection properties, etc.). In this sense images can be considered as an intermediate storage of unprocessed data. In terms of the previously presented filter concept, the sensor itself constitutes a rather broad filter, while the processing, analysis and interpretation of the imagery data constitute a much narrower filter.

The imagery data consist of contrast (brightness) values representing projections of a three-dimensional scene recorded by a (generalized) camera or imaging sensor. Several images from the same scene provide information in several spectral bands and polarizations and give multiple viewpoints in space (stereo) or in time (time sequence).

## *III. Image interpretation*

The goal of image interpretation is a concise description of the three-dimensional scene represented by the image. The kind and level of description is, however, dependent upon both the observer and the use of the information.

The information-carrying element in the image is its brightness (tone, contrast) and related spatial, spectral and temporal variations. For the sake of simplicity, the two latter co-ordinates have been disregarded and the spatial information alone has been considered.

The contrast of each individual picture element (pel) is determined by many physical factors of the corresponding (and other) scene points: incident illumination, reflecting properties of the surface, surface aspect (relative to the sensor and the illuminating source), absorption and diffusion properties of the intervening medium, and so on. In addition, the sensor itself influences the contrast value through the geometric and photometric properties of the sensor [1]

The information about the three-dimensional scene is ambiguous. First of all, the direct information about the three-dimensional structure is lost in the imaging process. Second, a wide variety of combinations of physical factors (illumination, reflection, diffusion, etc.) may result in the same brightness distribution. Therefore, an infinite number of different scenes may exhibit the same image. The more imagery data are collected, however, using different spectral bands, illumination, viewpoints, etc., the more the problem of this ambiguity is lessened. How is it at all possible then to interpret a single monochromatic image and come up with a correct result? The answer is that this is only possible with sufficient *a priori* information. The decoding of brightness value to recover scene description uses a considerable amount of *a priori* knowledge. Such knowledge may conveniently be described as models of the scene, of the illumination and of the imaging processes.

The task of image interpretation (and vision) is complicated and not completely understood. One fundamental part of this process is the detection task: detection of the presence of an object, a detail or a pattern. A basic part in this process of detection is the figure-ground concept (or segmentation as it is designated in computer processing). The basic signals in this process are contrast, mean contrast, texture, colour and line segments. On the next level, we may consider structure, shape, shadow, size, texture, shading, surface elements and surfaces as the building elements of objects. On an even higher level of interpretation, objects, relations between objects (context) and global arrangements are important. The analysis and interpretation of an image in this context is usually called image understanding, which is a subclass of a broader class of analysing perceptual and other kinds of data (signal understanding).

A helpful concept when discussing image processing and interpretation is to consider targets, objects or events in a hierarchical manner. It is obvious that the various tree structures for representing the image interpretation tasks will be both data- (object-kind) and goal-dependent.

Often, a summary description of image interpretation is made in terms of a few levels of description: detection, recognition and identification. Such a summary will always be arbitrary to some degree. Only when the kind of object of interest and the goal of interpretation are defined is it possible to quantify the various levels of description.

In order to handle, display, process and analyse images, the digital computer has become a very important tool. It is not, however, possible to have the computer perform the very difficult task of interpreting images from real-life 3D scenes. Images from flat scenes (of which Landsat images are examples due to the limited resolution) can, however, be analysed to a certain degree. It is possible, with the use of multispectral data, to analyse images by performing classification or segmentation. Computers can also be utilized in an interactive mode, in which case the interpreter is using the computer as a versatile instrument.

## *IV. Characteristics of imaging sensors*

In order to interpret images, a reliable model of the image generation process is a prerequisite. Such a model is needed for describing the physical characteristics of the image itself.

The following are some of the more important characteristics of an imaging system: geometric characteristics (scale, angular coverage, geometric distortions); resolution characteristics (including depth of focus); and radiometric characteristics (contrast, scale, dynamics, etc.). In order to simplify this discussion, only the following points will be taken into account: contrast quantization; spatial sampling; and spatial resolution.

Since the progress in several areas makes it possible both to digitize imagery and to build image sensors to deliver imagery data in digital form, the characteristic finite element images become more and more important. The digitization process involves a two-dimensional quantization (sampling) in the image plane as well as a quantization of the contrast values. These processes are well understood from a technical viewpoint. Their influence on image quality (perceptual quality) and on image interpretation is, however, not so easy to describe. The main reason for this is the lack of an image quality figure (one or many parametric) related to the physical characteristics of the image (resolution, contrast variation, noise, dynamics, etc.). (See [2] for a discussion of this problem.)

An image recorded with any sensor is always band-limited. Perfect reconstruction of such a signal is always possible provided that the sampling frequency is equal to or higher than the Nyquist frequency (equal to the highest frequency of the band-limited signal). If a shorter period (higher sampling frequency) is used, the original signal can still be recovered. In this case, redundant data are, however, recorded. If, on the other hand, lower sampling frequency is used, not only is information lost (bank-limitation) but distortion is also introduced due to the aliasing effect (spurious spatial frequency components).

An investigation of the perceptual effects of sampling distortions was recently made [3]. The main result is that the perceptual effect is a two-parameter effect. Also, the visual (perceptual) effect is due both to aliasing errors and band limitations. The conclusion therefore is that pre-filtering should be used so that sampling frequency coincides with the Nyquist frequency.

The quantization of image contrast is also simple to analyse from a mathematical point of view. Various non-linear schemes have been proposed and are often used (logarithmic to match the visual system and histogram-dependent to minimize rms errors, etc.) [4]. It is well known from picture processing that significant improvements in visual interpretation may be gained by simple-point operators (pseudo-colouring, contrast stretching, etc.) provided the operator is matched to the picture data and to the processing task.

An interesting result related to this problem is the following. If a signal consists of Gaussian noise of unknown spectral (energy) function, an estimation of this function can be made from the signal itself (using all amplitude information) or with almost the same accuracy, using only the polarity of the signal (i.e., using only the zero-crossings of the mean signal level).

In a study on image correlation, it was shown that provided quantization is adapted to the image data characteristics, correlation accuracy (e.g., for navigational purposes) did not deteriorate even when the image data was represented by two bits [5].

## *V. Effect of resolving power*

Any imaging system has the effect of smoothing luminance details. This effect is caused by the limiting resolving power of the instrument. The resolving power is conveniently described in terms of the image (signal) response the system gives when illuminated with a point source (at infinite distance). The resulting two-dimensional response function shall be denoted by $h$ (angular co-ordinates) (point-spread function). The system integrates [with weights given by $h(\cdot)$] the luminance distribution within the angular region given by $h(\cdot) \times 0$. Another way of describing the system response is by the (two-dimensional) Fourier transform (FT) of $h(\cdot)$ [or $H(\cdot)$]. This is the modulation transfer function (MTF) of the imaging system. In this way, the FT of the image contrast content is given by the product of the FT of the scene luminance distribution (observed with a sensor of infinite but good resolution) and $H(\cdot)$.

The resolution capacity of the sensor can be defined in terms of the accuracy with which we can determine the position of a point source. Accuracy is determined by the noise of the system (including visual noise) [6]. In this manner it is possible to formulate a detection theory based on a methodology determined from the theoretical field of information.

Experiments show that for detection a signal-to-noise ratio (contrast) of 3–5 is needed. Also, a few pels are needed in order to distinguish the 'object' from spurious single-noise pels (2–6 pels per target linear dimension). (See [7] for details on these problems.) Experiments with Landsat and other satellite images [8] indicate that for detection, the minimum target (linear) size is approximately 1.4–2 times the resolution cell size (pel size) (exact number dependent on contrast).

## *VI. Simulation of satellite imagery*

In order to study some of the effects produced when image resolution is decreased, the authors of this paper generated a series of photographic images of different resolution [9]. The original image used in this study was an air reconnaissance photograph recorded from 600-m altitude with a focal length of 15 cm (scale 1 : 4 500). This image was sampled at 25-$\mu$m intervals and converted to digital data using an Optronics C-4500 rotating scanner. The sample accuracy was 8 bits and the original 9.5″ film was converted to an image matrix of 7 000 × 7 000 pels (or approximately 400 megabits of digital data).

The digital image was then subject to computer processing using a DEC-10 GP computer system and a minicomputer image processing system (figure 6.1). In order to produce images of different resolution, repeated convolutions (linear filtering) were applied to the imagery data (see [9] for details). It can be shown that the application of repeated filtering results in data effectively filtered with a Gaussian-shaped filter. The effective image point-spread function is therefore also Gaussian. The effective size of the resolution function was both theoretically calculated and experimentally checked.

**Figure 6.1. Schematic description of the data flow in the simulation experiment**

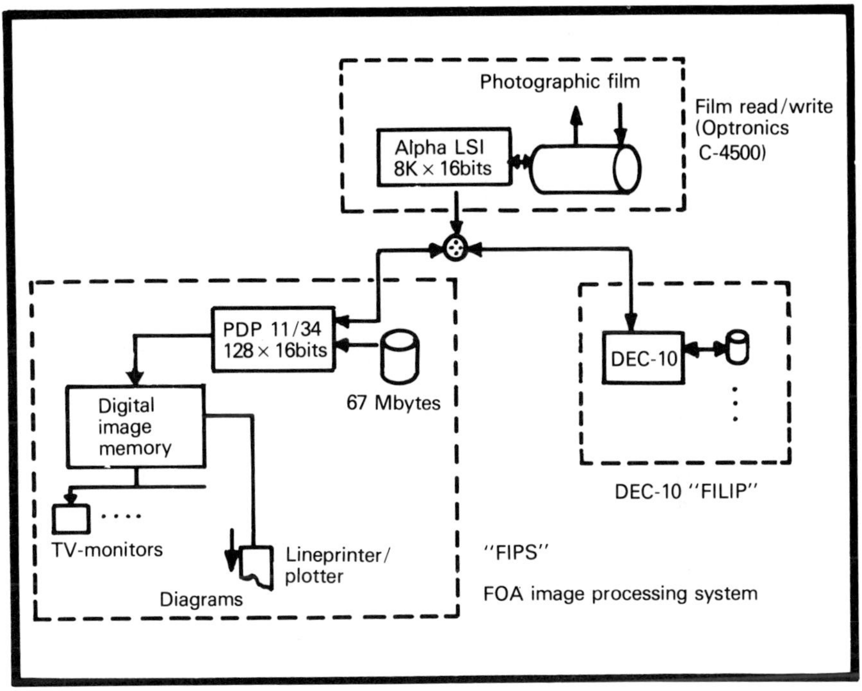


The filtering of the subsequent images was carried out to produce redundant data, that is, the resulting images were heavily oversampled. In this manner, photographic (and display) products could be made with 'photographic' quality (i.e., without the typical chequerboard pattern normally associated with digital imagery).

The original photographic image (the harbour of Gothenburg) was chosen so as to contain targets and patterns of various kinds and sizes (super-tanker, piers, boats, buildings, vehicles, etc.) and a detailed study of some of these targets was also conducted.

Some examples of simulated images having ground resolutions (defined as the width of the point-spread function between 37 per cent of its maximum value) of 1.8 m, 9 m, 24 m, 43 m, and 80 m are presented in figure 6.2. For a few objects (see table 6.1) detection tasks were performed for the following levels of detectability:

| | | |
|---|---|---|
| 0: | no detection | : object cannot be distinguished from background |
| 1: | detection | : object can be distinguished from background |
| 2: | recognition | : object can be recognized by overall shape |
| 3: | detailed recognition or description | : internal structure of the object can be recognized |

**Figure 6.2. Filtered image with ground resolution (pel size) of (*a*) 1.8 m, (*b*) 9 m, (*c*) 24 m, (*d*) 43 m, (*e*) 80 m**

(*a*)

(*b*)

(c)

(d)

(e)

**Table 6.1. List of objects used for interpretation tasks**

| Object number | Object type | Length × width (m × m) |
|---|---|---|
| 1 | Supertanker | 220 × 35 |
| 2 | Pier with boats | 140 × 16 |
| 3 | Narrow pier | 48 × 3 |
| 4 | Tow boats | 28 × 9 |
| 5 | Parked cars | 5 × 2 |

Figure 6.3 shows a graphical representation of levels of detectability obtained from simulated images by visual tests. Most of the objects used in this test were elongated objects (length/width = 2.5–15). It is therefore not surprising that the level of detection (level 1) approximately occurs for pel sizes three times object width, since integration is performed by the eye in the detection process.

**Figure 6.3. Graphical presentation of levels of detectability obtained through visual tests performed using the simulated images**

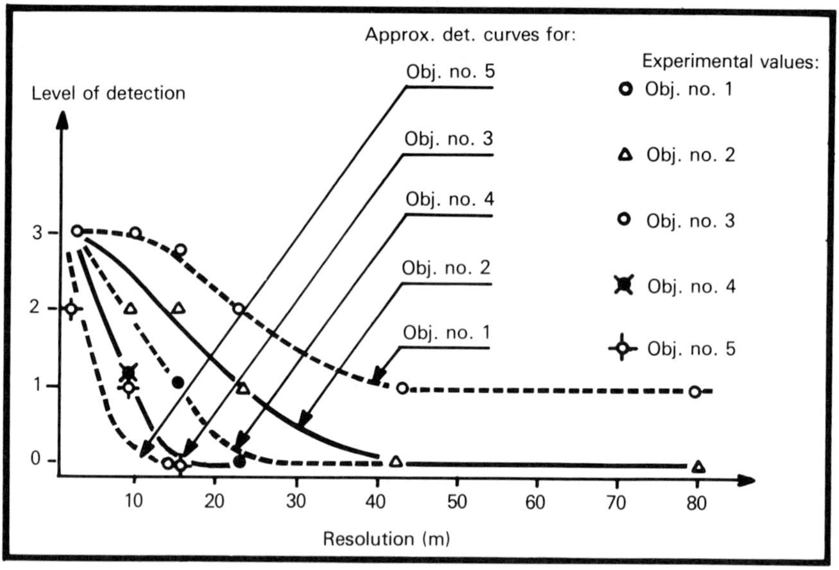


A comparison between the simulated images of 80-m and 34-m resolution can be made with Landsat images taken over the Gothenburg region (recording date not the same as for the photographic recording). The MSS (band 7) and RBV recordings are shown in figures 6.4 and 6.5. When comparing these Landsat recordings with the simulated images, one should keep in mind that band 7 gives very high contrast for water/land features. The monochromatic photographic image, on the other hand, gives less

**Figure 6.4. Part of Landsat scene MSS 7, 80-m ground resolution. Area covered by figure 6.2 as indicated**

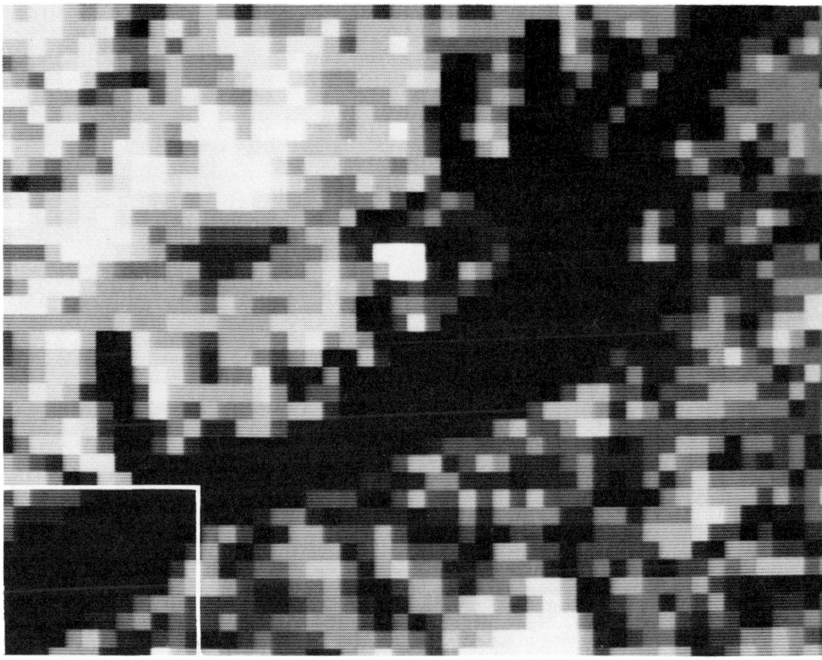

**Figure 6.5. Part of Landsat scene RBV, about 35-m ground resolution. Area covered by figure 6.2 as indicated**

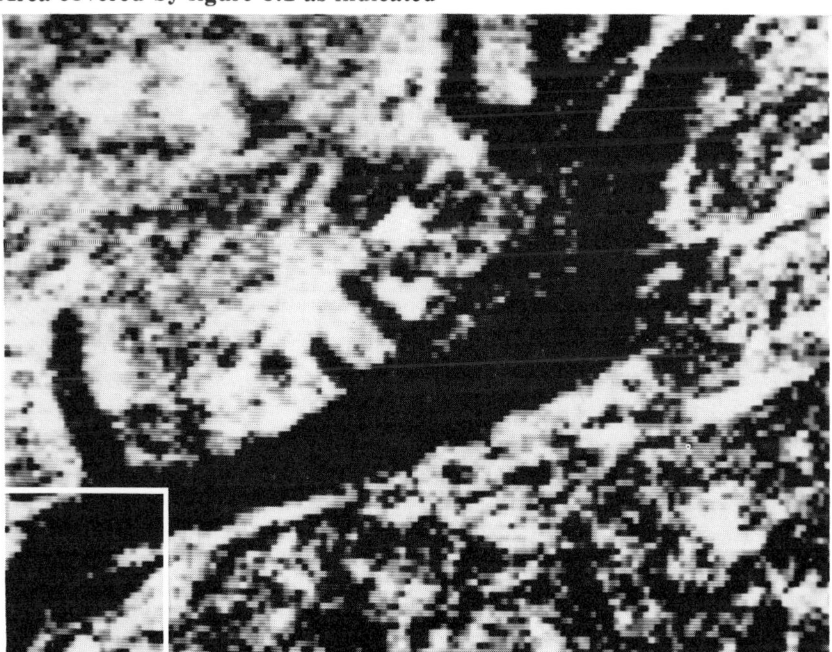

contrast. It is apparent that the RBV image (about 35-m resolution) gives much better resolution of the harbour details. Also, objects of the size of large ships are easily resolved (compare the broad pier with the narrow one in the photographic image).

Figure 6.6 shows a graph illustrating image resolutions for various satellite sensors (Landsat, the planned French SPOT, Skylab photographic camera). Also, the approximate range of resolutions for area surveillance and detailed reconnaissance is indicated. The turbulence of the atmosphere constitutes a resolution limitation of about 10 cm. Resolution for two photographic camera systems placed at various satellite orbital altitudes is also indicated. These cameras have film resolutions of 200 and 250 lines/mm and focal lengths of 0.50 and 4 m, respectively. The resolutions of the images produced in the simulation experiment dealt with in this paper are also indicated on the graph.

**Figure 6.6. Image resolutions for various satellite sensors**

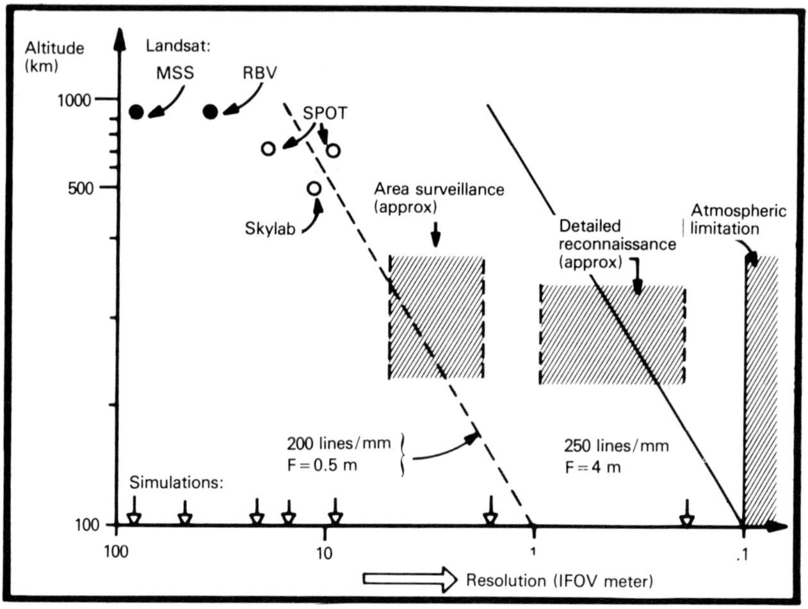


## *VII. Conclusions*

The digital computer can be used for many tasks in image analysis and photointerpretation. Therefore, the special effects produced by image digitization are important to understand. Also, the computer was used for simulating images having specific characteristics. This paper identifies one special application: the simulation of images having different ground resolutions.

Such images can be used to study integration performances of various spaceborne imaging sensors. Studies of this kind may be of benefit when discussing characteristics of sensors to be used in various monitoring tasks.

## References


1. Barrow, H.G. and Tenenbaum, J.M., 'Computational Vision', *Proceedings of the IEEE*, Vol. 69, No. 5, May 1981, pp. 571–595.
2. Marmolin, H. and Nyberg, S., 'Multidimensional scaling of subjective image quality', FOA Report, C 30039, November 1975.
3. Linde, L., Marmolin, H. and Nyberg, S., 'Visual effects of sampling in digital picture processing', *IEEE Transactions*, SMC-11, No. 3, 1981, pp. 201–207.
4. Rosenfeld, A. and Kak, A.C., 'Computer simulation of satellite imagery', FOA Report, C 30212, January 1981.
5. Könyves, P., 'Estimation of position by image correlation', FOA Report, D 30090, April 1978.
6. Orhaug, T., 'On the resolution of imaging systems', *Optica Acta*, Vol. 16, 1969, pp. 75–84.
7. Lloyd, J.M., *Thermal Imaging Systems*, (Plenum Press, London, 1975).
8. UN COPUOS, 'Characteristics and capabilities of sensors for earth resources surveys', UN Document A/AC.105/204/Add. 1, 21 December 1978.
9. Orhaug, T. and Forssell, G., 'Computer simulation of satellite imagery', FOA Report, C 30212, January 1981.

# Paper 7. The prospects for beam weapons


**A.M. Din***

*Institut de Physique Théorique, Université de Lausanne, Dorigny, CH-1015 Lausanne, Switzerland*


## *I. Introduction*

Recent history has presented numerous examples of gross underestimates concerning what is possible in science and technology, particularly where military application is concerned.

Before World War II few people would have given the nuclear bomb a practical chance of becoming a reality. Today it is the principal means of mass destruction and will probably remain so forever.

For some time the prominent scientist and military expert Vannevar Bush maintained that the intercontinental ballistic missile (ICBM) was a technical impossibility. Today it is the principal transportation vehicle of mutually assured destruction (MAD).

More recently it was said that a laser beam weapon (LBW) could do militarily significant damage only by being so big that one might as well drop it on the enemy. Today this is not necessarily considered to be the case, but the same opinion is now being expressed about particle beam weapons (PBWs). It might, however, only be a question of a few years before the story has to be revised again.

How can one understand the quest for exotic weapons like the LBW and PBW? It is important to realize that these weapons, as well as other new sophisticated military gadgets, are not supposed to be science fiction weapons, supplanting the basic mass destruction role of nuclear bombs. Rather, they are conceived as transforming advanced technology into devices able to produce a decisive advantage during a nuclear exchange by reducing (or ideally eliminating) the attack and defence capability of an adversary.

---

* The opinions expressed in this paper are those of the author and do not necessarily reflect the official policy of his professional affiliation.

Military planners see in the advent of LBWs and PBWs a way out, for the first time in the last two decades, of the no-exit dogma of MAD. The basic reason for such a presumption lies in the profound change in the technological time scale of warfare that these weapons would cause.

Typical current time scales are of the order of seconds in close exchanges with planes and missiles and of the order of several minutes for ICBMs to reach their targets. In contrast, during the time it takes the LBW photons, moving at a speed of 300 000 km/s, to reach a plane flying at mach 2 at a distance of 2 km, the plane would move only 4.5 mm, and a satellite circulating at 7 km/s at a distance of 500 km only 12 m. The same numbers apply in the case of particles of a PBW, which move very close to the speed of light. The time scale for target destruction has been reduced to a fraction of a second in all beam weapon (BW) applications.

Directed energy weapons (LBW or PBW) function by depositing on a target an energy of 1 kJ or more. An energy impact of this magnitude will in general create such serious secondary thermo-mechanical effects in the target material that the object will be destroyed or will malfunction. The physics of LBW and PBW is described in more detail below.

Which are the primary targets of directed energy weapons that come to mind? A BW placed in outer space on board a satellite could be used first of all in an anti-ballistic missile (ABM) mode by firing at missiles during their boost phase. In this phase the missile would be easy to locate and to track because of the limited angular movement. Secondly, satellite-based BWs may be deployed in an anti-satellite (ASAT) mode to attack other satellites or, in the only marginally different defence satellite (DSAT) mode, to defend against approaching space vehicles.

Another possibility, which may have some self-protective advantages, is to station BWs on board aeroplanes. The US Air Force has already deployed a Boeing NKC-135 airborne laser laboratory aircraft equipped with a carbon dioxide 400 kW laser. Such BWs could be used for tactical purposes, in an ABM mode for defending nuclear weapon bases and conceivably in an ASAT mode.

Earth-based BWs have an ABM as well as ASAT capability. Earth-based BWs may also have promising applications in conjunction with space-based relays. BWs, and in particular charged particle beam weapons (CPBWs), have further been considered as possible candidates for anti-ship missile defence (ASMD).

*A priori* the range of BW applications thus appears to be quite broad and the tactical and strategic implications promise to be more than spectacular. However, various physical constraints as well as policy considerations tend to somewhat bleach this rosy military picture, as will be discussed in more detail below.

Before going into these matters it is worth asking how seriously the great powers take the BW prospects by studying the amounts of money which have been and will be spent on BWs. Official figures only exist for the USA. It is estimated that up to 1980 about $1.5 billion have been spent on military-related laser research. For the next five years it is planned to spend an average of $200 million per year for this purpose. PBWs have received considerably less funding—about $60 million up to 1980—and projected near-term spending runs at about $35–45 million per year. These weapon progammes are administered at the Defense Advanced Research Projects Agency (DARPA) and, in fact, the LBW programme is the single largest technology-based programme sponsored by the Department of Defense (DoD). Other programmes which have closely related objectives dispose of separate budgets; thus, for instance, the DoD has projected about $175 million for ASAT weapons in the 1981 fiscal year.

Without access to the corresponding figures for the USSR, military trade journals have in recent times given the impression, based on more or less solid evidence, of the existence of certain technological gaps in the BW domain. Whether this is really so or yet another example of the machinations of the military-industrial complex referred to by President Eisenhower more than 20 years ago—or both—still remains to be seen.

It seems, however, that a BW race is well under way.

## *II. Laser beam weapons*

Since the discovery of the laser effect and its first experimental demonstration in 1960, tremendous progress has been made in pushing physical parameters like power output and single pulse energy to higher and higher levels. The current definition of a high-energy laser, according to the DoD, is one which has a single pulse energy of at least 30 kJ or an average power output of 20 kW.

Let us briefly recall the basic mechanism of the laser effect. Whenever the atoms of a certain medium are forced (pumped) into an excited energy state, such that a population inversion develops, it may be possible to stimulate coherent emission of a radiation which in general will be monochromatic and strongly collimated. The variety of lasers, such as solid, liquid, dye, gas, gas-dynamic and chemical lasers, is just a reflection of the many possible choices of medium and pumping mechanism to suit the desired output characteristics.

It was not until the invention in 1967 of the carbon-dioxide gas-dynamic laser (GDL), producing a beam of light of wavelength 10.6 $\mu$m in the far infra-red, that one could start thinking practically of generating sufficient energy output to have a weapon capability. A more recent device which may have interesting military applications is the free electron laser (FEL). The first operation of an FEL took place in 1975 at Stanford University and since then considerable effort has been put into developing the concept further.

The principle of the FEL is somewhat different from other lasers, since the lasing medium is made up of the free electrons of a mono-energetic, relativistic electron beam produced by linear (or storage ring) accelerators. In such a beam a population inversion exists with respect to the entire spectrum of lower energy and a magnetic (wiggler) modulation can generate a narrow band of coherent radiation with a tunable range of wavelengths from about 1 mm to 100 nm (in the far ultraviolet).

The efficiency and power output of an FEL can be very high. Studies at Lawrence Livermore Laboratories indicate that a peak power of $10^6$ MW at wavelengths about or below 300 nm could be attainable.

High-energy lasers have already been tested in a number of military field experiments in the USA. In 1973 the Air Force used a GDL to shoot down a winged drone at the Sandia Optical Range at Kirkland Air Force Base, New Mexico. In 1976 the Army, using an electron discharge laser in its Mobile Test Unit, successfully destroyed winged and helicopter drones at Redstone Arsenal, Alabama, and in 1978 the Navy, using a chemical laser, destroyed a TOW anti-tank missile in flight at San Juan Capistano, California. It may safely be supposed that the USSR has been able to perform similar tests.

To understand the possible scope of LBW applications, it is necessary to investigate closely a multitude of complicated physical problems involving operational characteristics and target impact of the laser beam. The propagation of a beam in vacuum is unobstructed and determined solely by the diffraction-limited field divergence, which can be made smaller by using large mirror apertures. It has therefore been generally acknowledged that LBWs are particularly suited to operate in outer space.

Applications wholly or partially within the atmosphere are, however, possible when due consideration is taken of the degrading effect of air. The propagation of a hole in the atmosphere is facilitated by operating the laser in a pulsed mode of nanosecond bursts. Also, particular choices of wavelength, determined by the resonant frequencies of air, will produce self-focusing beams (by a combination of diffractive and refractive processes) so as to avoid an undesired 'blooming' of the beam.

The above physical characteristics also tend to enhance the inflicted target damage. The pulsed mode operation produces complicated thermo-mechanical effects leading to quick rupture or 'through burning' of the target material. This effect is further enhanced by using small laser wavelengths (i.e., ultraviolet or ideally X-ray lasers).

A number of technical problems still remain to be solved before militarily effective BWs can be deployed. For deployment on satellites one of the problems is to ensure an adequate power supply for the laser and a virtually maintenance-free operation. These problems can be avoided by using Earth-based LBWs where the beam is relayed to the mirror of a satellite (possibly geostationary) in a near vertical position over the Earth station. These stations could with advantage be located close to the borders of enemy territory so that the satellites, knowing the location of ICBM silos, would be able to relay the beam in a not-too-oblique angle towards missiles in their boost phase, without demanding an excessive angular tracking capability of the laser mirrors. Hybrid inertial-optical trackers have achieved pointing accuracies of about 1 μrad and angular tracking rates of about 0.1 rad/s. For ASAT and boost phase ABM systems the latter figure may have to be improved by one or two orders of magnitude.

The angular tracking accuracy is one of the problems that make the use of LBWs less effective in connection with atmospheric close distance missions such as ASMD. For such purposes PBWs could be better suited.

Before deploying LBWs, military planners will require answers to a number of other questions, such as possible countermeasures and comparisons with other conceivable means of performing similar missions.

However, the recent rapid advances in LB technology make it seem likely that some kinds of LBW system could be deployed already around the mid- or late 1980s.

## *III. Particle beam weapons*

Two classes of PBWs exist: the charged particle beam weapons (CPBWs) with military applications mainly in the atmosphere and the neutral particle beam weapons (NPBWs) for use in outer space. The distinction in region of applicability appears for the following reasons. In outer space a CPB would suffer, over long distances, a dispersion due to electrostatic repulsion and a largely uncontrollable bending due to fluctuations in the geomagnetic field. Only in the atmosphere can a CPB have the chance to propagate over a distance of several kilometres thanks to self-focusing effects. An NPB on the other hand is unaffected by magnetic fields and the beam dispersion is determined only by the initial beam preparation. In the atmosphere an NPB would, however, quickly be degraded because of ionization.

The CPBs may consist of either electrons or protons which are brought up to near the velocity of light by more or less standard accelerator techniques. The NPB consists of hydrogen atoms which are produced by stripping off electrons from an accelerated beam of negatively charged hydrogen atoms. The CPBW and NPBW military programmes in the USA are run under the code names Chair Heritage and White Horse, respectively.

For military applications CPBs appear to have some advantages over LBs but numerous disadvantages, one of the latter being that the corresponding technological base is not very well developed. One of the advantages is the high target penetrability of charged particles which means that the beam energy will not be deposited only in a narrow surface layer. The impact effects include (as with LBW): hole boring leading to rupture due to internal pressure; generation of heat, X-rays and electric currents destroying sensitive electronics; and initiation of detonation of high explosives. In addition the radiation cone around the beam axis in the target may comprise neutrons which could lead to malfunctioning of nuclear warheads.

The main uncertainty about the possible use of CPBWs stems from the lack of precise knowledge about the characteristics of endo-atmospheric beam propagation. The simplest theoretical models are clearly inadequate and more realistic treatments quickly become extremely complex because of the interplay of a variety of physical effects.

Early accelerator experiments in the 1960s already indicated the presence of beam instabilities which would cause a rapid disruption of the beam, but also laid the basis for a better understanding of how a subtle preparation of beam characteristics like energy profile, pulse length and separation could enhance the self-focusing influence of the surrounding air plasma.

An experimental check on the viability of stable CPB propagation is scheduled to start in October 1982 when the Chair Heritage programme of DARPA will inaugurate the advanced test accelerator (ATA) at the Lawrence Livermore Laboratory near Berkeley, California. For a cost of about $40 million the ATA will be able to produce a beam of relativistic electrons with an energy of 50 MeV, a 10 kA current, a 70 nanosecond pulse length, a burst rate of 1 000 Hz and an average rate of 5 Hz. The beam energy is thus seen to be modest compared to what has been achieved in other research accelerators, which have in general, however, a beam current much too low for military applications.

The ATA is an induction-type accelerator of relatively traditional design. To improve the performance of CPBs a number of innovative accelerator designs have been devised, some of which, like the radial line accelerator, the auto-accelerator and the various electrostatic accelerators, may turn out to be of practical military interest.

An essential problem to be solved by new accelerator designs is making sure that the dimensions of the CPB accelerator will not be excessive, particularly if it is not going to be Earth-based. For linear accelerators the beam energy is, of course, directly determined by the accelerator length and the available accelerating gradients for which the average standard today is of the order of a few MeV/m. This quantity could most likely be increased one or several orders of magnitude by novel designs.

Another aspect of PBW design for non-Earth-based and, in particular, satellite-based accelerators is the problem of assuring a sufficient power supply for generating the accelerating fields. This, however, does not seem to be an insurmountable obstacle using currently available chemical explosives and detonation-coupled field generators. A more distant possibility is to use the power generated by fusion of deuterium–tritium pellets.

The tracking and pointing requirements of PBWs are roughly the same as for LBWs. In the case of PBWs faster angular tracking rates should be attainable because of the simpler mechanism of magnetic beam deflection. One of the basic purposes of the ATA is to test the performance of such steering nozzle devices.

The reasonable hopes of good performance in this respect and the hoped for stable atmospheric propagation, even under unfavourable air conditions, determine to a large extent the envisaged military applications of CPBWs. These include ASMD of large carriers against cruise missiles, where the warning distance and times could be as small as 1 km and 4 seconds, respectively. Also, point defence of missile silos and other ground installations against cruise missiles and re-entry vehicles could be envisaged.

The possible use of NPBWs in outer space is determined, in addition to the CPB base, by the development of an adequate electron-stripping mechanism with a beam dispersion limit of 1 $\mu$rad as well as the requirements of a self-sustained platform.

Before actual deployment of PBWs, an additional number of problems, most of which are common to LBW applications, have to be resolved. Among other things, one has to develop appropriate sensors for target acquisition and identification mechanisms for detecting miss vectors or hits. Another point to be considered is, again, possible countermeasures and alternative weapons for comparable missions.

The fact that PBW technology is still in a proof-of-principle stage makes it difficult to give predictions of future PBW deployment. It is not unimaginable, however, that the earliest systems applications might appear some time in the late 1980s or early 1990s.

## *IV. Tactical and strategic implications*

The distinction between tactical and strategic missions of BWs, as is also sometimes the case for other weapon systems, is not a very sharp one. There is a further ambiguity in BW missions since, from the description above of the functioning of BWs, it may seem superficially that they are designed solely to destroy small well-defined objects and therefore cannot be classified as a means of mass destruction.

There is however one (tactical) aspect of CPBW which could entail classification as a radiological weapon with a certain mass destruction capability: a CPB when propagating through air will generate a shower of energetic radiation around the beam axis and the associated radiological effects in human beings may be lethal or incapacitating up to distances of several tens of metres. When sweeping the CPB over the battlefield a non-localized anti-personnel effect results. Even if this aspect may not turn out to find significant military applications, it could be important for a characterization of BWs in connection with control measures, to be discussed in the next section.

Other tactical applications of BWs include deployment on tanks, aircraft and ships for ASMD. Such missions do not, however, seem to fall essentially outside the conceptual stage of other innovations in tactical warfare and will not be considered further here.

Almost all other applications of BWs appear to have conceptually new and potentially dangerous implications for the strategic balance of today. The most decisive application would be the ABM mode of satellite-based BWs for attacking either ground- or submarine-launched ICBMs in the boost phase. A couple of satellites in geostationary orbit or a couple of dozen in low orbit could ideally be sufficient to produce an ABM system with an ICBM intercept probably of nearly 100 per cent.

A second and third line ABM system could be formed by BW-equipped high-flying aircraft, such as the NKC-137 laser laboratory, and by Earth-based BWs situated around missile silos and other strategic centres.

These latter systems could also be effective against cruise missiles. However, the intercept rate is not likely to be very high under a massive attack. Thus, if one assumes the existence of an effective strategic bomber defence, a truly ironic hypothetical situation emerges: the only remaining decisive strategic delivery vehicle would be nothing else but the premonitor of the missile age—the German V-1 bomber!

Satellite-based BWs have important strategic implications when deployed as ASAT systems due to the increasing military reliance on outer space. Firstly, satellites have come to play a predominant role in both tactical and strategic communications and any attack on such satellites would automatically lead to a diminished warfare capability in all domains. Secondly, as already noted, satellites may acquire an important general ABM system capability.

A realistic military evaluation of the strategic impact of BW systems must include a number of elements. For instance, it is necessary to have reliable estimates of the minimal and maximal effectiveness of fully deployed systems in defence as well as in first-strike situations. But even before such questions become relevant it is vital to predict what could happen to the strategic balance during the time span between BW testing and actual deployment. In fact, even if only a fraction of the imaginable BW applications are ever realized, they will in all likelihood give rise to what may be called a technology-expectation-generated instability.

The point is that any weapon system which promises to nullify the military credibility of the MAD dogma will produce an instability which may be released through a variety of pre-emptive moves. For example, during a period of Earth-based prototype testing (which may be kept secret with some confidence) or of space-based systems development (which may be more difficult to hide) an adversary could feel forced to consider selective pre-emptive responses while the strategic balance had not yet been changed. This danger would of course be considerably magnified during the process of actual first weapon deployment. When making such considerations it should be recalled that military planners tend to put excessive emphasis on worst case scenarios and, as discussed, the many uncertainties surrounding BW effectiveness could only make the instability more acute.

Various scenarios related to the strategic balance may arise depending on which of the two great powers takes the lead and on the speed with which this takes place. With a lack of reliable comparative information on BW programmes in the USA and the USSR, it is difficult to make even vaguely accurate predictions in this respect.

As an educated guess, taking into account the traditional Soviet performance in the quantitative arms race contrasted with the US performance in the qualitative arms race, it may be reasonable to suppose that a US lead in the BW domain is most likely. With regard to the basic physics and technological development of BWs, one could probably expect the great powers to maintain a near parity. The above conclusion is based on the presumption that actual deployment of effective BWs will require the solution of a hitherto unseen collection of 'packaging' and logistics problems in which the US military-industrial complex should have a significant edge.

However, as discussed above, a US lead in BW systems will not necessarily imply that the strategic balance be changed in US favour. In fact, during a test and initial deployment phase, the USSR will have a number of options. On a relatively low level, the existing ASAT interceptor capability could be used in point pre-emptive strikes. And on the highest level there is, of course, the ultimate threat of an all-out nuclear pre-emptive strike.

It is also interesting to analyse a situation where both powers have attained some kind of parity in BW deployment. In such a case the strategic balance would superficially seem to have been maintained, but in reality the situation would be vastly different. This is related to the fact, alluded to in the introduction, that the characteristic technological time scale of warfare would change by several orders of magnitude. Today this time scale is, strategically, of the order of minutes, but in the above situation split second decisions would have to be taken. Thus, if the strategic balance today is not quite at the mercy of computers this would certainly be the case by then!

In conclusion, it seems hard, if not impossible, to envisage a situation where the deployment of BWs would enhance any national security, let alone world security.

## *V. Control measures*

The above observations on the development of BW systems were seen mainly from the perspective of military technology. Political perspectives must of course also be considered and, in fact, existing treaties interfere to some extent with possible BW applications. Furthermore, it is conceivable that the great powers, having noted some of the disquieting strategic implications, may reach a political consensus on new control measures in the not-too-distant future.

The most relevant treaty is the ABM Treaty of 1972 since, as discussed, BWs based on land, in the atmosphere and in outer space could have ABM capabilities. The ABM Treaty does not, however, specifically forbid BW deployment, but it has been interpreted as requiring that new ABM systems be submitted to discussions in accordance with Article XIII and that agreement be reached in accordance with Article XIV.

Two possibilities therefore seem to exist in this respect. Either the parties to the Treaty must, at a certain not very well-defined stage of weapon development, negotiate limitations (or abolition) of BW deployment, or perhaps due to optimistic military expectations, one of the parties may flatly terminate the Treaty.

Another treaty of less relevance in form, but not in spirit, is the Outer Space Treaty of 1967. Article IV of this Treaty obliges the parties not to place in orbit around the Earth nuclear weapons or any other kind of weapon of mass destruction. It is clear that BWs in general cannot be classified as such, but it is uncertain whether the aspect of PBWs, discussed in section III, concerning anti-personnel radiological effects, could be construed as infringing on the provisions of the Treaty.

While the spirit of the Treaty undoubtedly was to limit all weapon deployment in outer space, practically it leaves open all non-nuclear options. On the other hand, a recent Soviet draft proposal submitted by Foreign Minister Gromyko to the 36th Session of the UN General Assembly in August 1981 would demand that the parties "not put objects carrying any types of weapons into orbit around the earth". Even if this idea may appear sympathetic to arms controllers, probably realistic near-term treaties would have to provide for a more careful phrasing.

The point is that concepts such as defensive and offensive weapon systems and peaceful and military uses are becoming increasingly diffuse. Limitations on certain *bona fide* military space deployments may therefore easily be perceived as unduly restricting other activities, such as communication systems and reconnaissance capabilities, which could even have a purely civilian character. The central issue is how to define in a proper way the essential distinguishable elements of those space-based weapon systems that present the greatest dangers for the strategic balance. BWs certainly ought to be counted among these.

The problem of establishing treaties on weapon deployment which enhance security in an equitable way and which have a minimum of unforeseen technological loopholes is of course difficult to address. A few short remarks may therefore be in order.

It would seem that the majority of existing treaties prohibit a certain kind of activity. This automatically tends to imply that everything which is not prohibited is allowed—and should be undertaken. Such treaties can therefore only be effective during a rather limited time span. Future technological developments will tend to create uncontrollable instabilities. This could be said to be the case for the Outer Space Treaty.

The opposite approach is to provide for a treaty formulation by which a certain (existing) activity is allowed, with the understanding that anything else is prohibited. The ABM Treaty is of this nature although, as noted above, the prohibition clause is not clearly spelled out. Such formulations offer some advantage for the simple reason that it is easier to agree on maintaining and regulating something existing than to directly forbid a development on which it is difficult to be specific about possible future applications.

The abnormal situation which exists today concerning the activities of the great powers in outer space is that, although three out of four space missions have military purposes, this has never been admitted officially. The danger is that this situation will prevail until, thanks perhaps to BW deployment, a strategic instability becomes acute.

Following the above philosophy, a reasonable first step in controlling the activities in outer space would therefore be to acknowledge in treaty form the present state of affairs by establishing qualitative and quantitative guidelines for the existing military deployments. New technological developments would implicitly be forbidden or subjected to negotiations.

While such a scheme is far from satisfactory, it would at least save the world from an uncontrollable flourishing of 'technologically sweet' (in the words of Robert Oppenheimer) developments, which might turn out to be both strategically and politically sour.

# Paper 8. Orbital BMD and the space patrol

**P.J. Nahin***

*Kingsbury Hall/EE, University of New Hampshire, Durham, New Hampshire 03824, USA*

## *I. The present situation*

It was only a relatively short time ago that most people laughed at the idea of man in space, dismissing such talk as fantasy. Today, however, people are used to public debate about the pros and cons of war in space.

Operational satellite systems already in orbit make modern surface warfare as far advanced from that of World War II as that catastrophe was from the US Civil War. These satellites allow the command and control of ground, naval and air systems with far more efficiency and ease than was ever before possible. In military jargon, these satellites are called 'force multipliers'.

The next obvious step for the military is to move the weapons themselves (such as high acceleration anti-ballistic missiles (ABM) and ground bombardment rockets, laser projectors, particle beam generators, or ultra-high velocity electromagnetic rail guns) into space, too. The Soviet Union, with its well-publicized experimental programme for exploding killer satellites, has already made a significant move in that direction. In addition, the Soviets have every reason to question the military potential of the US space shuttle.

The realization that it is now technically possible to move men and the means for waging combat off the Earth and into space is, for many, exciting and hopeful. This is because the planet is presently littered with incredibly destructive weapons, and these weapons are deployed adjacent to where people live. It is one thing to live next door to a military depot that repairs broken Jeep engines. It is quite another to have ICBM silo fields underneath farms and next to elementary schools. The situation becomes especially

---

* The opinions expressed in this paper are those of the author and do not necessarily reflect the official policy of his professional affiliation.

ridiculous when such crazy schemes as 'shuttle-bug' and 'air-dropped' MX missiles are discussed. It is clearly more intelligent to base ballistic missiles at sea rather than on land where people live.

But why stop there—after all, the sea is full of living creatures too, including man. Why not move strategic weapons off the land and out of the sea and into space? Possibly we shouldn't dismiss this option too quickly in an automatic 'no arms in space no matter what' reaction.

## *II. War in space*

Violent struggle in space does seem to be, at least at first glance, a more attractive option than conducting a battle on Earth. Nuclear explosions in space may well spew much of their deadly radioactive debris away from Earth. The smallest of these particles, under the influence of the radiation pressure of sunlight, might tend to actually be blown out of the solar system. The targets in a space theatre of war will be property-intensive, with the resulting loss of life being fairly small (and limited to combatants, since it seems unlikely there will be innocent neutrals in space during the next 15 to 20 years or even longer).

The usual response to this argument is that it shows lack of understanding about the mechanisms of the arms race, or some such general dismissal. That is, once strategic weapons are in space the probability of a truly horrific global war will be greatly enhanced. Any clean and isolated battle that starts in space (and, according to those advocating a totally weapon-free sanctuary there, the very 'niceness' of space war will seduce an orbital conflict into existence) will not long remain in space. Such conflicts will, it is argued, inevitably spread downward to the surface and consume all of mankind in a raging Biblical Armageddon.

This point of view is compelling, and indeed, in the historical context of the nature of human affairs, its proponents may well be right. The concept of limiting a US–Soviet conflict that has degenerated to the level of direct lethal exchanges (as opposed to both sides using surrogate nations as battlefields, the traditional technique) strikes many as naive and wishful thinking and there are many proposals for keeping space the sanctuary the 1967 Outer Space Treaty claims it should be.

Still, in the spirit of promoting a wide spectrum of thought and debate, an idea is presented here that rejects a black-and-white casting of the role of weapons in space. There may be a way to place strategic weapons in orbit that would decrease the probability of world-wide disaster. This is a radical concept, but just as in science new ground is not broken by constantly repeating old ways, so it may be with arms control and the striving for some sort of stability in world military forces. We should give careful consideration to new concepts that could capture the imagination of the world.

One such concept is that of an orbital BMD operated by what can only be called a modern-day space patrol.

## III. *Orbital BMD*

The concept of a space patrol may bring a smile to many, but actually it's just a (giant) mutation in the evolutionary course that ballistic missile defence (BMD) has been following for years. In extremely broad terms, we can describe the history of the US BMD programme in three phases. First, in the mid-1950s, came the sophisticated radar controlled Nike-Zeus, itself the direct descendant of the Nike-Ajax and Nike-Hercules air defence weapons. Nike-Zeus moved the point of target engagement upward from those altitudes at which manned, air-breathing machines operated.

Next, in the early 1960s, came Nike-X with vastly improved radars, computers, and the ability to handle the increasingly sophisticated penetration aids known to be under development by the Soviets. Direct by-products of Nike-X were the ground-based ABM systems of the 1960s (Sentinel and Safeguard). These systems let incoming warheads penetrate deeply into the atmosphere to allow air drag effects to help distinguish light decoys from the more massive, armed re-entry vehicles. They were opposed, and finally defeated, (and the ABM Treaty signed) for all sorts of reasons, some of which were:

1. Some regions of the country would receive more protection than others.
2. Rocket ABM systems can always be defeated by the offender launching more missiles than the defender has interceptors.
3. There would be local devastation produced by one's own exploding defensive missiles.
4. It would be like 'shooting a bullet to hit a bullet', and 'everybody' knows that cannot be done.[1]
5. The existence of an ABM would make the defence less vulnerable to a counter-attack after launching a first-strike, thus increasing the chances of general war.
6. The technology couldn't be tested except (and too late) by an actual attack (e.g., the performance of the acquisition and tracking radars would remain an unknown, especially in the face of such strange and exotic phenomena as powerful EMP pulses).

But the ABM issue didn't die. The continuing technological war between defence and offence has raged on, despite the 1972 ABM Treaty; and the SS-17, SS-18 and SS-19 silos are more blast-resistant than the Minuteman silos, which can withstand 2 000 psi! [2]. So the missiles have

[1] But Herman Kahn, in an interview, said to the contrary: "It's really quite easy" [1].

become more accurate in response. This again makes an ABM capability at least an option to consider for the defence.

Thus, in the 1970s, people began thinking about using non-nuclear explosive, terminal homing interceptors to avoid both the radioactive explosion debris and the radar issues.

Now, in the 1980s, we find articles appearing describing 'layered' defences, with the first phase of interception occurring far above the atmosphere (new identification technology, such as high resolution, real-time radar imaging, may not need to use air drag). A new concept called 'leakiness' has been introduced to explain how a lower altitude, second phase of interception will eliminate those re-entry vehicles that survive the first layer of defence.

Clearly, BMD thinking is evolving away from near-Earth nuclear ABM rockets and pushing the interception of re-entry vehicles up to higher altitudes. What an orbital BMD does is just take this process to its natural completion in one fell swoop.

With this motivation as the driving force, I suggest that it may well be worth a look at what might be the world's best military use of space—the deployment of appropriate ABM weapons in orbit under the control of an international space patrol.

## *IV. The space patrol*

This weapon system in orbit would have as its mission the following tasks: (*a*) to engage and destroy thermo-nuclear ballistic missiles during the initial boost and mid-course phases; (*b*) to destroy warplanes in flight; (*c*) to strike terror into naval surface forces; and (*d*) to defend itself from any attack.

The technology used to perform these awesome objectives would be a broad-based mix, including, for example:

(*a*) ejected clouds of debris (e.g., nails); (*b*) laser projectors; (*c*) particle beam accelerators; (*d*) ultra-high velocity electromagnetic rail guns; (*e*) high acceleration ABM rockets; and (*f*) ground bombardment rockets.

The construction and arming of the orbiting platforms (there would have to be several, possibly many, separate stations) would be based on the space programme knowledge and skills of the USA and the USSR, but, and most importantly, they would be operated by elite crews selected from all the nations of the world. These men and women would constitute the overseers of the space patrol.

These crews could either be on board or in control from the surface via wide-band jam-proof radio links. If control were from the ground, the system could be one with a spatially distributed, netted command structure to prevent the terrorist take-over of an individual facility resulting in the control of the entire system. The 1967 Outer Space Treaty presently bans weapons of mass destruction from space. This treaty would require a modification to ban such weapons from any individual sovereign nation, allowing only those of the space patrol to be in orbit. Individual nations would still retain the right to maintain defence forces, armed with tactical weapons, if they so felt the need. All nations would have the right to petition the space patrol for aid in repelling invasion.

Orbital BMD would provide both the USA and the USSR with protection against the one threat neither can now defeat—an unsophisticated missile attack from a third country. Both nations now look with fear at the possibility of a massive first-strike wave of enemy ICBMs. But what should the response be to a single incoming missile?

These are, admittedly, ideas that have a strong idealistic flavour. They presuppose that both the USA and the USSR really do want a stable world and desire the elimination of so-called police action flare-ups and limited wars like Korea, Viet Nam and Afghanistan. This is a pleasant belief (and one would like to say, even rational), but it is not universally held.

There are observers of Soviet behaviour who assert that the USSR actually has a plan that requires the continual creation of global disruption and revolution (particularly in the emerging Third World nations). They point to Soviet literature and propaganda for support, for example, the following ideological jingle:

> Eat pineapple, gorge on grouse,
> Your last day is coming, bourgeois louse.

On the other hand, the Soviets have an easy target in the rich and powerful US Department of Defense. How can the United States survive economically, they ask, without a huge war-making machine? As one who has worked for the US military from time to time (indeed, as I write this my students are US Naval Officers), I must admit that the concerns of the USSR are not more easily dismissed than those of the USA.

## *V. Can it work?*

This is a complex question with enormously difficult technical and political aspects. On the technical side the debate on space-based BMDs or ABMs has ranged over the entire spectrum from dismissal as the mumblings of idiots, to the strident urgings of 'let's get our stuff into orbit before the other side does'. Some examples include:

1. USAF Lieutenant-General Kelly H. Burke (Deputy Chief of Staff of Research, Development and Acquisition) was recently quoted as saying, "In this country we don't know how to build a space-based laser that would be an effective anti-ballistic missile system" [3].

2. *Aviation Week and Space Technology* carried a short note

describing development work at Los Alamos National Laboratory on an X-ray laser weapon that would "produce a high-intensity beam of several hundred terawatts" that would be placed on "about 20 to 30 laser battle stations . . . for space-based ballistic missile defense" [4].

3. Richard Garwin, a highly respected scientist and arms control analyst, recently presented a paper that attempts to show the enormous technical problems in the construction of space-based weapons [5]. If he is right, the whole issue of orbital BMD is a moot one. If he is wrong then a unilateral weapons-in-space presence would be a stupendous technological surprise on the side left behind down on the surface of the planet.

4. Beginning in 1977 *Aviation Week and Space Technology* has published impressively detailed articles on directed-energy weapon developments in both the USA and the USSR. Enormous amounts of resources are being poured, by both countries, into this work, according to the magazine.

So how is the average person on the street to assess the technical question of 'can it be done?' It is very difficult to sort out the truth from speculation and deliberate misinformation. However, I have seen nothing to cause me to doubt that, on a theoretical basis, these devices can indeed perform as claimed.

But these are mere technical issues; much harder are the political issues.

Does the concept of a space patrol mean world government? Would the acceptance of such a concept, one creating an organization literally with world-wide military authority, mean that all nations would effectively be abrogating their sovereignty? Would it be the United Nations with teeth that bite instead of gums that sputter? These questions are not easily answered.

The space patrol is an idealistic concept. There will always be those who will argue that only a fool gives up power, and that safety is won only from a position of military superiority. Even military parity is said to be totally unacceptable (the USA is particularly hard-nosed on this point). This gloomy assessment does not even invoke the argument of the frightful wrath that could surely blast forth from the existing military services if they were to see their traditional roles and missions either radically altered or made obsolete.

Given the present state of affairs, it is obvious that these objections are perfectly valid. And yet, I still feel that the concept of an orbital BMD operated by the space patrol should be pursued. I feel this way because of one reason, and a non-technical one, at that.

I believe that the arms control people, on both sides, are talking to no one but themselves. Their sophisticated gyrations, their delicate slicing of the precise meaning of each and every word they use, have all taken on the musty flavour of the medieval debate as to how many angels can dance on the head of a pin. Quite frankly, with the exception of the participants themselves (and there is doubt, here, too), nobody really knows what SALT and similar such arms limitation efforts mean. Of course people understand the words they use, but the arms control debates have long-range implications that are often beyond our abilities to accurately predict. Most people find the evening sports page more relevant to their immediate lives.

I believe we need a new, imaginative arms control goal around which the people of the world can enthusiastically gather, and not just a small, highly elite group of technical experts. In the 1950s there was a popular juvenile science fiction programme on US television called 'Space Patrol'. It caught the fancy of adults as well as of children. In 1975, the Apollo–Soyuz link-up showed that Americans and Soviets can work together in space if it is considered to be in their joint interest.

In the 1980s and 1990s, the USA and the USSR could make what was once just a television fantasy, a space patrol guarding world peace, actually come true. What could be more in their joint interest than that?

## References


1. Kahn, H., interview in *US News and World Report*, Vol. XCI, No. 12, 21 September 1981, pp. 53–54.
2. *Strategic Survey 1980–81*, (IISS, London).
3. Burke, K.H., in *Military Electronics/Countermeasures*, May 1981.
4. Industry Observer, *Aviation Week and Space Technology*, Vol. 115, No. 10, 7 September 1981, p. 15.
5. Garwin, R., 'Are we on the verge of an arms race in space?', *The Bulletin of the Atomic Scientists*, Vol. 37, No. 5, May 1981, pp. 48–53.

# Paper 9. Some remarks on US and Soviet strategies concerning manned activities in outer space

**H. Kautzleben***

*Akademie der Wissenschaften der DDR, Zentralinstitut für Physik der Erde, DDR-1500 Potsdam, GDR*

## *I. Introduction*

There is serious concern over the re-emergence of the US political strategy of peace through strength. According to official statements this policy aims at gaining total technological and military superiority over the USSR. Such efforts to change the existing military equilibrium between the Soviet Union and the United States, between the Warsaw Treaty Organization and NATO, would inevitably cause a new round of the arms race. In their statements the leading Soviet representatives have made no secret of their response to these efforts on the part of the USA, and to the measures which the USSR is taking due to this development.

## *II. The US space programme*

Space technology plays an important part as far as the efforts to attain military superiority are concerned. Different systems of military satellites have been used for years and are being systematically completed: (*a*) three systems of satellites for military communications; (*b*) a system of satellites for navigation; (*c*) military meteorological satellites; and (*d*) systems of military reconnaissance and control.

According to the report on US space activities submitted by the President to the Congress in 1979, funds for space activities granted by the Department of Defense (DoD), rose from $3.2 billion in fiscal year 1979 to

* The opinions expressed in this paper are those of the author and do not necessarily reflect the official policy of his professional affiliation.

$4.0 billion in 1980 and to $4.9 billion in 1981. These funds constitute 43, 45 and 49 per cent of the total expenditure on space activities allocated by the US federal government in the respective fiscal years. For space activities with direct or indirect military purposes the other participating federal agencies also dispose of noteworthy funds. The same applies to the US National Aeronautics and Space Administration (NASA). In 1979 its share of total space expenditure amounted to 54 per cent, in 1980—53 per cent, and in 1981—50 per cent. About 50 per cent of this sum was spent on the development and testing of the space shuttle. US estimates indicate that over and above the total funds in these years another $2.1-, $2.3- and $2.5-billion, respectively, had been spent on secret military space programmes.

The employment of artificial satellites has turned out to be more advantageous than the earlier techniques used for weather forecasting, navigation and communications. In addition to this, especially in the USA, the conventional non-satellite-based techniques have been neglected to such an extent that it would be extremely difficult to return to using them.

Along with the process of introducing satellite systems in military practice, there emerged simultaneously a considerable dependence on them. Thus it is obvious why such an effort is being made to secure them against enemy attacks and disturbances, and why there are disputes about the so-called anti-satellite systems.

**Manned space activities**

Whereas automatic satellites have been used extensively for military purposes, a clearly military use of manned space flights can be related only to the beginning of the space shuttle programme. Though manned space flights were carried out from the very beginning by the armed forces, the only earlier project with possible military aims was the Manned Orbiting Laboratory (MOL). This project had been managed in close connection with the Gemini project by the US Air Force in the mid-1960s but ended abruptly for various reasons.

Manned space travel has so far been carried out exclusively by the USSR and the USA. An examination of the historical development of manned space flights indicates that there have been quite different strategic aims and stages.

The history of US manned space activities has included six projects with different goals. Project Mercury—undertaken by six astronauts from May 1961 to May 1963—was dictated by the ambition of the USA to keep abreast of the pioneer activities of the USSR in this area. Project Gemini—with 10 manned space flights from March 1965 to November 1966—showed intensive preparations for manned lunar-landing space flights. In his message of 25 May 1961 to the US Congress, President Kennedy declared that the USA should be the first nation on the Moon. This task was fulfilled within the Apollo programme which consisted of 11 successful manned flights with 33 astronauts between October 1968 and December 1972—the fifth flight landed on the moon. The Skylab programme, implemented in three space flights with nine astronauts from May 1973 to February 1974, used the technology of the Apollo programme and involved world records in long-endurance flights for the US astronauts. The Apollo–Soyuz test project in July 1975, with three US astronauts and two Soviet cosmonauts participating, was the last manned space activity by the USA for over six years.

A decisive argument for giving up US manned space activities was that they would serve mainly civil missions whereas military goals could be achieved more effectively and economically by other means. The same argument was applied to the MOL project which was terminated at the end of the 1960s.

In the space shuttle programme, US military men and politicians discovered a better way to achieve military goals and, at the same time, make the best possible use of the capabilities of man in space activities. This programme has continued despite the escalating costs.

**The space shuttle**

The space shuttle programme was discussed by the US Congress in the summer of 1971, and the go-ahead was given by President Nixon in January 1972. This project aims at developing a re-usable transport system with the required manoeuvrability, high payloads and a great number of applications. Due to technical problems the first test flight of the prototype 'Columbia' took place only in April 1981, after several delays. Depending on the chosen orbit the space shuttle can carry into a near-Earth orbit up to heights of about 600 km a load ranging from 10 000 kg to about 30 000 kg with a load capacity of 4.6 m by 18.3 m. The crew consists of three to seven men. By order of the DoD a booster to the space shuttle vehicle has already been developed that can carry payloads up to 2 200 kg into higher orbits (up to and including geostationary ones). Meanwhile the logistical preparations for using the space shuttle have been met. Space ports have been built at Cape Kennedy, Florida, and at Vandenberg, California, and some additional landing runways are available.

Three other models of the space shuttle called Challenger (1982), Discoverer (1983) and Atlantis (1984) are now in production or have already been completed.

The space shuttle is a space transport system which is more economical, more effective and more flexible in its applications than the former systems. Flights aboard the space shuttle are less strenuous for the crew, thus enabling a broader spectrum of tasks to be accomplished, ranging from carrying out scientific and technical experiments to installing large-scale constructions in space. The space shuttle will also enable inspections of satellites in orbit.

Approximately one-third of the 487 experiments planned for future space shuttle flights (until 1994) will be conducted with military aims.

Among the measures planned are: transport of communications, reconnaissance and hunter-killer satellites into outer space as well as experiments on board the space shuttle aiming at destroying enemy satellites, among other means by using laser weapons and other radiation armaments. First tests with laser weapons are to take place in 1982. It would also be possible to carry nuclear weapons into space on the space shuttle.

The confirmed flight programme for the space shuttle envisages for 1983 the transport into orbit of the Spacelab developed by the European Space Agency (ESA). By September 1981, 110 experiments had been scheduled for the mission of the Spacelab. They are to be carried out by the astronauts from the countries belonging to the ESA. In 1981 the necessary funds were drastically reduced by NASA in favour of military-based projects, in spite of protests on the part of the ESA.

The DoD also co-operates very closely with NASA in selecting and preparing astronauts for their missions aboard the space shuttle. In 1977, out of 35 candidates nominated, 21 were delegates from the DoD. All of them passed the preparatory training successfully.

It is worth mentioning that only US citizens are considered as possible participants in the space shuttle programmes. This is another substantial difference between the Soviet and US programmes for manned space flights, but certainly not the most significant one.

## *III. Soviet manned space flights*

The development of Soviet manned space activities is characterized by a number of closely interlinked stages. The main purpose of this development is the systematic construction of space stations in near-Earth orbits (with an altitude of 200 to 400 km and an inclination of 50°). These stations are served by satellites of the model Soyuz or Progress, which transfer people and goods.

1. The pioneer stage included eight space flights aboard Vostok 1–6 from April 1961 to June 1963 and aboard Voskhod 1 and 2 from October 1964 to March 1965. Altogether, there were 11 cosmonauts on board, among them the first, and up to now only, spacewoman.

2. In April 1967 a new stage commenced using the Soyuz-type space ships. These space ships made eight manned space flights in preparation for the placing of space stations in near-Earth orbit. Thirteen cosmonauts participated in these flights, two of them twice.

3. In April 1971 the activities on board space stations started with the launching of the space station Salyut 1 into orbit and the space flight of Soyuz 10. Two three-man crews of cosmonauts were on board Salyut 1; two two-man crews on Salyut 2; three two-man crews on Salyut 3; three two-man crews on Salyut 4; and four two-man crews on Salyut 5. At this stage there

were further flights of Soyuz space ships, fulfilling such other tasks as testing improved prototypes of space ships or newly developed installations. In some of the space flights the planned docking with the space station failed. At this stage the joint Soviet–US Soyuz–Apollo test project, including a preparatory space flight, also took place.

4. The current stage of Soviet manned space activities commenced with the launching of Salyut 6 in September 1977 and Soyuz 25 in October 1977. This stage incorporates several phases in which new prototypes of space ships have been tested. The five two-man crews of cosmonauts on board Salyut 6 remained in orbit for longer than any previous crews had done, thus setting new endurance records. Between April 1971 and May 1981 there were 33 manned space missions including one that was manned only on the return flight. Altogether 59 cosmonauts participated in this period, 13 twice and 6 three times. Nine of the 59 were non-Soviet members of the Intercosmos organization.

About 200 scientific or scientific-technical experiments have so far been carried out on board the orbital station Salyut 6 within the framework of a long-term programme. This programme is mainly concentrated on the following activities: (*a*) to carry out medical check-ups; (*b*) to observe the Earth's surface using elementary instruments as well as high-precision cameras (such as the MFK 6 multispectrum camera produced by Carl Zeiss in the German Democratic Republic); (*c*) to conduct material research under the conditions of outer space; and (*d*) to investigate near-Earth space as well as to conduct astrophysical observations.

## The Intercosmos programme

The orbital space station Salyut 6 served as a basis for conducting the first Intercosmos programme for manned space flights. The participating countries signed the Intercosmos agreement on 14 September 1976; since then, space flights envisaged by this agreement have been performed by nine international crews of cosmonauts. Work aboard Salyut 6 was done jointly by crews which consisted of a Soviet cosmonaut as commander, a flight engineer from one of the other countries participating in Intercosmos, who arrived in specially launched space ships, and the skeleton crew of the orbital station. Their research work was concentrated in four main areas: (*a*) remote sensing of the Earth; (*b*) material processing in outer space; (*c*) biological/medical experiments; and (*d*) astronomical and astro-physical observations.

The cosmonauts from each participating country carried out completely new experiments in orbit according to their specific scientific traditions and industrial structure.

The participation of cosmonauts in space flights from nine other Socialist countries—Poland, Czechoslovakia, the German Democratic Republic (GDR), Bulgaria, Hungary, Viet Nam, Cuba, Romania and Mongolia—made an important contribution to international knowledge about manned space flights. By realizing the ambitious experimental programmes, many significant results have been achieved, although the possibilities for using them for economic development have not yet been fully exploited. For example, the GDR cosmonaut, S. Jaehn, aboard the Soviet space station Salyut 6 in September 1978, conducted or jointly carried out about 20 scientific-technical experiments with the three Soviet cosmonauts on board. The GDR provided such scientific equipment for the space station Salyut 6 as, for example, photographic and medical instruments. According to the Intercosmos programme these instruments may be used by the other participating countries as well. Moreover, the subsequent crews continued the scientific-technical experiments started by the GDR cosmonaut.

### Co-operation for peaceful purposes

The successful work of the international crews of cosmonauts demonstrates that international co-operation in exploring and using outer space for peaceful purposes has become an important factor in the development of science and technology.

Moreover, the collaboration of international crews aboard the Soviet orbital station demonstrates the Soviet approach to exploration and use of outer space. Countries have the opportunity to make use of the generous offer of the Soviet government in the Intercosmos Agreement to conduct experiments on their own or jointly on Soviet space ships, without having to contribute towards expenses. This also applies to the use of the respective systems for transport and data transmission. This Agreement is open to all countries interested in co-operation for peaceful purposes. France, India and Sweden have participated in outer space experiments aboard Soviet spacecraft, and French and Indian citizens are preparing at the Soviet training centre for participation in a manned space flight on board a Soviet space station in the near future.

## *IV. Conclusions*

1982 will mark the 25th anniversary of the start of the space age. Since then the exploration of outer space has brought considerable results for communications, exploration of natural resources of the Earth, meteorology, navigation and other areas. At the very beginning of the space age the USSR submitted a proposal to the United Nations to prohibit the exploration of outer space for military aims, on the basis of strict observance of the principle of equal security and renunciation of military advantages for any country. In the following years Soviet efforts were consistently concentrated on making outer space a region of peaceful co-operation. In this respect many positive results could be achieved.

Present international treaties do not exclude the possibility of deploying in outer space weapons that are not categorized as weapons of mass destruction. Consequently, the danger of the militarization of outer space continues to exist and is even growing.

For this reason the Soviet Union has demanded that at the 36th Session of the UN General Assembly discussions be held on a treaty banning the deployment of weapons of any kind in outer space. The USSR has submitted a draft treaty. The conclusion of such a treaty would be a further important step towards the use of outer space for exclusively peaceful purposes and would help to prevent the use of manned space flights for military aims.

# Paper 10. Recent advances in the use of space for military purposes and on second generation nuclear weapons

**M. Felden***

*Laboratoire de Physique des Milieux Ionisés, Université de Nancy I, C.O. 140, 54037 Nancy Cedex, France*

## *I. Introduction*

In the last 10 years or so, the greatly enhanced role of artificial satellites has caused fundamental changes in the strategic possibilities relative to the conduct of a future war. Nuclear weapons, defence capabilities and the means to paralyse an enemy are also evolving; thus, it is difficult, in the period of rapid changes we are going through, to sum up globally this crucial question. We shall nevertheless proceed to certain reflections on these subjects.

## *II. The nervous system of modern armed forces*

Today the credibility of any military force depends first of all upon its ability to collect, transmit and process all forms of data and then upon its ability to intervene in, and adapt and react to, any situation it finds itself confronted with.

To that end, it must possess a highly efficient communications system, one that is also very well protected. Thus the USA and the USSR have developed a vast world-wide network of communications which enables them to observe each other continuously and, at the same time, to remain in contact with all their units disseminated across the globe.

What has therefore been established is a world-wide space network based upon two main links: the artificial satellite and the Earth-based relay

* The opinions expressed in this paper are those of the author and do not necessarily reflect the official policy of his professional affiliation.

station. Once such a network is disrupted, the whole military complex becomes blind and thus ineffectual. In this sense it can be said that the real nervous system of any modern armed force is its means of communication, of which the satellite is one of the major king-pins. Of course, there are fall-back capabilities, such as, the means to multiply chains of transmission and to render them self-sufficient. But as long as Hertzian wavelengths are used there is the problem of distortion preventing decoding.

In particular, the role and efficiency of missile-launching nuclear submarines are dependent upon their links to the centre of decision, and this poses a large number of problems that have not yet been completely resolved. In this case, where the delivery vehicles represent a decisive element, the artificial satellite plays a determining role. The difficulty of submarine detection is partially offset by the vulnerability of its communications.

## *III. Disrupting communications*

### The electromagnetic pulse weapon: mechanism and effects

The effects of a high altitude nuclear explosion are as follows: the gamma rays produced travel until they collide with electrons in the air molecules of the upper atmosphere. This gives rise to a flash of Compton electrons that are accelerated and deflected by the Earth's magnetic field, generating a transverse electric field. For a 10-megaton burst at an altitude of 400 km, the effect of this field extends over an area with a diameter of 3 000 km. The current produced sets up a rapidly rising electromagnetic pulse (EMP), radiating towards an even larger area of the Earth's surface. A few bombs (perhaps three) would be enough to cover the entire territory of the USA, USSR or Europe.

The power delivered, in a few nanoseconds, tails off in about one microsecond. The peak electric field may be as high as several kilovolts per metre (between 10 and 100 kV/m). Radio-frequency signals are generated over a broad band of frequencies (several kilohertz to several thousand megahertz). The instantaneous peak power density can reach several megawatts per square metre but, due to the pulses' short duration, the energy received is only a fraction of a joule per square metre. However, because any conductor acts as an antenna, the power is propagated through the electrical network within milliseconds and can knock out the system as a whole. Long-distance electrical transmission lines may receive peak powers as high as several tens of thousands of megawatts; the power induced is of the order of that occurring with the largest lightning bolts. However, the rise time of the current is considerably faster and is induced simultaneously in the entire electrical grid. Thus the various standard protection circuits will be ineffective. Similarly, long-distance communications and telephone links would be put out of order.

At the same time, the high frequency signals can be picked up by components of basic electronic circuits and destroy them. For example, energy of less than one microjoule would be sufficient to produce serious malfunctions in an integrated circuit. Therefore the airborne command posts and ground-based satellite communications systems are potentially vulnerable and very difficult to protect. The main significance is: (*a*) a possible disruption of military and civilian communications networks; (*b*) irreversible effects on all kinds of electronic components and computers; and (*c*) a possible complete disruption of the electrical power grid. The EMP could also damage electronic control of emergency cooling systems of nuclear reactors leading to serious accidents.

Protection against EMP is possible but not at a reasonable cost; nor is any technology available which could reliably protect all the components of a very large system. In the long run communications systems must be changed and replaced by other less vulnerable ones, such as fibre optics and laser communications.

In conclusion, it appears that the EMP bomb is a new class of weapon in the same vein as the neutron bomb, in that it enhances one of the effects of a nuclear explosion. EMP, however, can effectively cripple military and civilian systems almost instantly. This and other not yet well-known effects raise serious questions about nuclear strategy. In particular, they invalidate the previous exclusive reliance on massive retaliation, although several experts contest that point of view, arguing that military systems *can* be protected against EMP. Certainly, the strategic implications of EMP are one of the major elements determining the necessity to revise current defence policies.

## Other possibilities

Another possibility, in the same vein, would be to create anomalies in the wave-carrying environment. A very powerful bomb that explodes relatively slowly could be detonated within the ionosphere; this would sufficiently modify the environment to create serious disruptions in the transmission of electromagnetic waves. Furthermore, either through the creation of plasma or by movements in the ionospheric layer, powerful interference could be generated causing a background noise which would prohibit detection of signals.

From a satellite it is also possible to project powerful pulsed rays of charged particles in such a way that they are trapped by the Earth's magnetic field—the Van Allen Belt—and remain long enough to seriously disrupt communications. In this case the attacking party is subjected to the same constraints as the enemy and must adapt his strategy accordingly.

Yet another way of disrupting communications would be to install powerful electromagnetic generators of the klystron or magnetron type in satellites. If the latter can be precisely directed, they can beam down onto the target country to produce effects analogous to those of the previous example.

The question which remains is that of the energy supply of such stations. Several solutions are feasible, but none seem to be perfectly mastered. It is certain that for chemical explosives the limits have already been reached. The development of an onboard nuclear generator seems to be the best solution, and in this much work is being done, although a few more years are likely to be needed to reach the operational phase. The space shuttle could make the building of such a generator in space possible. A solution could also be found by placing large efficient solar panels on satellites as visualized in the E. GASER project, but they are highly vulnerable. Perhaps the technology of transmitting high-powered electromagnetic beams from the Earth's surface, using a satellite for relay, may be mastered first.

Another method of disrupting communications is to use hunter-killer satellites to destroy other satellites, either by ejection of pellets or by explosion. Lastly, the space shuttle should permit the capture of certain satellites if their altitude is low enough and if their configuration permits such captures.

## *IV. High-energy beam weapons*

High-energy beam weapons use high-energy lasers and particle beams (the particles being neutral in space and charged in the atmosphere). The USA is pushing forward an ambitious programme of high-energy lasers for utilization in space—carbon monoxide, iodine, hydrogen fluoride, halogen and excimetric lasers—which will allow coverage of a wide range of frequencies (from infra-red to near ultraviolet). Research is also being done to explore the possibility of Xaser generation—a coherent beam of electromagnetic radiation at X-ray frequencies. For the moment, the aim is to achieve a range of 5 000 km and an energy equivalent of 30 kJ per impulse. The aim for 1990 or thereabouts is to attain a 40 000 km range with 100 kJ per impulse energy equivalent. The purpose is to disturb the electronics of enemy satellites, in order to render them ineffectual. It is known that the USSR is running similar programmes and may even be ahead in the experimental field.

What is foreseen at a later stage is the destruction of intercontinental ballistic missiles with nuclear warheads, using satellite-borne high-energy beam weapons, but, in addition to questions of energy supply, there are problems regarding targeting, evaluation of firing inaccuracies and identification of the induced effects in case of non-explosion. Furthermore, there is the problem of overcoming possible countermeasures. These weapons call into question the global strategy developed over the last 30 years. Not only is the viability of this classic strategy being questioned but nuclear weapons and their delivery vehicles, including nuclear submarines, could become obsolete in the long run. If attention is not paid to this, the absolute supremacy of the two great powers would be almost definitely assured; it is hard to conceive what other nations could do simply to ensure their own security.

## Towards second generation nuclear weapons

There is an evolutionary logic in the development of nuclear weapons. It is well known that a nuclear explosion has essentially three types of effect: (*a*) blast, (*b*) heat flash, and (*c*) intense radiation and particle emission, to which must be added the radioactive fall-out. The central idea since the development of these weapons has been to make the most of each of these effects and, if possible, to enhance one of them in relation to the others. It is this process which has led to the development of weapons with enhanced radiation (the neutron bomb) and to research into an electromagnetic pulse bomb, not to mention other possibilities.

The same basic idea lies behind high-energy beam weapons, which, however, use different technology. Thus, instead of transporting destructive energy in concentrated form (as chemical or nuclear charge) towards the objective, which necessitates a vector, it is possible to increase considerably the kinetic energy of particles, as well as their number, and to bombard the targets with them. They can be fired towards the target at a speed approaching the speed of light. If the number of impacts is sufficient and if the energy is transported to the material in less time than it takes for the effects to be dissipated, there are local physical effects which are possibly as destructive as those caused by fusion. The quantity of energy absorbed depends on the total energy of the particles as well as the nature of the target (which determines the degree of penetration). There are also associated effects; for instance, in certain circumstances, there appears a strong beam of gamma rays, projected forwards, which can damage or destroy electronic components. It is therefore possible either to cause the explosion of in-flight nuclear warheads or else to destroy the electronics which guide and detonate them. There are a great number of technological difficulties to resolve, in order to perfect these types of weapon. Nevertheless, it should be noted that there are no physical principles to oppose such developments. It is therefore essentially a question of time and means, both human and financial. The use of these weapons is feasible, but their ideal environment would be in space, either on board artificial satellites, or on board manned space stations. There are possible countermeasures, but it can be envisaged that such countermeasures could be warded off, depending on the type of beam weapon.

In spite of the considerable number of difficulties, this panoply of arms should constitute a second generation of nuclear weapons, and they could see the light of day before the end of this century. It should be noted, however, that the character of these is essentially defensive, for it would be difficult to envisage offensive use of such weapons at this stage, given the power at their disposal.

## *V. A new conception of space in military strategy*

The real issue is that a second phase of this new generation of nuclear weapons would result in a whole new conception of nuclear war. High-energy beam weapons could so undermine the offensive potential of today's nuclear weapons that it would be considered necessary to enhance the war-fighting character of the beam weapons. One would, in fact, have to pass from defence to attack, and go on towards the destruction of the enemy, including the civilian population. The following scenario is conceivable: a vast portion of territory (several thousand square kilometres) would be scanned, in about the same way as the screen is in a television set, by an adequately equipped space platform; for example, one with powerful, high performance generator(s). Thus, at ground level, a very high level of radiation would be created; depending on their nature, the particles would cause very considerable damage by their direct as well as by their secondary effects, notably with a flow of penetrating gamma rays. The effects would be analogous to those of a super neutron bomb, though greatly amplified and extended in time. Furthermore, if the energy source were sufficient, there would be no limit to the destructive possibilities of these weapons. It follows, therefore, that the reinforced radiation weapons, and in the same way the EMP weapon, constitute an intermediate step in the evolutionary development of nuclear weapons. And this hypothesis is backed up by the fact that certain US and Soviet projects foresee the construction of space stations weighing around 100 000 kg as the first stage and over 1 000 000 kg before the end of the century.

It is true that present-day technology does not yet permit the implementation of these new concepts. But the space shuttle development is a step in this direction, and the development of its successor (the heavy-lift vehicle) can be interpreted in the same way; for these developments will make it possible to send into space large masses, including nuclear power generators, to realize the concept of beam weapons in space. Technological progress should allow these weapons to become a reality by the beginning of the next century.

Lastly, it should be noted that it is possible to increase remarkably the performance of these weapons, by using other types of particles—neutrons, gamma photons, mesons, and, in the near future, beams of radioactive nuclei, not to mention other, even more efficient possibilities.

## VI. *Conclusions*

Space has become a major issue in the arms race, and it is necessary to emphasize that its exploitation is only just beginning. Furthermore, since many of its possibilities remain to be realized, great technological progress can be foreseen by the end of the century. Thus the problem of nuclear warfare, in which space will play a determining role, will be placed on a new footing, among other reasons because new developments will enable one side to completely disrupt communications and paralyse the enemy. In any case, the development of a new generation of nuclear weapons is not contrary to any of the laws of physics, and therefore it can be reasonably supposed that they will end up being developed and perfected. It is therefore necessary, first of all, that the governments of those nations which do not have the means to perfect these weapons (there are hardly five or six in the world who can—and at what price?) become aware of this fact. Secondly, it is necessary that the public does likewise. Lastly, it is essential that a collective reaction occur. This would entail common sense, for what is at stake is nothing less than the future of the human race, and, in this domain, everyone is concerned.

# Paper 11. Use of satellites in crisis monitoring

**K. Santhanam***

*ISRO Headquarters, Dist. Office Road, Bangalore 560009, India*

## *I. Introduction*

In analysing the use of reconnaissance satellites performing imaging and ferret functions, the most important aspects of concern are that these are inseparable parts and important 'assets' of the military establishment of the USA and the USSR, whose self-justifying, self-perpetuating need for information on global developments is a contemporary phenomenon. Such satellites would not have a continuing lease of life if they were civilian and if the information provided by them were not integrated with their strategic defence apparatus and their military/political interests in alliances/spheres of influence. It is relevant to note that the ill-fated U-2 flight in 1960 of Gary Powers from Pakistan over Soviet territory brought an end to illegal reconnaissance over the USSR, and during the same year the first successful photo-reconnaissance satellites were launched by the USA. Interestingly, the USSR made a proposal in 1962 to the United Nations for a ban on military activities in outer space, including reconnaissance [1a]. It then reversed its position in 1963, presumably because its own space programme had begun to serve defence interests. Inevitably, outer space became a logical extension of the reconnaissance apparatus of the military systems of the USA and the USSR, and the way was paved to claim 'customary usage'. Subsequently, military reconnaissance satellites have been implicitly characterized as being peaceful and baptized under SALT as verification progeny.

In dealing with reconnaissance satellites, it may be useful to consider space reconnaissance essentially as an extension of aerial reconnaissance. This paper attempts such an exercise in a limited way and highlights some of the crises in which such satellites have provided information of importance

---

* The opinions expressed in this paper are those of the author and do not necessarily reflect the official policy of his professional affiliation.

to the space powers. The effectiveness of satellite reconnaissance in the past and in the future is also addressed.

## *II. Aerial reconnaissance and space reconnaissance*

The background to military reconnaissance from outer space clearly indicates that its development was perceived and promoted as a method for providing photographic (and electronic) intelligence over areas in which such acquisition by aircraft was inconvenient, illegal, provocative, risky and necessitated air bases in other countries. With advances in launcher technology and payload stabilization systems, the limitations in scale and quality of photography from space have been overcome by the space powers to a level where these photographs have been described (even as early as 1963) as being as good as aircraft reconnaissance photographs [2]. Ground resolution in the range 1–3 m for area surveillance and of 0.3 m for close-look satellites have been reported, based on unclassified analyses [1b]. A ground resolution of 0.4 m is mentioned in reference [3] from an altitude of 100 miles for a camera with a focal length of 2.4 m and a film resolution of 180 lines per mm. A recent review article on verification of the SALT II Treaty says "existing reconnaissance satellites are presumed to have 15 cm or better ground resolution" [4a]. US photo-reconnaissance satellites are also credited with a ground resolution limited only by atmospheric effects, which would be of the order of a few centimetres from this altitude. It is to be presumed that the Soviets have similar capabilities.

In addition to yielding photographic information equal to that of high altitude aircraft, reconnaissance satellites have other advantages which have made them attractive to military establishments. These include: (*a*) repetitive coverage of target areas subject to satellite energetics and orbital decay (and film inventory in the case of cameras); (*b*) legal permissibility based on customary usage and peaceful (meaning 'non-aggressive') purposes and a sanctified role under SALT as part of 'national technical means of verification'; (*c*) the capability to alter orbital parameters to provide multiple passes in a day over areas in which military events occur rapidly; and (*d*) the capability to utilize interesting area surveillance information for subsequent close-look imagery.

There are, however, some limitations in the coverage of target areas which are common to both satellite and aircraft reconnaissance. For instance, night-time coverage is poor compared to day-time coverage. Depending on latitude, season and local weather conditions, target areas may be under cloud cover. These two limitations could obviously be reduced (though not overcome) by infra-red (IR) coverage and radar imaging at lower resolutions. Further, real-time processing of weather satellite information could be used to perform reconnaissance when a break in cloud cover is observed.

Another problem could arise from camouflage and deception techniques implemented at the target site. These could include procedures like disruptive painting, tonal blending and installation of dummy facilities and equipment. That camouflage is indeed an area of concern in reconnaissance from outer space may be inferred from Article V.3 of SALT I which says: "Each party undertakes not to use deliberate concealment measures which impede verification by national technical means of compliance with the provisions of this Interim Agreement. This obligation shall not require changes in current construction, assembly, conversion, or overhaul practices."

Camouflage detection has been possible, for instance, by using the difference in spectral reflectance characteristics of foliage and paint [5]. Vegetation has high reflectance in the IR region beyond 700 nm and has an absorption (the 'chlorophyll dip') in the range of 670–680 nm. False-colour camouflage detection films utilizing this IR reflectance difference have been used in aerial photography. Multiband photographs and multispectral imagery combined with colour-additive viewing could provide information for camouflage detection, though there is no published information in this regard. Analysis of target-associated features and cultural characteristics of the area in the course of photo-interpretation could also be expected to assist camouflage detection.

It is also important to note that information on military exercises and manoeuvres from reconnaissance imagery would not reveal military intent, and a training exercise could, in fact, be a cover for invasion (see, for example, [6] regarding Egyptian tactics in the Yom Kippur War). This is why other intelligence available to the military establishment should be integrated with image analysis in the evolution of policies consistent with the 'national interests' of the powers carrying out the reconnaissance. Theoretically, it may be possible to conceive of military manoeuvres planned to avoid satellite detection, though it is difficult to say whether such dodging would be effective or worthwhile.

In comparing aerial and space reconnaissance for the military establishment, it is reasonable to assume that the aims are not different. As any textbook on aerial photography reveals [7], aerial reconnaissance in peace-time can provide advance knowledge of a hostile nation's reactions to various political, economic and military crises and it can remove the advantage of surprise in the initiation of hostilities. During war it can assist the military establishment in achieving early victory by providing up-to-date information on force deployment, terrain factors, communication systems, weapon installations, etc. So, both in peace and in war, it is unlikely that the aims of space reconnaissance would be any different from those of aerial reconnaissance. Under weather conditions that reduce scene availability for satellite imagery or for more rapid higher resolution imagery over targets ill-protected by SAMs, recourse may also be made to high-flying aircraft to supplement satellite imagery. The linkage between aerial and space reconnaissance would then be direct.

Though it has been claimed that space reconnaissance has promoted disarmament/arms control, the military use of photographic reconnaissance has an intent which is rather aggressive. While foreknowledge of a hostile nation's military preparations may deter the outbreak of hostilities, the equally important function of photographic reconnaissance in enabling victory by appropriate aggressive military action needs to be emphasized.

In this connection, it is worth recalling that military action does have functions in the following order of importance: (*a*) destruction of military forces and installations with retaliatory capacity (missile-launching sites, anti-aircraft emplacements, bomber bases and military airfields, shipyards and submarine bases, military ports and troop concentrations, etc.); (*b*) destruction of military production facilities and supporting industries (armament manufacturing plants, aircraft factories, ship construction and repair facilities, vehicle assembly plants, lubricant production, oil refining and storage facilities, steel mills, electricity-generating stations, chemical industries, etc.); and (*c*) destruction of transportation and communications networks (railways, roads, shipping lanes, radio and radar installations, etc.). Like the impossibility of inventing a gun that would fire only in one direction, it would be equally impossible to imagine information from military reconnaissance satellites being used exclusively for disarmament/arms control measures.

A question which frequently arises regarding reconnaissance from space is whether tactical and strategic targets of importance in military crises can be identified from satellite altitudes which typically are about 10 times higher than aircraft altitudes. As mentioned earlier, it is inferred that ground resolution lower than 1 m for close looks and in the range 1–3 m for area surveillance functions have been achieved.

The ground resolution needed in military photo-interpretation for five-task levels is given in table 4.1 of Part I. It would appear that the surveillance resolution number of 3 m would be able to detect, and in most cases identify, military targets. Further, any changes from the normal, as measured against earlier records of the target area, could be easily monitored.

Ground resolutions in the range of a fraction of 1 m would appear to be of special importance for detailed analysis of weapon systems which have previously been precisely identified, mainly to study in detail externally observable improvements or changes like, say, the upgrading of a tank's gun. Most of the targets listed in table 4.1 of Part I would appear to render themselves to precise identification by imaging systems having a resolution in the range 1–5 m. The exceptions are radar, radio systems, supply dumps and vehicles which need a resolution of about 0.3 m and artillery/rockets which need a resolution of 0.15 m for precise identification. The higher level tasks of target description and analysis generally need higher resolution. These tasks presumably lead to, for instance, information on 'functionally related observable differences' (FROD) and 'externally observable design features' (EODF) of strategic arms under limitation [4b]. In other words, area surveillance imagery in resolution range 1–3 m may be inferred as capable of providing precise or near-precise identification of most military targets, while close-look imagery in the resolution range of 0.1–0.5 m would enable target description and analysis.

In image analysis, ground resolution, unlike ripeness, is not all. Spectral information and spectral resolution also have importance. Computerized image-processing techniques, like density slicing, contrast stretching and automated feature extraction, may be expected to assist the photo-interpreter troubled by purely geometric limitations in the acquired imagery.

Interpretation of satellite imagery in the context of providing information on military crises may be expected to be a sophisticated and time-consuming job. It is reasonable to assume that the talents and skills of military photo-interpreters, suitably trained in computerized image analysis, would be used in this task. They would need, additionally, a vast amount of information on the geology and meteorology of the target country as well as on its military equipment table, the strength and structure of its armed forces, its military installations and periodic military exercises. The image acquisition or targeting process would need timely information on political and other tensions so that satellite coverage could be initiated appropriately. These are systemic requirements, and image acquisition by satellites is just one element in a long chain.

## *III. Inferred instances of crisis monitoring by satellites*

There have been a number of military crises over the last two decades which may have been monitored by military reconnaissance satellites. It is also known that such satellites have enabled the conclusion of bilateral limitation of strategic arms by providing 'non-intrusive' verification of compliance. The short survey of the limited literature attempted here is confined to crisis monitoring by satellites. The instances cited should be accepted as indicative only because it is reasonable to assume that the great powers have used their satellites to monitor every international crisis to test and improve their image acquisition and interpretation systems for privileged information in line with their national and international interests.

Neither the USA nor the USSR has directly admitted that its satellites have monitored one crisis or another. Analysts have therefore looked for 'clues' in parameters like orbital inclination, perigee, ground track and orbital manoeuvres to identify intended target coverage and to correlate them with crisis development. Analysis of telemetry signals provides additional information on some functions. Most analysts have drawn inferences based on ground-track calculations utilizing published data from satellite-tracking networks. This survey, admittedly incomplete, draws on published information which seems to have focused more on Soviet crisis monitoring than US.

Cosmos 281, 283, 286 and 289 launched on 13 May, 20 May, 15 June and 10 July 1969, respectively, at an inclination of 65° may have covered the Sino-Russian border conflict [8]. Photographic targets may have been at 55°N from an altitude of 210 km, twice a day during the daytime. It is likely that the satellites would have monitored incidents on the Tasta river in the Semipalatinsk region and Goldinsky Island in the Amur river. US satellites 19A, 26A, 39A, 41A and 50A may also be presumed to have monitored the area.

As far as the Indo–Pakistani crisis of 1971 leading to the formation of Bangladesh is concerned, Cosmos 463 launched on 6 December 1971 and Cosmos 464 launched on 10 December are reported to have provided photographic coverage of the target area [9]. Cosmos 463 was recovered on 11 December and Cosmos 464 on 16 December. Cosmos 465 which was launched on 16 December remained in orbit for 11 days, while the ceasefire went into effect on 17 December. A Big Bird satellite (1971–56A) may also have covered the crisis.

Orbit analyses of Cosmos 596–600 and 602 [10] indicate the use of recoverable Cosmos satellites to monitor the Arab–Israeli war which broke out on 6 October 1973. Cosmos 596 launched on 3 October 1976 was of the non-manoeuvrable, area-surveillance type. Cosmos 597, a high resolution manoeuvrable satellite, was launched within an hour of the outbreak of hostilities on 6 October 1973 with its perigee over the Middle East on the northbound pass. Cosmos 598 was launched on 10 October enabling a photographic pass over the target area at the end of the first day's mission itself. For stabilizing ground tracks, Cosmos satellites apparently execute a two-burn manoeuvre. In the case of Cosmos 598, the first burn lowered the perigee to about 175 km on 12 October to provide optimum coverage at the desired altitude on the northbound pass. It was recovered on 16 October. Cosmos 599 was launched on 15 October and did not have the Middle East as a primary target, providing battle-area coverage only during the last three days of its mission. Cosmos 600 launched on 16 October provided coverage on the first day of its mission as Cosmos 597 had done. Its apogee was lowered on 18 October to produce an eastward shift of the ground-track over the battle area. The satellite's perigee was shifted to the southbound pass while preserving an altitude of 215 km on the northbound pass. The apogee was raised on 20 October to restore the normal westward drift of ground-track over the battle area. It was recovered after a flight of seven days. Cosmos 602 was launched on 20 October and recovered on 29 October with a repetitively stabilized ground-track. Cease-fire was called by the Security Council in this period. Cosmos 603 launched on 27 October was recovered after a 13-day flight, its perigee being lowered to 175 km only on 1 November and track stabilization being achieved on 5 November. The peace agreement between Egypt and Israel was signed on 11 November. It is likely that Cosmos 607, launched on 12 October, and its successors, Cosmos 609, 612, 616 and 625, were used to monitor the cease-fire. Cosmos 630 launched on 30 January might have observed the redeployment of Israeli forces after their withdrawal from the Suez Canal.

The Cyprus crisis in 1974 invited satellite observation by the USA and the USSR [11]. On the day of the army coup (15 July) Cosmos 666 was over western Cyprus in the morning, and over eastern Cyprus in the afternoon of 22 July. The Big Bird (1974-20A) made two passes over Cyprus, one on 20 July when the Turkish invasion began and another on 24 July. Another US satellite, 1974-65A, made two passes over Cyprus on 22 August. Cosmos 667 launched on 25 July was manoeuvred on 2 August to provide a stabilized track over Cyprus, making two passes during the daytime on 7 August. Cosmos 670, launched on 6 August and recovered on 9 August, has also been mentioned [12].

Reconnaissance satellites have also received publicity in relation to a possible South African nuclear test in 1977 [13a] and observation of the rocket-launching range of the West German firm, OTRAG, in Zaire [13b]. Analyses of the ground-tracks of the Big Bird satellite (1977-56A) launched on 27 June 1977 and two Cosmos satellites, 922 launched on 30 June and 932 launched on 20 July, have indicated photographic coverage of the Kalahari desert in South Africa. It has been reported that the USSR detected preparations for a nuclear test in this desert and informed the USA and other countries. It has been inferred that the same satellites covered OTRAG's rocket range in the Shaba province. Cosmos 653 launched on 15 May 1974 and the Big Bird satellite (1974-20A) launched on 10 April 1974 may have observed the crater formed by India's peaceful nuclear explosion (conducted on 18 May 1974) on 20 May and 22 May, respectively. Though these three instances do not obviously constitute crises, they are indicative of the privileged photographic information of various territories which the great powers derive from their military reconnaissance satellite programmes.

## *IV. Effectiveness of satellite monitoring of crises*

The instances cited above on inferred satellite monitoring in some crises would be incomplete without a discussion of satellite effectiveness in crisis monitoring and management. Though photographic information had been obtained by the USA and the USSR on these military crises, the question arises as to whether such information has enabled the prevention of the outbreak of hostilities, the identification of the aggressor, or just management and resolution of the conflicts.

There is an argument that reconnaissance satellites would deter one state from attacking another because advance information on military preparations would be available. A question that needs to be answered is: how many military conflicts have been deterred or prevented because of the existence of the military reconnaissance satellites of the USA and the USSR? In the period 1960–80, more than a dozen wars have broken out in the Third World affecting international peace. There is no information to indicate that satellite reconnaissance has been a factor inhibiting the outbreak of hostilities.

In the case of a dispute over an unsettled border, the involved parties are likely to know each other's military deployment and manoeuvres before war breaks out. There may be surprise in timing, but it is unlikely that military manoeuvres would be unknown to each of the parties. Would military reconnaissance satellites be able to cover the start of a military crisis involving surprise in time of attack? Even with continuous monitoring this is likely to be difficult. And with discontinuous monitoring imposed by resources and priorities, such photographic coverage would occur, if at all, only by chance. Assuming that photographic coverage once or twice a day has been initiated because of tension indicators, it is possible to identify the massing of troops and military equipment across a border. However, would such imagery be able to identify who fired the first shot and when? Would such imagery prove who committed aggression first? When conflicting political claims in regard to alignment of borders exist, what would be the weight indeed of proof of aggression provided, if at all, by a reconnaissance satellite?

Photographic coverage after war has broken out would be easier because the battle theatres and targets of aerial/naval attack would be known. A daily pass by the satellites would give the satellite owner independent information on the military progress of the conflict, consolidation and occupation of territories. The first advantage to the owner would be hard information on the progress of the war, independent of bulletins issued by the parties in conflict which often may have roots in propaganda. The second advantage would be the use of such information for military offensive/defensive plans and targeting. Information on, say, the tank concentrations of one party or vulnerable gaps in army formations would be considered militarily very useful by the second party in the conflict. If the second party is involved in a strategic/special relationship with the satellite owner, the possibility of information sharing cannot be dismissed out of hand. A question then arises: is post-attack information provided by reconnaissance satellites useful in crisis management and resolution? Given the possibility of information sharing with 'partners' who could use it for offensive (or defensive) operations, would it not lead to conflict aggravation, and imply conflict enlargement with the power carrying out the reconnaissance getting embroiled in the proxy war?

The questions raised regarding effectiveness of satellites in crisis monitoring and management would appear to reveal the rather unsatisfactory current situation of satellite surveillance. As long as reconnaissance satellites form part of the military establishments of the space powers, the availability of independent, non-partisan satellite-sensed information may be ruled out. The use of this non-intrusive tool of surveillance in reducing military tension and promoting peace and stability in the international community remains largely theoretical. As far as military conflicts in the Third World are concerned, it may be reasonably argued that many forces of instability are external in origin and have roots in a system of economic, political and security management pursued by major powers. To the extent that conflicts are influenced by these powers, it is highly unlikely that prevention of war or proof of aggression based on satellite data would ever occur in a non-partisan manner.

## *V. Crisis monitoring and the Third World*

A theory which may be accepted without debate is that the Third World, with its need for economic development, is keenly interested in peace and stability. In an interdependent world, political and military upheavals tend to cast long shadows. Obviously, problems disturbing peace in the Third World need political solutions, just as they do for developed states. Many Third World countries have internal elements of instability which are serious enough without the involvement of external influences, but these influences form a disturbing dimension because Third World countries get drawn into great power rivalry and relationships with strategic overtones.

Among adversary states in the Third World, would non-partisan information from reconnaissance satellites available from an agency independent of the great powers contribute to confidence building and tension reduction? The theoretical answer to this question is likely to be an optimistic 'yes', though the achievement of a system providing such information would face more political than financial and technical problems. The adaptation of social and political systems to technical advances may be expected to be slow. However, the international community (and the Third World in particular) will have to think of information management when five or six additional countries become high resolution sensing states. Could this proliferation lead to the availability of independent, non-partisan and timely dissemination of information on crises? Despite the limitations in crisis monitoring satellites mentioned earlier, the overall effect of an independent information system may be expected to be positive.

The phrase 'information system' here would of course imply an institutionalized way of acquiring, processing and analyzing satellite imagery (with other open information), preferably in an open manner, to meet agreed dissemination patterns. Though satellite reconnaissance is essentially a new dimension to the ancient craft of espionage, its potential to promote confidence building, to reduce tension and to maintain peace between nations is strong. This potential should not be oversold, nor should it wither for lack of support. While promoting confidence between nations in the Third World, this potential is bound to inform countries about military aspects of great power rivalry, an equally important function.

## References


1. Scoville, H., Jr, 'The case of verification and detection: the contribution of satellite technology to the verification of arms control agreements and the prevention of conflicts', Paper presented at *Colloque Science et Désarmement*, Paris, 15–17 January 1981.
   (a) —, pp. 2–3.
   (b) —, p. 6.
2. Gatland, K.W., 'Espionage from orbit, part two', *Flight International*, 17 April 1969, pp. 642–645.
3. Greenwood, T., 'Reconnaissance and arms control', *Scientific American*, Vol. 228, No. 2, February 1973, pp. 14–25.
4. SIPRI, *World Armaments and Disarmament, SIPRI Yearbook 1980* (Taylor & Francis, London, 1980).
   (a) —, chapter 7.
   (b) —, pp. 164–174.
5. *Manual of Remote Sensing*, Vol. 1, ed. R.G. Reeves (American Society of Photogrammetry, Falls Church, Virginia, USA, 1975), p. 36.
6. *Aviation Week & Space Technology*, 31 March 1975, p. 7.
7. Avery, T.E., *Interpretation of Aerial Photographs*, 2nd Edition (Burgess Publishing Co., Minneapolis, Minnesota, USA, 1968).
8. Perry, G.E., 'Cosmos observation', *Flight International*, July 1971, pp. 29–32.
9. Perry, G.E., 'Cosmos coverage of the Indo-Pakistani war', *Spaceflight*, September 1972, p. 350.
10. Perry, G.E., 'Looking down on the Middle East war', *Flight International*, 21 February 1974, pp. 240–245.
11. SIPRI, *World Armaments and Disarmament, SIPRI Yearbook 1975* (Almqvist & Wiksell, Stockholm, 1975), pp. 385–391.
12. Perry, G.E., 'Observing the Cyprus War', *Flight International*, 5 September 1974, p. 270.
13. SIPRI, *World Armaments and Disarmament, SIPRI Yearbook 1978* (Taylor & Francis, London, 1978).
   (a) —, pp. 73–79.
   (b) —, pp. 79–80.

# Paper 12. Verification using satellites, feasibility of an international or multinational agency

**M. Abdel-Hady***

*Remote Sensing Centre, Academy of Scientific Research, 101 Kasr El-Eini Street, Cairo, Egypt*

and

**A. Sadek***

*National Committee on Peaceful Uses of Outer Space, Cairo, Egypt*

## *I. Introduction*

The growing use of outer space has brought into focus new technology and applications that were not generally known until recently. In view of the prospects of this rapid technological development and the resultant monopoly on its use by the space powers, many developing countries have become increasingly concerned about their own interests and security. It has also become evident that international space law cannot cope with such fast technological progress. This situation is causing a continuously increasing gap between the space powers and the non-space powers and creating many problems in the regulation of those outer space activities that would benefit the international community.

In view of the rapid development of space technology, it seems unfair that legal rights have been established allowing the space powers to practice certain space reconnaissance activities without somehow protecting the rights of the other states. In international legal practice, a precedent can be created by permitting the continued practice of certain activities by space powers. The establishment of a legal basis and rights for the space powers is more likely if other countries do not object or declare their positions in regard to such activities, either because of their unawareness of the legal implications or their lack of understanding of the technology involved.

Although the technological and economic capabilities of developing countries may not enable them to carry out space activities or to take active part in their development, they can play an important role in establishing proper legal practice for such activities and in the use of data collected. In addition to insuring their security and national interests, they would enhance world peace

* The opinions expressed in this paper are those of the authors and do not necessarily reflect the official policy of their professional affiliations.

by serving the entire international community on a non-discriminatory basis. One of the most significant means by which developing countries can accomplish this goal is through the establishment of an international or multinational agency for verification of certain parts or types of arms control and disarmament treaties and for monitoring potential crisis areas. The international or multinational participation would reduce to a great extent the present risks of confining this practice of military reconnaissance to only a very few space powers.

## *II. Military reconnaissance by satellites*

The USA and the USSR rely to a great extent on military reconnaissance by satellites for verification of compliance with disarmament/arms control agreements, and for monitoring crisis areas around the world.

In view of this situation, the position taken by many developing countries that do not possess space capabilities of their own can be summarized as follows:

1. The inability of developing countries to control such activities over their territories or to use effective concealment and camouflage for their sensitive targets against military reconnaissance by satellites launched by the space powers has caused such countries to be increasingly concerned about the implications of these activities for their national security and interests.

2. The non-defined status, in international law, for the sovereignty of countries over the space above their territories, has made it possible for other countries possessing satellite reconnaissance capabilities to use this space freely and at progressively lower altitudes, allowing the gathering of information of significant strategic and tactical value. Since developing countries are not informed about satellite data information collected over their territories and have no control over dissemination of such data to other parties, they feel their national security can be dangerously compromised.

3. The general change in international law regarding the tendency to legalize the unilateral use of outer space for military reconnaissance is also of major concern to many developing countries.

## *III. Consequences of continuing present military reconnaissance activities*

The continued practice of current military space reconnaissance activities without official protests from non-space countries and the lack of definite international provisions regarding the status of these activities has led many officials and scholars, especially in Western countries, to consider them as legal. Hence, the space powers consider such reconnaissance over national territories of other countries as legal activities.

The continuation of such activity, while its status remains undecided, can lead to some critical international situations. This has been proven by historic precedent with the incident of the US U-2 plane in 1960 which caused a severe political crisis between the USA and the Soviet Union.

If the space powers continue to affirm the legality of military space reconnaissance, in the absence of participation from the rest of the international community (possibly through an international agency), violation of the independence and sovereignty of the non-space countries may result in the long run. The threats which these countries would then face might increase with the expected rise in the number of space countries. It is expected that about 10 to 15 states will possess space surveillance capabilities by the end of this century.

In view of this situation, the non-space countries, in order to maintain their interests, may have to pursue other alternatives, such as:

1. The creation of an international satellite monitoring agency (ISMA). The acquisition of military space reconnaissance information would thus be controlled by an international agency insuring the interests of the entire international community on a non-discriminatory basis, and would be conducted according to acceptable international legal regimes.

2. Seeking, by all available means, to halt attempts aimed at converting present practices for unilateral acquisition and discriminatory dissemination of military satellite reconnaissance data or information into an international legal customary rule. Actions which non-space countries may pursue in this respect are: (*a*) presenting official protests, and (*b*) arguing the illegality of acquiring military information by satellites, without prior permission of the subjacent state, through the rules of international law, in general, and through the 1967 'Space Treaty' in particular.

## *IV. The establishment of an ISMA*

At the first special session of the General Assembly devoted to Disarmament in May 1978, the French delegation proposed the establishment of an ISMA. The General Assembly noted the proposal in the final document at the end of the special session.

At its 33rd regular session, the General Assembly requested the Secretary-General to undertake, with the assistance of qualified governmental experts, an in-depth study of the technical, legal and financial implications of establishing such an agency.

In pursuance of this resolution, and upon nomination by their governments, the Secretary-General appointed a group of experts which met several times in Geneva and prepared in February 1981 a comprehensive report outlining the technical, legal and financial implications of establishing an ISMA. This report will be submitted by the Secretary-General of the UN to the General Assembly's Second Special Session devoted to Disarmament in 1982.

The objective of the Agency, as recommended by this committee, is to use satellite technology to gather information for the purpose of: (*a*) verification of compliance with international arms control and disarmament agreements, both those already in force and those to be concluded in the future; and (*b*) investigating international crises.

The Governmental Expert Group, in their report, outlined an evolutionary process for the establishment of an ISMA, whereby in the first phase the agency would operate an international data processing and interpretation facility and disseminate resulting data/information to member states. In this initial phase ISMA would depend on states which operate surveillance satellites. In the second phase ISMA would have ground stations working with national satellites. These stations would also be needed in the third or final phase of the agency's development where an ISMA would operate its own satellites.

### Legal issues

During the Expert Group's discussions, legal issues associated with the establishment of an ISMA and the legal regimes governing its activities raised several controversies over their significance and implications for the national security of member states. Therefore, the group proposed several alternative options to be further investigated by member states to decide on the best course to suit their various national interests and that of the international community. A general outline will be given in the following paragraphs focusing on these options and the various arguments for or against each of them.

### International legal status of an ISMA

The following alternatives were presented as proposals for the ISMA operation. Each alternative has its own advantages and disadvantages, which would have certain implications for the operation of the Agency after its establishment: (*a*) as a specialized agency of the United Nations—in some respects this may be similar to the IAEA; (*b*) as a subsidiary organ of the General Assembly; (*c*) as a subsidiary organ of the Security Council; and (*d*) as an independent organization (outside the UN system). This last option was proposed, but not strongly supported by most members of the Group.

## Initiation of ISMA action

This issue also evoked serious discussions between members of the Group, and several options were presented for further research and study by member states and by the international community. Since an ISMA's mission is not to be considered as a general routine surveillance organ for the world, it is essential to identify with precision the conditions under which it will spring into action, which could be upon request by one or more of the following: (*a*) a principal organ of the United Nations; (*b*) an intergovernmental organization; (*c*) a member of an ISMA (for its own territory, for the territory of [an]other state[s] or for areas not subject to the sovereignty of any state); (*d*) non-member state(s); and (*e*) states parties to an agreement providing for international monitoring.

## Acquisition, use and dissemination by ISMA of data and/or information

The complexity of an ISMA, as dealing primarily with the collection and dissemination of data and information, was reflected in the lengthy discussions by the Expert Group preceding the adoption of their report.

Taking into account that technical developments may make it possible in the future to transmit from satellites not only raw primary data to the Earth but also partly analysed imagery data, the Group concluded that a more complete definition of the terms 'data' and 'information' would have to be clarified. For the purpose of their report the Group used the following understanding of these terms:

1. The term 'data' refers to 'raw' (primary imagery) data, either photographic or digital, obtained by satellites.
2. The term 'information' refers to information obtained by analysing raw primary satellite data.

## Question of confidentiality of data processing, formating and archiving

This issue was also one of the elaborate ones in the Group discussions before preparing its report. The Group discussed arguments for and against the confidential handling of primary data obtained from satellites in the Agency. The Group was not able to reach unanimous decision on this issue and did not consider it to be its task to recommend a particular solution to this problem. The Group recognized that this aspect would be of great importance for states considering their participation in the establishment of an ISMA [1].

Arguments in favour of openness were that any misuse of the Agency could be controlled and, more important, the end-product of the work of the Agency, a factual report, could be scrutinized in the way which would provide a guarantee for the impartial operation of the Agency. The adoption of open working methods should not, however, mean that outsiders would be able to request data in a way that would prevent the Agency from working efficiently. It would rather mean that the material used for the report would be available, at least after completion of the report. The details for such a working method would have to be subject to provisions in the Constitutive Act of the Agency and possibly to internal regulations.

On the other hand, the arguments for confidentiality were that, in view of the sensitive nature of some of the data obtained by the Agency, appropriate measures might have to be considered to safeguard their confidentiality during transmission, processing, analysing and archiving. For instance, though weather satellite and most Earth resources satellite data would be handled without restriction, the data required by an ISMA may be considered sensitive by supplier states in Phases I and II, and probably by others in Phase III. It was further noted that the required confidentiality may not be assured by legal, technical or administrative constraints, but the Constitutive Act may, nevertheless, have to provide appropriate provisions in this regard.

**Questions of dissemination of data and/or information by an ISMA**

Considerable discussion by the Expert Group was devoted to the form in which an ISMA report should be released. The mere dissemination of data, including auxiliary data, without any interpretation by the Agency, would tend to promote confidence in the accuracy and impartiality of the findings, because no human evaluation would be involved. However, the adoption of this format for the ISMA reports would produce unintelligible information for those users who do not possess appropriate technology and skills to do their own interpretation. This method would clearly discriminate in favour of the technologically more advanced states. It therefore seems that an ISMA's role would be to provide a factual report based on the processing and analysis of the data available to it. The Group also recognized that inconclusive or contradictory interpretation could emerge in the course of analysis of data by teams at the Image Processing and Interpretation Centre (IPIC) and was of the view that in such cases it might be necessary to provide the users with more than one analysis together with data used for such analysis.

**Access to ISMA data and reports**

While the existing international law contains no specific prohibition against unrestricted dissemination of data and information collected by satellites, regardless of the nature of such data/information, the widely differing views of states on this matter will have to be taken into account in the establishment of an ISMA.

With regard to the question of who should have access to ISMA reports, various possibilities can be identified, among them: (*a*) all members of the

United Nations; (*b*) only members of the Agency; (*c*) the Security Council, the executive organ of the Agency and states directly concerned (e.g., parties to a dispute, parties to a relevant treaty); and (*d*) only states directly concerned and the executive organ of the Agency. These options were examined by the Group and arguments for and against each of them were presented in the report adopted by the Group. It should, however, be emphasized that application of each option will also depend upon the relevant provisions of the Agency's Consultative Act.

## *V. Technical implications*

During the past decade, increasing reliance on military satellites for area surveillance and crisis monitoring has taken place. The advantages of the high speed of a satellite (30 000 km/h), and its capability of repetitive coverage and synoptic view, made satellite reconnaissance a very powerful tool in the hands of space powers for verification of arms control agreements and identification of military targets. At the same time civilian reconnaissance satellites, while primarily used for mapping and surveying natural resources and with their sensors having much less resolution capability compared to military satellites, could be also useful in the future for military reconnaissance. Within the next few years the great improvement in resolution of sensors used in the civilian satellites and the greatly increased volume of data to be collected by these sensors will be noteworthy.

The only two countries which today possess the necessary space technology for gathering data pertaining to military installations and activities relevant to the verification of arms control agreements and crisis monitoring are the United States and the Soviet Union, with a number of other countries presently developing capabilities in this area (China, France, India and Japan).

There are primarily two types of reconnaissance satellite—photographic and electronic. An international monitoring agency, such as the proposed ISMA, would mainly use the first type, that is, photographic. Instruments used in photographic (or imaging) satellites are sensitive to electromagnetic radiation with a wavelength between 0.4 and 0.7 $\mu$m and in the infra-red region (wavelength between 0.7 and 300 $\mu$m).

Figure 12.1 shows the relationship of ground resolution as a function of focal length for a satellite altitude of 150 km (the altitude at which most US military reconnaissance satellites orbit). From these relationships, it is reasonable to assume that it would be possible, with long focal-length cameras and with image resolution of about 100 lines per millimetre, for a ground resolution of 25 cm to be achieved using military reconnaissance satellites. Ground resolution could therefore be improved considerably if the focal length of the camera were increased. However, it is also known that the diameter of the optical system increases with increased focal length. The maximum size of the diameter that can be used is limited, however, by practical considerations and by the diffraction effect within the optical system. But even taking this into account, the diameter required to achieve a focal length of 600 cm is not outside the bounds of practicability. With a ground resolution as high as 15 cm, it is not surprising to find the use of reconnaissance satellites extended to activities other than merely providing assurances of compliance to arms control treaties.

**Figure 12.1. Ground resolution as a function of focal length (satellite altitude = 150 km)**

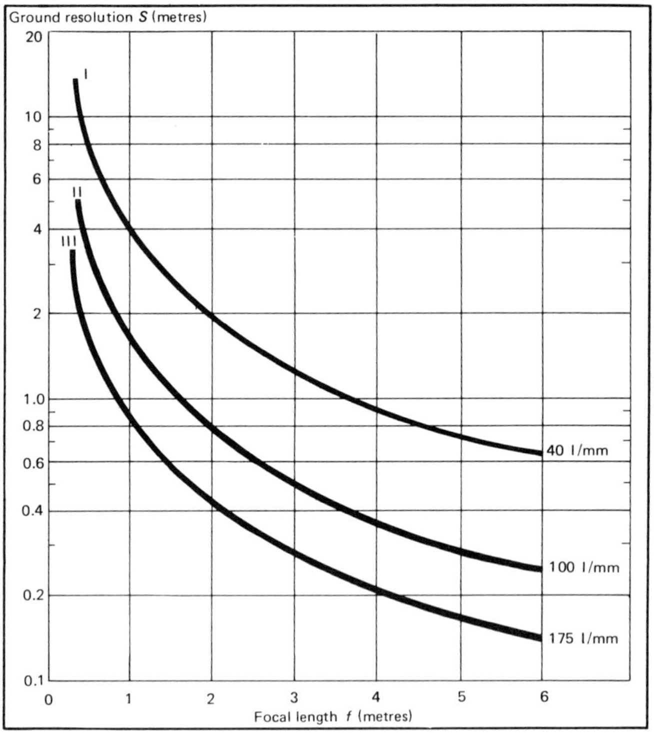


*Source:* Reference [7]

For Earth observation from space, a satellite's orbit and its sensor payload are selected depending on the type of information needed. Sensors observe and record images of the Earth's surface and the atmosphere by detecting radiated or reflected electromagnetic radiation. The data collected by the sensors are either recorded on films or by tape recorders. The exposed film can either be brought back to Earth, be developed and scanned on board the satellite by electronic devices or be directly transmitted to Earth via telemetry links. In the latter case, the resulting electrical signals are transmitted by radio to a receiving ground station when the satellite is within radio-visibility. A satellite data receiving station thus comprises antennae, receivers, data processing, storage, and normally data analysis equipment.

Resolutions required of a sensor to detect, recognize, identify and describe various targets of military significance are indicated in table 4.1 of Part I. Two types of mission are used for military surveillance: (*a*) *area surveillance* using a resolution in the range of 3–5 metres and (*b*) *close-look* using a resolution of 0.5–2 metres. However, even better resolutions (of the order of 0.3 m) are obtainable from sensors on military reconnaissance satellites, which suggests that these satellites are capable of performing most of the tasks indicated in Table 4.1 of Part I [1].

Military surveillance satellites are not only capable of contributing to the verification process regarding the implementation of several arms control and disarmament agreements, but they have also become an important source of information for this process, thus providing a means of inspection which is not based on the idea of 'on-site inspection'. The use of satellites has already become embodied in several major arms control treaties such as SALT I and II.

It is worth noting that during the past decade, major efforts have been made and considerable progress has been achieved in improving the performance of military observation satellites. The performance of civilian satellites has also improved considerably. For example, the return-beam vidicon cameras (RBV) on board the US Landsat-1 and -2 satellites had resolutions of 80-m IFOV, while that on the Landsat-3 has a resolution of 40-m IFOV. While all three Landsat satellites carried multispectral scanners (MSSs) with 80-m IFOV, an MSS on the proposed French SPOT satellite is expected to have a 20-m IFOV. A camera on this satellite operating in panchromatic mode is expected to have a 10-m IFOV. The new Landsat-D satellite is expected to carry a new sensor called a thematic mapper with 30-m IFOV. A synthetic radar (SAR) on board the US ocean surveillance satellite Seasat-1 had a resolution of about 25-m IFOV. On the other hand, an Earth terrain camera on board the US Skylab manned satellite had a resolution of between 15 and 30 m. It is interesting to note that if some of these sensors were to orbit at an altitude of 200 km rather than 440 km (Skylab) or 900 km (Landsat), the resolution would approach that of a camera on board a military area surveillance satellite.

It is also interesting to note that in Landsat images, even with present resolution limitations, one can identify major roads, airports and to some extent their runways. Image quality, however, could be considerably improved with the aid of computer enhancement techniques resulting in better delineation of such target areas. For example, figure 12.2 shows a Landsat scene of the Nile Delta in band 7 (near infra-red) to a scale of 1 : 1 250 000. Several sectors of this image were separated from the scene and with computer enhancement techniques and image enlargement to a scale of 1 : 312 500, several major military and civilian air bases become very clearly distinguishable. Figure 12.3a shows the enhanced sector for the city of Cairo, where major roads and intersections are clearly visible, as well as Cairo International Airport with its runways and another small military airport southwest of the international one. Figures 12.3b, c and d show other military airports along the eastern and western fringes of the Delta which otherwise would not be clearly identified from the original 1 : 1 250 000 image if it were not digitally and photographically enhanced.

The same thing is noticed in figure 12.4, which shows a Landsat image 1 : 1 250 000 of the northeastern Sinai Peninsula, where digital enhancement and enlargement of the bottom-centre sector shows very clearly the development of a large military area with an air strip, figure 12.5. Another interesting feature in the same image is the distinct tonal contrast along the imaginary line on the ground delineating the pre-1967 political borders between Egypt and Israel, a phenomenon that is worth investigating.

As for the US space shuttle, the first launching took place in April 1981 and the second in November 1981. Many future missions are planned, including the one to carry into space the manned space laboratory developed by the European Space Agency (ESA) which will orbit at an altitude of 300 km. The contribution of the space shuttle to remote sensing will be derived from its capacity to carry into space a wide variety of sensors, particularly large format high resolution (10 m) cameras, stereo cameras, radar, scanning devices and infra-red thermal scanners.

However, for purposes of routine military reconnaissance, especially for day-to-day requirements of an ISMA, the space shuttle could be of limited value because of its short flight duration, its high orbital inclination, and difficulty with repetitive coverage of exact locations.

The few scenes released by NASA taken from the first space shuttle mission show little use for either military or even civilian reconnaissance. In many respects they are similar to earlier space photographs taken by the astronauts during the Gemini and Apollo flights in the sixties. They are highly oblique, with a poor resolution and with segments of the space shuttle obscuring parts of the scene. Figures 12.6a and b show pictures taken during the first space shuttle mission over Egypt; the former shows primarily the Sinai Peninsula, and the latter shows parts of the Nile Valley in Upper Egypt with parts of the western and eastern deserts.

The technical missions which an international or multinational monitoring agency would be required to perform could be of two main types: (*a*) monitoring of compliance with disarmament/arms control agreements; and (*b*) monitoring of crises.

The requirements concerning resolutions, area coverage, frequency of coverage, response time and auxiliary data would be different for these two types of mission.

During discussions of the Governmental Expert Group on the implications of establishing an ISMA, it became evident that the following phases would appear to be feasible:

*Phase I:* Development of an IPIC using imagery data available from existing and future civilian satellite systems for training; also using other imagery data made available to ISMA from national (non-civilian) satellite systems.

**Figure 12.2. Landsat-1 digitally processed image of the Nile Delta in Egypt, scale 1 : 1 250 000**

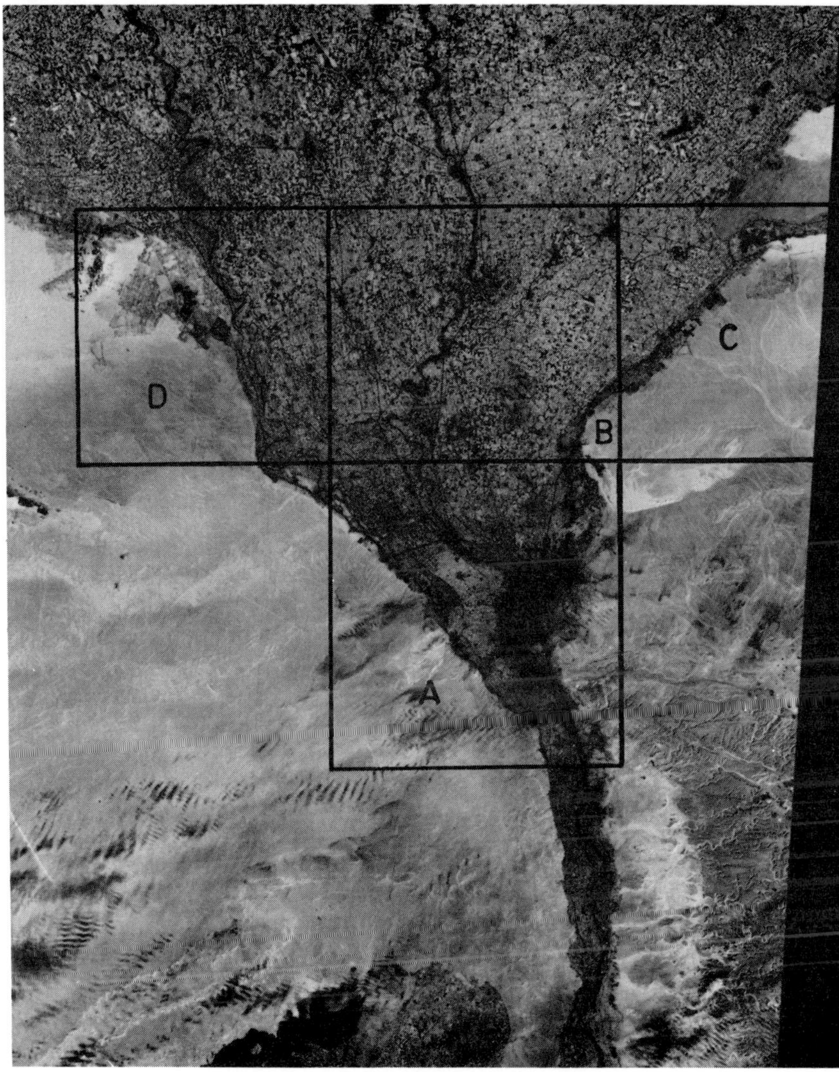


*Phase II:* Development of ground stations for receiving data acquired by relevant civilian satellite systems and by non-civilian satellite systems, as available.

*Phase III:* Development of ISMA space segments for operational monitoring satellites.

It should be pointed out, however, that other evolutionary phases would be conceivable. One option would be to start with Phase I and then to implement Phase II and Phase III at the same time.

Starting with Phase I, the establishment of an IPIC might be considered

Figure 12.3a. Computer enhancement and photo-enlargement for part of sector A in figure 12.2, scale 1 : 312 500

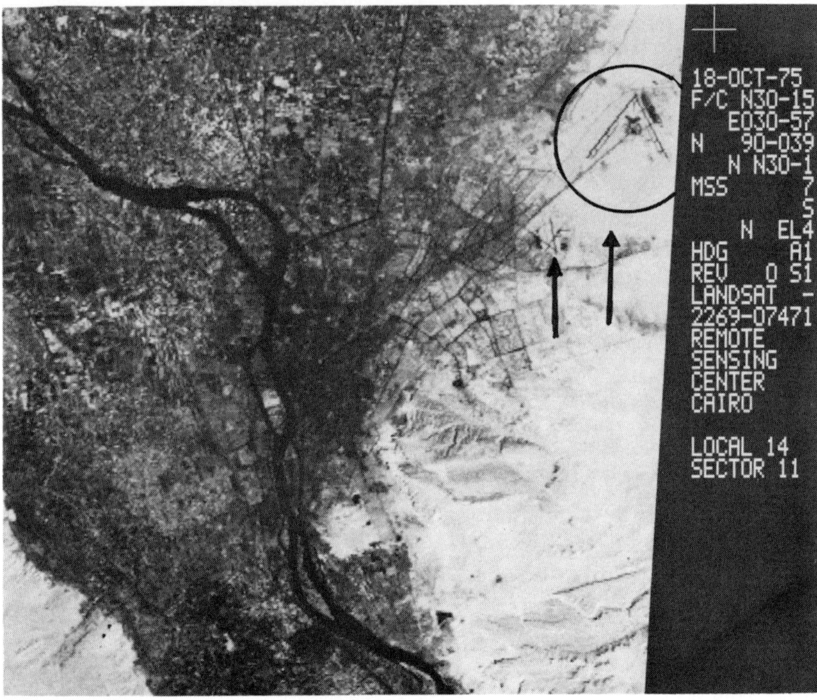


Figure 12.3b. Computer enhancement and photo-enlargement for part of sector B in figure 12.2, scale 1 : 312 500

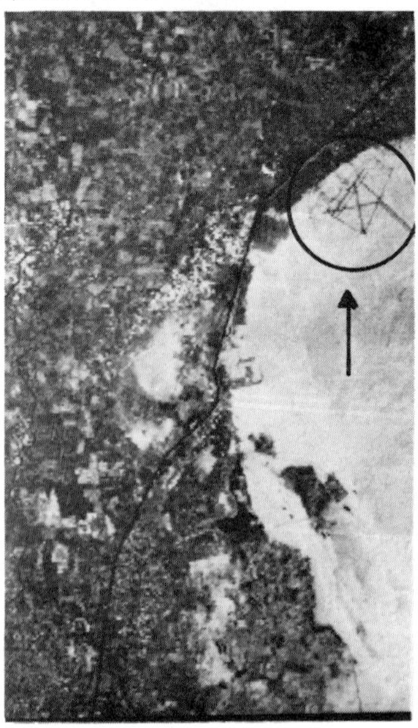

**Figure 12.3c. Computer enhancement and photo-enlargement for part of sector C in figure 12.2, scale 1 : 312 500**

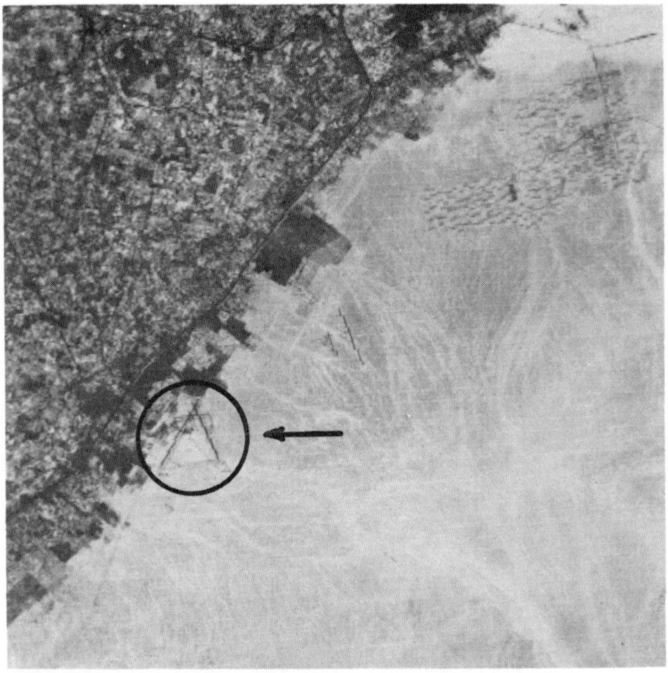

**Figure 12.3d. Computer enhancement and photo-enlargement for part of sector D in figure 12.2, scale 1 : 312 500**

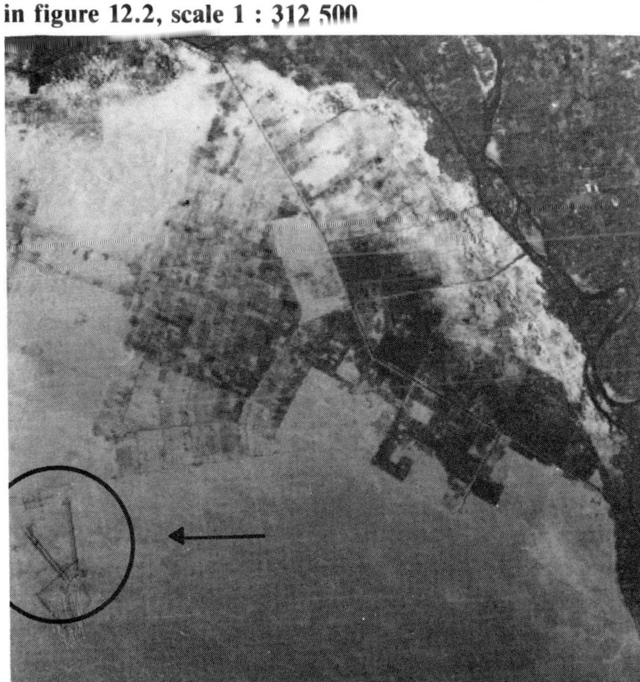

**Figure 12.4. Landsat-2 digitally processed image of northeastern Sinai Peninsula and S.E. Mediterranean coast, scale 1 : 1 250 000**

as a natural way of starting the activities of an ISMA. This phase could also be viewed as a necessary training period for interpreters and analysts and for establishing the required procedures, methods and standards.

Data availability for an ISMA, especially during its early stages of operation, would be dependent upon data delivery from other national satellite systems of co-operating space powers. Such data sources may be considered as follows: (*a*) data from existing meteorological satellites; (*b*) data

**Figure 12.5. Computer enhancement and enlargement of part of sector A in figure 12.4, scale 1 : 312 500**

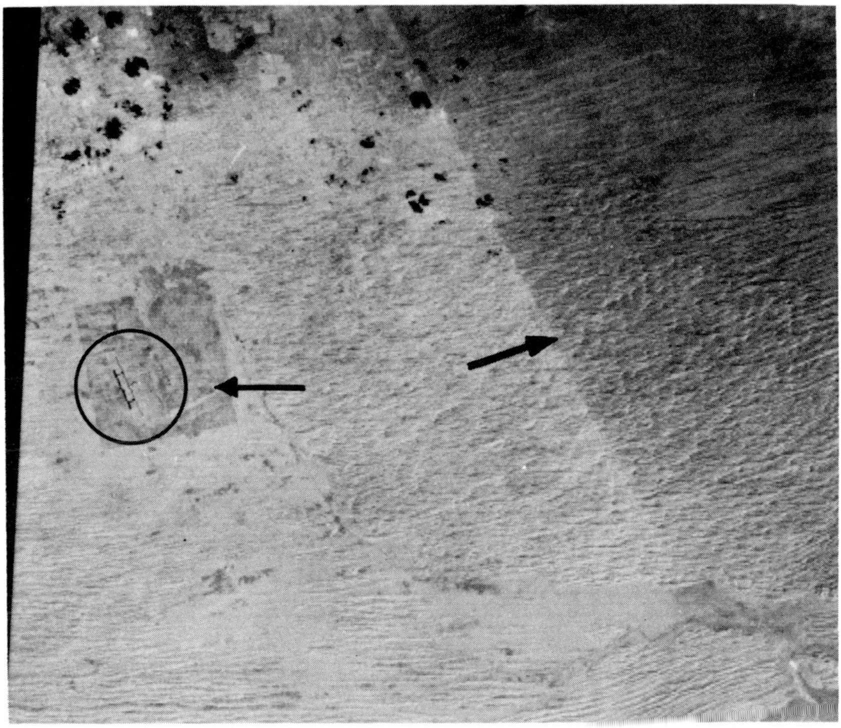

from existing and planned Earth resources satellites—USA (Landsat), USSR, ESA, France (SPOT), and others; and (*c*) previously recorded satellite data (Skylab, etc.). In all likelihood this data would be acquired by an ISMA without difficulty and could be used for the purpose of training, and for establishing procedures and standards. The data would, however, be limited to an IFOV in the range of 10–20 m and such data, therefore, would be of limited interest for the monitoring and verification functions of an ISMA.

On the other hand, national systems in which fine/high resolution data are generated could be as follows: (*a*) US military reconnaissance satellite data (area surveillance [resolution 3–5 m], close-look [fine/high resolution, about 0.5 m], electronic surveillance, high resolution radar data [from 1985?] and IR data [high resolution, from 1990?]); (*b*) Soviet military reconnaissance satellite data (area surveillance [medium resolution], close-look [high resolution] and radar surveillance); and (*c*) other (improved SPOT [radar and optical, from 1985?]).

From a technical point of view, some data could be delivered to an ISMA which would be just adequate to its functions. The delivery of the data from national (and other) organizations in this phase could be made in the

**Figure 12.6a. Photograph of Sinai Peninsula and Gulf of Suez area taken during first US space shuttle mission (1981)**

**Figure 12.6b. Photograph of Upper Nile Valley and parts of the Eastern and Western deserts taken during the first US space shuttle mission (1981)**

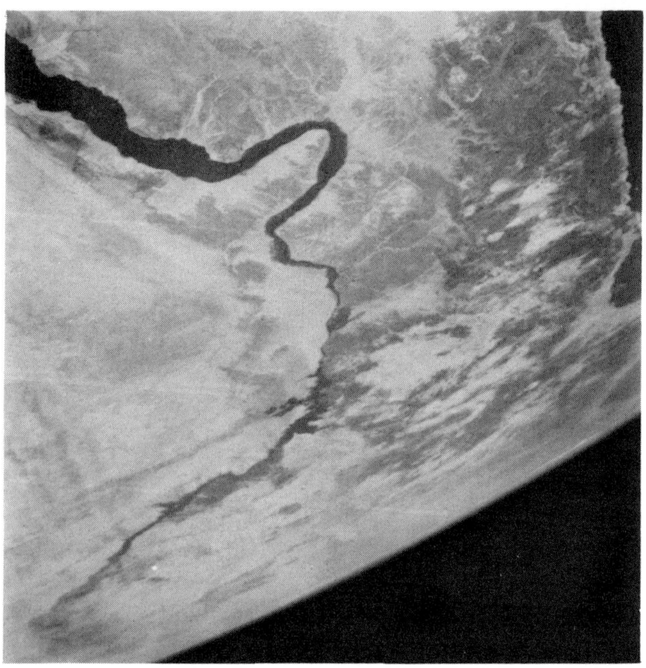

following ways: (*a*) physical delivery of photographic images; (*b*) physical delivery of imagery data stored on magnetic tape; and (*c*) transmission via data network, including satellite (INTELSAT) network.

The technical section of the report prepared by the governmental experts on the implications of establishing an ISMA (SMA/2—April 1981) [1], has dealt in a fair amount of detail with other technical facilities needed for an ISMA, including an IPIC, the ground segment, the space segment and technical aspects of the evolution of an ISMA, with requirements for various phases of evolution.

## *VI. Financial implications of an ISMA*

### General

Since a variety of technical options, possibilities and combinations of options are feasible when implementing various phases of an ISMA, the Expert Group found that precise or quantitative cost estimates were difficult to make. Such a variety of technical possibilities when viewed in conjunction with different policies, possible reactions and co-operative attitudes of states towards the Agency, would necessarily lead to a broad range of cost estimates. Several financial estimates of relevance to the development of an ISMA have been made by various groups [2–6]. Any cost estimates for a project like that of establishing an ISMA are difficult to make since they would have to be based on different assumptions and methods. To make assumptions or choices as a basis for a more elaborate analysis of the costs would, in fact, amount to a design of what the Agency should look like. The Group neither found it possible, nor considered it its task to make such choices. However, in an attempt to give some idea of what financial implications the establishment of an ISMA and the implementation of the different phases could entail, some approximate costs are indicated below. The cost figures presented here are related to 1980 values [1]. Should the international community decide to proceed with the establishment of an ISMA, one of the first steps recommended would be to commission a detailed engineering and related cost study which should be prepared by a specialized professional organization or establishment.

### Phase I

In this phase, an ISMA would function as an IPIC. Since the task of the ISMA's IPIC would be quite different from other existing national or regional centres for processing satellite data for natural resource assessments, utilization by an ISMA of such national or regional centres would be rather limited. It would therefore be highly desirable, in the initial phases of the ISMA's operation, to acquire and operate its own IPIC. The Agency would thus acquire and operate its own (*a*) data processing subsystem (data formating, correction and pre-processing), (*b*) data management subsystem (reproduction, storage, archiving, cataloguing and dissemination) and (*c*) data analysis and interpretation subsystem (including computer- and machine-aided techniques, as well as human deductive and inductive analyses). Capital investment for acquiring such systems and subsystems (exclusive of space and building costs) may amount to approximately eight million dollars (1980 prices). The operational costs for Phase I of an ISMA based on some rough estimates of data volumes to be processed and analysed as well as on technical and manpower resources needed, would be approximately in the range of 25 to 30 million dollars per year. The fees that would have to be paid, if any, to the states providing data would also have to be taken into consideration.

**Phase II**

In this phase, the cost estimate is based on the assumption that the ISMA would acquire and operate its own ground receiving stations for receiving data. Most estimates give a figure of anywhere between 6 and 8 million dollars as the capital cost per station, and approximately 2 million dollars per station annually as the operational cost. In case an ISMA has to set up and operate a global system of 10 stations with data links, the initial investment cost could be in the range of 60 to 80 million dollars. The annual operational costs for these 10 stations would therefore be about 20 million dollars. However, these cost figures can be significantly reduced if member states possessing space systems and/or receiving stations would be willing to make them available to the Agency as and when needed. The fees to be paid to states permitting access to their satellites would also have to be taken into account.

**Phase III**

Figures given here are based on the assumption that an ISMA may have to establish its own space segment as a supplement to national systems. In such a case, the range of cost figures would be approximately as follows:

1. The total cost of launching a single satellite for area monitoring (including cost of R&D) varies between 300 and 400 million dollars (spread over a period of four to six years). To ensure adequate area coverage, the Agency may typically have to launch a three-satellite system carrying combinations of optical, infra-red and radar-type satellites. Cost figures could then be in the range of 0.9 to 1.2 billion dollars. Cost of satellite renewal (assuming one launching every two to three years) and of systems management (e.g.,

operations personnel, R&D personnel, software development contracts, equipment maintenance, in-house R&D) is estimated to vary in the range of 50–200 million dollars per year.

2. Cost estimates for launching and operating a single low altitude manoeuvrable satellite for close-look, including R&D and other facilities needed for command and control of the satellite, may be approximately 1.5 billion dollars, spread over a 10-year period. Satellite renewal and systems management, based on a launching every 2 years, would cost approximately 120 million dollars per year. As far as less expensive alternatives to the manoeuvrable close-look satellite are concerned (e.g., a 'launch on request' system to cover specific targets with very high resolution), the Group felt that precise cost estimates would need a detailed engineering examination by specialists who would have to consider the relative merits of all systems from the cost-benefit angle.

3. In all of the above options for a space segment, development and operational costs can be substantially reduced (by several hundred million dollars) if the ISMA is allowed to use instruments, equipment, satellite platforms, etc., developed under a national programme.

## *VII. Attitudes of major states concerning the establishment of an ISMA*

When the French proposal for the establishment of this agency was discussed at the 33rd Regular Session of the General Assembly of the UN, it was evident that most member states were convinced of the important contribution such technology can make to the solution of monitoring problems, considering the need for international non-discriminatory measures without interference in the internal affairs of states. However, neither the USA nor the USSR supported the proposal. This is thought to be because they were both concerned about such an ISMA acquiring military reconnaissance data/information over territories of many countries, a matter which would be of disadvantage to both countries as it would affect their present monopoly on such kind of activity. Neither country participated in the Governmental Expert Group that prepared the implications of establishing an ISMA.

It is believed that the US opposition to creating an ISMA is based on two main reasons:

1. After World War II, it appeared that there was an intelligence gap between US and Soviet intelligence systems, as the latter was superior in traditional espionage operations. The United States attempted to close this gap by using advanced technological means, such as, high altitude reconnaissance aeroplanes (the U-2) and later military reconnaissance satellites. It was therefore logical for the USA to feel that the establishment of an ISMA

could affect the present balance of information-gathering systems with the USSR. This is in addition to the possible risks involved in the transfer and control of such vital security-data gathering and dissemination systems from purely US hands to an international organization.

2. The USA is concerned that some vital classified data regarding their space reconnaissance systems and capabilities might be uncovered.

In spite of these concerns that most probably account for the absence of US participation in the preliminary stages, it appears that there are some signs of possible change in this attitude in the future, as the establishment of this agency may not be totally against US interests for several reasons, for example:

1. It may be in the strategic interest of the USA that information regarding some critical areas in the world, particularly along the Soviet–Chinese border, become available to other countries through an ISMA.

2. It does not appear in the interests of the USA that such an ISMA, if it comes to reality (a matter which appears to be feasible with the support of the international community and present French and European technological capabilities), be dominated by French as well as European technology.

3. It is also evident to the USA that the monopoly between the two major space powers over military reconnaissance by satellites will not continue, as it is expected that several other countries will possess such a capability within the next decade. An ISMA may then prove to be the most logical alternative to conflicts resulting from such a large number of national satellites and the possible consequences of violations to international legal order. Therefore, it may be concluded that the US attitude is not that of total opposition to the establishment of such an agency, but is rather a 'wait and see' attitude.

The Soviet attitude towards the establishment of such an agency appears to be a strongly negative one, a position which is shared by several other Eastern Block countries. Although the Soviet position is considered to be generally for the same reasons as that of the USA, it appears to be much stronger.

Many of the developing non-space countries, Egypt included, consider that there are many factors in their interest which can be served by establishing such an ISMA. For example, Egypt's security interests would be well served, as a country in the Middle East which is an area of increasing local and international conflicts, if the area were continuously monitored by an ISMA affiliated with an international body, such as the UN.

## *VIII. Conclusions*

The "Group of Governmental Experts on the Implications of Establishing an International Monitoring Agency" reached the following conclusions at the end of their deliberations:

1. The Group recognized the valuable contribution which monitoring by satellites could make to the verification of compliance with certain arms control and disarmament agreements. It further recognized the positive role that satellite monitoring could play in preventing or settling international crises and thus contributing to confidence building among nations.

2. The Group also took note that from a technical point of view observations from satellites for the purpose of information gathering related to verification of compliance with treaties and for crisis monitoring are both possible and feasible. The technical facilities for an ISMA, including the satellites necessary to carry out the needed missions, could be acquired in stages: for instance, Phase I could comprise only an image processing and interpretation centre, Phase II could comprise data-receiving stations that could receive appropriate data from observation satellites of various states and in Phase III the Agency could have its own space segment comprising a number of satellites.

3. From a legal point of view, the Group found no provision in international law, including space law, that would prohibit an international governmental organization, such as an ISMA, from carrying out monitoring activities by satellites.

4. As regards the financial implications, a variety of technical options were found to be possible, leading to a broad range in cost estimates. Whatever the assumptions on which the estimates were based, even in Phase III, which is the most complete and most expensive phase, an ISMA would cost the international community each year well under one per cent of the total annual expenditure on armaments.

**References**


1. 'Group of governmental experts on the implications of establishing an International Satellite Monitoring Agency', UN Report, SM/2, GE. 81–61130, April 1981.
2. 'On the establishment of an International Satellite Monitoring Agency', French document, SMA/WP/3, Group of Governmental Experts for Establishing an ISMA, 11 June 1979.
3. 'Contribution to the technical study concerning an International Satellite Monitoring Agency', French document SMA/WP/10, Group of Governmental Experts for Establishing an ISMA, 12 May 1980.
4. 'An international agency for the use of satellite observation data for security purposes', Paper presented at 34th Pugwash Symposium, Avignon, France, 14–17 April 1980.
5. Tsipis, K. 'Technical, operational and policy considerations and alternatives for the use of satellite observation data for security purposes', Paper presented at 34th Pugwash Symposium, Avignon, France, 14–17 April 1980.
6. Perry, R., 'Prospects for an International Satellite Monitoring Agency', Paper presented to the United Nations Experts Group on ISMA, 27 May 1980.
7. SIPRI, *Outer Space—Battlefield of the Future?*, (Taylor & Francis, London, 1978).

# Paper 13. What additional arms control measures related to outer space could be proposed?

**D. Goedhuis**

*Flat 4, 37 Cadogan Square, London SW1, UK*

## *I. Divergent interpretations of existing measures*

Before considering the possible approaches that might lead to an acceptance of the measures referred to in the title of this paper, attention should be drawn to the way the rules adopted so far, aimed at limiting the use of outer space for military purposes, have been interpreted.

An analysis of the statements on the meaning and content of these rules, made by the representatives of several countries as well as by a number of jurists concerned with space law, shows considerable divergencies. Little attention has been given to the harmful consequences of these dissensions which—though they have, up till now, not given rise to serious political confrontations—will, unless halted, have a damaging effect on international space relations.

The scope of this paper does not allow a survey of the many controversial opinions on the meaning and effect of the present rules, but some of the recent views on the extent to which the most important of these rules constrain the military uses of outer space will be evaluated.

So far four treaties have been adopted which contain provisions aimed at some form of arms control in space.

1. The Moscow Treaty of 1963 on the banning of nuclear weapon tests in the atmosphere, in outer space and under water. In Article I of the Treaty, outer space is mentioned as one of the environments where such tests are prohibited.

2. The Treaty adopted by the USA and the USSR in 1972 on the Limitation of Antiballistic Missile Systems. Article VI prohibits the development, testing or deployment of ABM systems which are sea-based, air-based, space-based, or mobile land-based.

3. The Treaty on Principles Governing the Activities of States in the Exploration and Use of Outer Space, including the Moon and other Celestial

Bodies, of 1967. Article IV of this Treaty, which can be considered as the Charter of Outer Space, provides:

> States Parties to the Treaty undertake not to place in orbit around the earth any objects carrying nuclear weapons or any other kinds of weapons of mass destruction, install such weapons on celestial bodies, or station such weapons in outer space in any other manner. The moon and other celestial bodies shall be used by all States Parties to the Treaty exclusively for peaceful purposes. The establishment of military bases, installations and fortifications, the testing of any type of weapons and the conduct of military manoeuvres on celestial bodies shall be forbidden. The use of military personnel for scientific research or for any other peaceful purposes shall not be prohibited. The use of any equipment or facility necessary for peaceful exploration of the moon and other celestial bodies shall also not be prohibited.

As will be seen below, it is the interpretation of this article which has given rise to the most serious controversies.

4. The Moon Treaty: Agreement Governing the Activities of States on the Moon and other Celestial Bodies within the Solar System other than the Earth, of 1979. Article III of this Treaty provides:

> 1. The moon shall be used by all States Parties exclusively for peaceful purposes.
> 2. Any threat or use of force or any other hostile act or threat of hostile act on the moon is prohibited. It is likewise prohibited to use the moon in order to commit any such act or to engage in any such threat in relation to the earth, the moon, spacecraft, the personnel of spacecraft or man-made space objects.
> 3. States Parties shall not place into orbit around or other trajectory to or around the moon objects carrying nuclear weapons or any other kinds of weapons of mass destruction or place or use such weapons on or in the moon.
> 4. The establishment of military bases, installations and fortifications, the testing of any type of weapons and the conduct of military manoeuvres on the moon shall be forbidden. The use of military personnel for scientific research or for any other peaceful purposes shall not be prohibited. The use of any equipment or facility necessary for peaceful exploration and use of the moon shall also not be prohibited.

Reference should also be made to the Convention on Registration of Objects Launched into Outer Space (1975) which, although it does not contain any specific arms control measures, could—provided it were interpreted in the right way—play a confidence-building role in the military sphere.

Article IV of this Convention requires states launching space objects to provide the Secretary-General of the United Nations information on a number of data, including "the general function of the space object". In this context it should be noted that, notwithstanding the fact that much more than half of the US and Soviet satellites launched serve military purposes, not one of the launchings registered has so far been described as having a military function.

There have been, in particular, three statements made in the last few years which demonstrate the present misconceptions regarding the legal content of the rules aimed at a limitation of the military uses in space.

1. During the discussions in the UN Committee on the Peaceful Uses of Outer Space (COPUOS), both the Swedish and Chilean delegates expressed the view that, under the terms of the Outer Space Treaty of 1967, outer space has been established as the common heritage of mankind.[1]

2. In two articles in *The Times* in February 1981 by the Defence and Science correspondent, the Space Treaty was described as having declared outer space as "a zone of peace". It was further suggested that the Treaty expressly forbade the destruction of satellites by any kind of weapon [3].

3. In the most recent and well-documented treatise on international space law, the author, Professor Marcoff, asserted that all military activities in space are illegal [4].

## Space as a common heritage of mankind

The above statements call for the following comments. Insofar as the contention that the Outer Space Treaty has established outer space as the common heritage of mankind is concerned: although the debates both in the Law of the Sea Conferences and in COPUOS have shown that the import of this term is far from agreed upon, it is generally recognized that any area which is the "common heritage of mankind" must be dedicated to exclusively peaceful purposes. As will be seen, the Space Treaty has not assigned the use of the whole of outer space to exclusively peaceful purposes and moreover does not satisfy several of the other basic implications of the concept [5].

## Are all military activities in space illegal?

As to the erroneous belief that the Space Treaty has prohibited the use of all military activities in this area, the advocates of this idea refer to two provisions in the Treaty. One is the provision contained in the Preamble by which the common interest of all mankind in the progress of the exploration and use of outer space for peaceful purposes is recognized. But a provision of this kind in the Preamble of the Treaty does not create any legal obligation. Moreover in the first three articles of the operative text of the Space Treaty, in which the leading principles governing space activities have been laid down, no mention of the use of the whole of outer space for peaceful purposes has been made. It is only as regards the Moon and other celestial bodies that this concept has been accepted.

The second provision which has led to the assertion that all military activities in outer space shall be carried out for the benefit and interest of all countries. The question of the meaning and effect of the benefit clause has

---

[1] See the statement made by the Swedish delegate on 30 March 1979 [1] and that made by the Chilean delegate on 10 July 1979 [2].

been ardently debated for many years but the views expressed on this clause will not be considered here. Suffice it to say that a purely grammatical interpretation of this provision without considering its meaning in the context of the Treaty as a whole would lead to totally unacceptable consequences [6].

*What are peaceful purposes?*
In the context of Article 3 of the Moon Treaty crucial questions arise, the answers to which have wide implications regarding the effect of the present rules aimed at a prohibition of certain military uses in space.

Paragraph 1 provides that the Moon and other celestial bodies shall be used "exclusively for peaceful purposes". Does this provision mean that the Moon and other celestial bodies are completely demilitarized? Since the conclusion of the Space Treaty, the interpretation of the term "peaceful purposes" has given rise to fundamental controversies. As a great lack of awareness appears to exist on the practical consequences of the conflicting views, it may be useful to make the following comments. Two fundamentally different interpretations of the term "peaceful purposes" have come to the fore. Under one, this term means "non-military", while under the other it means "non-aggressive". The latter interpretation is being followed by the United States.

In an article published in 1968 attention was drawn to the damaging consequences of this interpretation and it was pointed out that during the deliberations in COPUOS, prior to the conclusion of the Space Treaty, the great majority of delegates insisted that the term "peaceful" should be interpreted in the sense of "non-military" [7]. They were thus inspired by the Antarctic Treaty, which starts from the premise that military purposes, defensive as well as offensive, are not peaceful purposes. The consequences of the interpretation that peaceful purposes are non-aggressive become particularly apparent in the context of the Moon Treaty of 1979.

Article 3.4 of this Treaty, reiterating Article IV (2) of the Space Treaty, provides that the use of military personnel for scientific research or for any other peaceful purposes shall not be prohibited and that the use of any equipment or facility necessary for the peaceful exploration of the Moon and other celestial bodies shall also not be prohibited. By virtue of Article 9 of the Moon Treaty, states may establish manned and unmanned stations on the Moon. As the use of military personnel on these stations for peaceful purposes is not prohibited, the interpretation of this term as "non-aggressive" would mean that the stations could be used for all kinds of military purpose so long as they could not be considered to be aggressive. But if one followed this line of reasoning one would necessarily come into conflict with the provision laid down in the first sentence of Article 3.4, by which the establishment of military bases has been forbidden.

Another example may be given of the dissensions which would result from permitting the use of the Moon for non-aggressive purposes. It has been submitted that since defensive and deterrent capabilities serve the cause of peace, it is only when such devices are intentionally used for aggressive purposes that they lose their peaceful status. As all arms have deterrent capabilities, states—on the basis of such a contention—would be able to claim that any arms on the Moon constitute a use of the Moon for peaceful purposes.

What about the position of the Soviet Union on this issue? In the treatise *International Space Law*, edited by Professor A.S. Piradov, who acted as Soviet representative in several meetings of COPUOS, the following statement has been made: "The principle of the partial demilitarization of outer space and the *total demilitarization of celestial bodies* [emphasis added] is formulated in Article IV of the Space Treaty" [8]. Another authoritative Soviet expert on space law, Professor G.P. Zhukov, in his lectures to the Académie de Droit International in 1978 remarked in the same sense: "The Outer Space Treaty of 1967 establishes for the moon and other celestial bodies a regime of total demilitarization" [9].

From these statements the conclusion might be drawn that an important gap exists between the position of the two major powers on this issue, the Soviet Union interpreting the term peaceful as "non-military".

However, as mentioned above, the Soviet Union—like the United States —pretends that all its satellites serve peaceful purposes. Apparently it considers that none of its present military activities can be considered as "non-peaceful".

*Weapons of mass destruction*

A question which arises in the context of Article IV (2) of the Space Treaty and Article 3.3 of the Moon Treaty concerns the interpretation of the term "weapons of mass destruction". This notion, which also appears in the Treaty on the Prohibitions of the Emplacement of Nuclear Weapons and other Weapons of Mass Destruction on the Seabed, the Ocean Floor and the Subsoil thereof (which was "commended" by the UN General Assembly on 7 December 1970), was discussed in the Conference of the Committee on Disarmament. At this Conference it became clear that the great powers understood certain weapons to be covered by the term, including biological, chemical and radiation weapons, but speculation was extended to lasers, weather modifiers and anti-satellite devices [10].

As the meaning and the content of this prohibition is elusive and may lead to disputes, it is important—especially in view of the uncertainty whether the emplacement of anti-satellite devices in outer space is covered by the term—that an attempt be made to clarify precisely what the notion "weapons of mass destruction" implies. Regarding the suggestion that anti-satellite weapons may be considered to be included in the prohibition, the negotiations between the United States and the Soviet Union on this matter [11] indicate that they reject the view that, under the terms of either the Space Treaty or the Moon Treaty, the emplacement of anti-satellite devices in outer space is prohibited. This attitude is further emphasized by the Soviet proposal for banning all weapons in space. From the fact that, during the discussions in the last session of COPUOS, a considerable number of delegates made a strong appeal to the two great powers to resume without delay their negotiations on a ban of anti-satellite weapons, the conclusion can be drawn that they also consider the term in question as not covering the emplacement in outer space of such devices.

On the basis of the above considerations, it is submitted that the need to clarify the meaning and the content of the two terms referred to can hardly be denied.

## *II. The appropriate medium to study arms control measures in outer space*

Logically, COPUOS—the medium through which both the Space and Moon Treaties have been prepared—should be given the task of studying arms control measures in outer space to examine the possibility of arriving at a qualification and clarification of the ambiguous provisions in these treaties. However the question arises whether the Committee could achieve a general agreement on putting the consideration of this issue on its agenda.

As the possible implementation of the rules referred to is closely related to the wider issue of the prevention or limitation of an arms race in outer space, attention should be given to the opinions expressed in COPUOS on the body which would be the most appropriate to devise rules on this issue.

Whereas numerous delegates expressed the view that the Committee should be considered the most appropriate forum,[2] the American delegate stated that, as arms control in outer space was inseparable from the complex question of security on Earth and arms control in general, the issue was a subject which went well beyond the expertise and mandate of COPUOS [12].

The suggestion that the Committee would not possess the necessary expertise to study this matter is difficult to accept. However that may be, when one gives thought to the forum which would be the most effective in dealing with space-related military matters, the following considerations should be taken into account.

Reference has been made to the treaties containing provisions aimed at a limitation of military uses in outer space which were prepared by separate international bodies. Attention should also be drawn to the debates during the special session of the UN General Assembly devoted to disarmament. The Italian delegation at this session sponsored a provision which was eventually incorporated in the Final Document, reading as follows: "In order to prevent an arms race in outer space, further measures should be taken and

---

[2] During the Twenty-fourth Session of COPUOS held in New York from 22 June–2 July 1981, the delegates of Austria, Brazil, Canada, Chile, Egypt, India, Romania and Sweden expressed the opinion that it would be the duty and responsibility of COPUOS to prevent an arms race in space.

appropriate international negotiations be held, in accordance with the spirit of the 'Treaty on Principles Governing the Activities in the Exploration and use of Outer Space, including the Moon and other Celestial Bodies'" [13]. As a follow-up to this provision, the Italian government introduced in the Committee on Disarmament a proposal for the elaboration of an additional protocol to the Space Treaty [14], on which some comments will be made below.

As there is at present a complete lack of co-ordination between the various organizations involved in space-related military matters, there appears to be a strong need to study the possibility of creating, within the framework of the UN, a special forum consisting of representatives of the Committees which at present are attempting to limit the military uses of outer space.

## *III. The task of the special forum*

If an agreement on the creation of such a forum could be reached, the question arises as to what kind of arms control measures might offer a reasonable chance of achieving positive results and should therefore be discussed. In this context some comments should first be made on a suggestion which, in the view of the author of this paper, is unrealistic. During the debates in the twenty-fourth session of COPUOS, the Soviet delegate, Professor Piradov, expressed the opinion that the time had come to think of the development of legal norms banning the stationing of weapons of any kind in outer space and he referred in this context to the following statement made by Mr Brezhnev on 17 April 1981: "May the shoreless ocean of outer space be clean and free of any kind of weaponry. We are in favour of deploying joint efforts to achieve that great humanitarian aim: 'the prevention of militarization in outer space'" [15].

The idea of prohibiting the installation of any kind of weapon in outer space was supported by the delegates of the German Democratic Republic, Mongolia and Bulgaria.[3]

### Complete demilitarization?

Do these statements imply the belief that a complete demilitarization of outer space could be achieved independently of disarmament in a wider

[3] See also the following statement made by the Canadian delegate: "Of increasing concern to my country is the growing use, or the potential for use, of outer space for other than purely peaceful purposes. We are deeply committed to the peaceful uses of outer space and strongly believe that all Governments should exert the utmost efforts towards this objective, both within the Committee and outside it" [16].

field? As is well known, efforts to arrive at such demilitarization were made in the first year after the launching of the first Sputnik, but it soon became recognized that this aim could not be achieved without a general disarmament in the world. It is submitted that the arguments on which this recognition was based have not lost any of their validity. Efforts to arrive at a complete demilitarization of space can only lead to a delay in achieving the limited arms control measures which might be considered feasible in the present phase of international relations.

In this context some comments should be made on the Italian proposal to the Committee on Disarmament and referred to above. At the 193rd meeting of COPUOS, the Italian delegate specified that the purpose of this proposal was:

> to ensure that outer space, including the moon and other celestial bodies, be used for peaceful purposes only and that States refrain from engaging in, encouraging or authorizing, directly or indirectly or in any way participating *in any measure of a military or other hostile nature* [emphasis added], such as the establishment of military bases, installations or fortifications, the stationing of devices having the same effect, the launching into orbit or beyond of objects carrying weapons of mass destruction or any other type of devices *designed for offensive purposes* [emphasis added], the conduct of military manoeuvres as well as the testing of any kind of weapons. [17].

When one analyses this proposal, the following question arises. On the one hand, the proposal is directed towards a prohibition of any measure of a military or other hostile nature in outer space, but on the other hand, in its second part, it refers to a prohibition of launching types of devices for 'offensive' purposes. Does this mean that the launchings in outer space including the Moon of any devices for 'defensive' purposes are to be allowed? Does this not contradict the proposal to prohibit any measure of a military nature?

In his explanation of the proposal, the Italian delegate stated: "Of course, the use of reconnaissance, surveillance and communications satellites and, indeed, of any space system which would reinforce strategic stability by ensuring, *inter alia*, the verification of disarmament and other limitation agreements, will not be prejudiced."

Does the statement "any space system which would reinforce strategic stability" imply that by virtue of the proposal all measures in outer space, including on the Moon, which could be considered as reinforcing such stability would be permitted?

Here again one must ask how such a proposition can be reconciled with the proposed prohibition of any measures of a military nature in space. The inconsistencies in the Italian proposal clearly demonstrate the insurmountable obstacles with which the advocates of a complete demilitarization of outer space are faced. But what are the measures of arms control in space which, in the contemporary international political situation, might be considered as not being outside the bounds of feasibility?

As was mentioned above, an increasing number of states have stressed the crucial and urgent need to halt and reverse the deployment of anti-satellite devices. The prospect for an international anti-satellite treaty will not be discussed here. Suffice it to say that it might be decided by an international body (to be established), either to create a special anti-satellite treaty or to extend the arms provisions at present laid down in the Space and Moon Treaties.

**Defining terms**

Reverting to the lack of awareness of the harmful consequences which are likely to arise from the controversial interpretations of the term "peaceful purposes" the question arises whether it would be possible to arrive at an agreement by which the concept would be given legal content. In this context, it seems useful to give attention to the debates on this notion as laid down in the Declaration of Principles Governing the Seabed and Ocean Floor and the Subsoil thereof, beyond the Limits of National Jurisdiction (paragraphs 5 and 8), as well as in various provisions of the negotiating texts of the Law of the Sea Conference. Professor Tullio Treves has given a thorough analysis of the changing official attitudes on the interpretation of this concept [18a]. As the author remarked, originally numerous states, including the Soviet Union, favoured demilitarization of the sea-bed, and consequently maintained that no military use of the sea-bed could be considered "peaceful". During the Third United Nations Law of the Sea Conference the Soviet Union changed its attitude. "The Soviet delegate did not state his country's support for the demilitarization of ocean space. Rather he emphasized that the problem of strengthening peace on the seas was beyond the scope of the work facing the Conference on the Law of the Sea. In this, the Soviet position came very close to that of the United States."

On the basis of the debates on this concept Professor Treves concluded that since the earliest mention of it in the context of the Law of the Sea, the expression "peaceful purposes" had undergone an evolution which, though never reaching total clarification, had considerably narrowed its scope.

When one compares the trend towards the interpretation of this term in the context of the Law of the Sea with the trend on this issue in outer space law, one notices a certain parallel.

Insofar as the US attitude is concerned, it is noteworthy that President Johnson, in his United Nations' Day Proclamation (August 1967), described the meaning and the purpose of the Space Treaty in the following words: "The Treaty, *following the precedent of the Antarctic Treaty* [emphasis added] reserves an unspoilt area for strictly peaceful purposes to benefit all mankind." However, since the time of President Johnson's declaration, the US government in several statements has continuously interpreted the term "peaceful" as "non-aggressive".[4] Reference has already been made to the ambivalence in the Soviet attitude.

When one takes account of the dominant influence of the two great powers on the framing of the rules of space law, one may have strong doubts about the feasibility of arriving in the near future at an agreement on a universally acceptable precise definition of the concept "peaceful purposes".

However, as the present dissensions on the meaning of this term—in particular as regards the question whether any military activities on the Moon are permitted—are bound to lead to serious conflicts, the need to agree on which activities can be considered *not* conducted for peaceful purposes appears to be of crucial importance.

In this context consideration should also be given to the desirability of greater disclosure of the nature and location of space-related military activities in outer space.

In this paper's introductory remarks attention was drawn to the fact that not one of all the launchings registered so far has been described as having a military function. As was stressed in the publication of the Stanley Foundation entitled *Cooperation or Confrontation in Outer Space*, all parties to the Space Treaty and the Convention on Registration should be encouraged to increase the content and timeliness of disclosure. "Formalized extension of disclosure of military activities through unilateral initiatives or international agreements could make a positive contribution to peace and security" [20].

Another issue to which attention needs to be given concerns the interpretation of the notion "weapons of mass destruction". Although the term cannot be considered to cover anti-satellite devices, speculations have been made as to whether it includes lasers. During the US Senate Hearings on the Seabed Arms Control Treaty the US negotiator gave the following definition:

> It has . . . the meaning of embracing nuclear weapons, embracing also chemical and biological weapons and then being open-ended . . . in order to take care of developments which one cannot specify at the present time, some form of weapon which might be invented or developed in the future, which would have devastating effects comparable to those of nuclear or chemical or biological weapons which one cannot simply describe at the present time. [18b].

## *IV. Final observations*

Contrary to the school of thought which holds that the conquest of space has led to a fundamental change in international relations, it should be recognized that this conquest has not brought—and, in the presently prevailing

[4] In the meeting of COPUOS on 16 July 1979 the US delegate, Mr. Hosenball, declared: "Art. III (of the Moon Treaty) is a clear statement that celestial bodies and those orbits around them are to be only for peaceful—that is non-aggressive—purposes" [19].

political situation, cannot bring—any basic transformation in the power structure of international society [21].

As the USA and the USSR govern all major space activities and will continue to govern these activities in the foreseeable future, it is clear that the success of efforts to restrain the arms race in outer space depends decisively on their attitude. In this context attention may be drawn to the tendency of both space powers to shy away from publicly giving any information on the aim and results of their present military activities in space.

A typical example of the reluctance of both powers in this respect is provided by the statements made in the paper prepared by the Secretariat of COPUOS and published on 15 May 1978. This contains the views expressed by a number of countries on the experience gained in the application of the Space Treaty. The US statement is worded as follows:

> Without the establishment of the principles of the law of outer space, it is open to doubt whether the benefits of all mankind from the past decade of space exploration and use would have been fully realized. Unlike the sixteenth and seventeenth century explorations from Western Europe, with which man's ventures into space has frequently been compared, the movement into space *has not been a race of conquest* [emphasis added] or a type of colonial or territorial expansion.

The Soviet Union, for its part, stated that it had consistently held the position that the investigation of outer space must be a basis for co-operation and not a field for competition and rivalry. Reference was made to a statement made by Mr. Brezhnev:

"We in the Soviet Union do not regard our space research as an end in itself, *as some kind of 'race'*"" [emphasis added] [22].

These statements can hardly be considered to represent an accurate picture of the existing position. Outer space activities have proceeded and are proceeding in the strongest rivalry for political ends.[5]

It is interesting to note that the great powers' statements on the present situation in outer space were made the same year that the recognition of the serious dangers of the deployment of anti-satellite weapons led to their bilateral talks aimed at achieving a ban on these weapons.

As to the views of the countries which at present do not play a significant role in space developments, it has only been in the course of the more recent deliberations in COPUOS that several of the less powerful nations stressed the importance of gaining some perspective, in the words of the Brazilian delegate, "of the seldom discussed but essential problems of the present and future characteristics of the distribution of power in the use of outer space" [24].

Insofar as the suggestions aimed at arms control in outer space are concerned, it was submitted above that the proposals aspiring to a complete

---

[5] See *inter alia* the article by Sir Bernard Lovell [23], and the statement by Professor Piradov in the Soviet treatise, International Space Law [8b]: "Contemporary Space Law is the result not only of cooperation but also of the most acute struggle between States."

demilitarization of outer space should be considered to be ineffective and even counter-productive. Such an aim could become merely an excuse for delaying the urgently needed curbs on anti-satellite weapons.

Earlier, reference was made to the strong appeal by many states to the two space powers to resume without delay their negotiations on a prohibition of such weapons. Insofar as the chances of arriving at a consensus on a ban are concerned, there is one important factor which might exercise a favourable influence. Whereas it is generally conceded that the Soviet Union has been ahead of the United States in the deployment of anti-satellite systems, the successful flight of the space shuttle—a device which can also be used for anti-satellite purposes—has resulted in a transformation of the previously existing imbalance between the two powers. Equivalence of capabilities in this field should be considered as the most important prerequisite for the acceptance of a ban on the weapons in question.

Reverting to the harmful consequences of the divergencies in the interpretation of the term "peaceful purposes" as laid down both in the Space Treaty and in the Moon Treaty, it has been suggested that it is unlikely that an agreement could be reached on a universally acceptable precise definition of this term. There is an urgent need—in particular in relation to the question of whether any military activities on the Moon are permissible—to make an effort to identify activities which can be considered *not* to be conducted for peaceful purposes.[6]

Closely related to the need for such identification is the need to lay greater emphasis on the disclosure of the mission of satellites launched. Efforts should be made to arrive at mandatory reporting procedures by which the nature of the mission can be verified.

Attention has also been drawn to certain problems which have arisen in the interpretation of the term "weapons of mass destruction". This term certainly cannot be interpreted as covering anti-satellite weapons as they are not capable of "destroying masses".

Earlier in this paper consideration was given to the US statement that the subject of arms control in outer space is inseparable from the complex question of security on Earth and arms control in general. Does this statement imply that, in the opinion of the US government, negotiations on measures aimed at a prevention of an escalation of military uses of outer space can only be held in the context of negotiations embracing all aspects of strategic arms limitations in the world?

There is in particular one reason why an inclusion of the issue of arms

---

[6] See Stephen Gorove [25]: "It would be better to identify those activities which constitute minimal threats to national security and permit those regardless of the ultimate purpose for such activities or their conduct by military personnel." The author of this paper, Professor Goedhuis, expressed doubts in an article on the possibility of states agreeing on a definition of what activities constitute "a minimal threat to their security". He then received a letter from Professor Gorove clarifying that he did not propose a definition of the term "minimal threat" but only an identification of the type of activities that are permissible, presumably because they would be regarded as "minimal threats to security".

limitations in space in the discussions on arms limitation in general appears to be undesirable. Although, as noted, with the deployment of anti-satellite devices a movement has started which could result in the growth of military confrontation in space, this medium is at present still mainly free from kill-mechanisms. It can therefore be assumed that the problems which have to be overcome in arriving at a consensus on certain arms limitations in outer space are comparatively less arduous than those which arise in other fields where an unbridled military competition has already taken place.

Strong arguments are militating in favour of establishing a special forum to deal with space-related matters, composed of representatives of the various bodies which are concerned with reducing the risk of war in outer space. It is obvious that the military competition in space poses dangers to the security of all states and that progress towards arms limitation in this medium requires a unified approach which would avoid arms control being seen solely as an expression of a US–Soviet condominium in space. Among the other subjects to be discussed by a multilateral forum to be instituted, would be the possible creation of an international satellite monitoring agency as proposed by France, by which adequate attention would be given to verification of the purposes of space missions.

Finally, there is an extraordinary lack of awareness in the world population at large of what is at stake in outer space developments and in particular of the disastrous consequences which would result from an intensification of the arms race in this area. The possibilities of achieving arms limitations in space on a broad scale will, to a considerable degree, depend on a widespread public consciousness of the crucial need to reach an agreement on such limitations.

The paramount importance of arousing and orienting world public opinion on this matter need therefore hardly be stressed.

## References


1. UN Document A/AC.105/C.2/SR 314, p. 4.
2. UN Document A/AC.105/PV.197, p. 6.
3. *The Times*, 5 February 1981.
4. Marcoff, M., *Traite de Droit International Public de l'Espace* (Editions Universitaires Fribourg Suisse, 1973), pp. 357 and 679.
5. Goedhuis, D., 'Some recent trends in the interpretation and implementation of the rules of international space law', *Columbia Journal of Transnational Law*, Vol. 19, No. 2, 1981, p. 219.
6. Goedhuis, D., 'Some substantive and procedural issues presently at stake in space legislation', *Zeitschrift für Luft- und Weltraumrecht*, No. 3, 1976, p. 199.
7. Goedhuis, D., 'An evaluation of the leading principles of the treaty of outer space', *Netherlands International Law Review*, 1968, p. 25.
8. Piradov, A.S., *International Space Law* (Profrers Publishers, Moscow, 1976).
   (a) —, p. 91.
   (b) —, p. 259.

9. Zhukov, G.R., 'Tendances contemporaines du développement du droit spatial international', *Recueil des Cours de l'Académie de Droit International*, Vol. 3, 1978, p. 257.
10. O'Connell, D., *The Influence of Law on Sea Power* (Manchester University Press, 1975), p. 156.
11. Asbeck, F., *The Militarisation of Space*, Armament and Disarmament Information Unit, Vol. 2, No. 2, April/May 1980.
12. UN Document A/AC.105/PV.220, p. 27.
13. General Assembly Resolutions S.10/2 para. 80.
14. UN Document CD/9, 26 March 1979.
15. UN Document A/AC.105/PV.216, p. 53.
16. UN Document A/AC.105/PV.217, p. 42.
17. UN Document A/AC.105/PV.193, p. 8.
18. Treves, T., 'Military installations, structures, and devices on the seabed', *American Journal of International Law*, October 1980.
    (a) —, p. 808.
    (b) —, Note 82.
19. UN Document A/AC.105/RV.203, p. 22.
20. '*Cooperation on Confrontation in Outer Space*', The Thirteenth Conference on the United Nations of the Next Decade (sponsored by the Stanley Foundation), July 1978, p. 25.
21. Goedhuis, D., 'The present state of space law', *Centenary Volume of the International Law Association* (Kluwer, The Netherlands, 1973), p. 201.
22. UN Document A/AC.105/PV.219, 1978.
23. Lovell, B., 'The great competition in space', *Foreign Affairs*, No. 124, 1972.
15. UN Document A/AC.105/PV.197, p. 22.
25. Gorove, S., *Studies in Space Law: Its Challenges and Prospects* (Sijthoff & Noordhoff, Alphen aan den Rijn, The Netherlands, 1977), p. 91.

# Paper 14. Anti-satellite weapons: the prospects for arms control

**D.L. Hafner***

*Boston College, Department of Political Science, Chestnut Hill, Massachusetts 02167, USA*

## *I. Introduction*

In June 1979, when the Soviet Union and the United States completed their third round of negotiations on arms control measures for anti-satellite (ASAT) weapons, the prospects for reaching agreement on ASAT limitations seemed good—but not overwhelmingly so. The Soviet Union had been reluctant to label the first round of talks (8–16 June 1978) as anything more than 'preliminary consultations', reflecting their uncertainty about what the USA had had in mind when proposing the talks in March 1977. But at the second round (23 January–19 February 1979), the participants made the symbolic transition to 'negotiations seeking concrete results'. Enough progress was made to encourage the two parties to the negotiations to schedule a third round, convened in Vienna just prior to the SALT II summit (23 April–17 June 1979), in the hope that an ASAT accord might be drafted and signed in conjunction with the SALT Treaty.

The events of December 1979 and the usual diplomatic hiatus imposed by the US presidential elections have blocked further negotiations on ASAT arms control. Nonetheless, in the intervening period, neither the United States nor the Soviet Union has demonstrably advanced its ASAT weapon programme to a significant degree, so the prospects for ASAT arms control appear to remain good, although still not overwhelmingly so.

* The opinions expressed in this paper are those of the author and do not necessarily reflect the official policy of his professional affiliation.

## II. *Current treaties protecting satellites*

In discussing these prospects, it is important first to note that the United States and the Soviet Union are already parties to a body of international obligations that ban certain weapons from outer space and offer some protection to satellites. These include, *inter alia*:

1. The Limited Test Ban Treaty (1963), which bans nuclear detonations in outer space;
2. The Outer Space Treaty (1967), which bans nuclear weapons and other weapons of mass destruction from orbit and obliges the signatories to explore and to use outer space "for the benefit and in the interests of all countries";
3. The International Telecommunications Convention and related agreements, which establish general regulations to minimize radio-frequency interference with satellite systems;
4. The 'Hot Line' Agreement (1963) and 'Hot Line' Modernization Agreement (1971), which oblige the United States and the Soviet Union to maintain two direct communication links at all times using satellite communication networks;
5. The 'Accidents Measures' Agreement (1971), in conjunction with the Prevention of Nuclear War Agreement (1974), which together oblige the Soviet Union and the United States to refrain from interfering with the attack early warning systems of either side, which would include satellites that are components of such warning systems; and
6. The SALT I ABM (anti-ballistic missile) Treaty and Interim Agreement (1972), which prohibit the United States and the Soviet Union from interfering with the national technical means of verification of the other party, which would include satellites that are components of such verification systems. The ABM Treaty also bans the development, testing or deployment of any space-based ABM systems.

Although the obligations of these and other related agreements are substantial, they alone cannot stem a competition in ASAT weaponry by the two great powers. These accords place some limits on the use of ASAT weapons, but their crucial deficiency is that they do not fully limit the testing, acquisition and deployment of such weapons. The recent Soviet proposal in the UN General Assembly, to ban all weapons from orbit, is a proposal worthy of consideration, but it too suffers this same defect. If ASAT weapons can be freely tested or deployed, each side must anticipate that the weapons will be used—and must take appropriate measures. In the absence of limits on testing and deployment, it is not surprising that the USA and the USSR have asserted that they retain their rights of self-protection, including the right to use ASAT systems, both against satellites engaged in hostile acts and as a general act of retaliation against other nations.

## *III. The incentives for ASAT competition*

The incentives to acquire ASAT weapons are considerable because satellites are such inviting military targets. Increasingly, satellite systems are being used as the eyes, ears and voices of modern military forces—and as with the eyes and ears of men, they are few in number and their loss can be crippling. In the 1980s the USA and the USSR together could have only about 200 active satellites in orbit, assisting in tactical and strategic communications, early warning of attack, photographic and electronic surveillance, meteorology, navigation, and enemy target location (as well as a broad variety of non-military tasks, including resource location, arms control verification, and so on). Many of these tasks could be performed without satellites (and indeed were in the past), though less reliably or inexpensively. But at least in the case of the United States, the Earth-based alternatives were abandoned or curtailed as satellites became available. Hence, the military dependence upon relatively few satellites is high. If one side should gain an advantage in blinding or muting the other side's satellites with ASAT weaponry, military planners argue, this could be translated into an important military advantage during conflict.

The motivation to acquire ASAT systems is heightened by the current vulnerability of satellites to an array of potential ASAT weapons. Nuclear warheads, directed-energy systems (e.g., lasers, particle beams and radio-frequency transmitters, whether ground-based or in orbit), shrapnel or projectiles driven by high explosives, co-orbital or direct-ascent interceptors —each of these could disable, disorient or destroy a satellite. Although a clever defender can hope to devise countermeasures against these weapons (e.g., heat- and radiation-resistant coatings, armour cladding, 'hardened' or jam-resistant electronics, engines for manoeuvring out of an attacker's range, and so on), a satellite equipped with all of them would be costly and cumbersome indeed. The modern naval warship or main battle tank provides ample evidence of the defence necessary against precision guided munitions. In any case, satellites in current use have few such self-protective devices and, given the high cost of satellites and the low rates at which they are ordinarily replaced, most satellites will remain relatively unprotected into the future. Military planners will be tempted to argue, therefore, that the only way they can protect their own satellites is to acquire an ASAT weapon and threaten retaliation against an attacker's satellites. A 'defence' of this sort, of course, will naturally appear as a provocative 'offence' to the other side.

Perhaps the strongest incentive for acquiring ASAT weapons, however, is that satellites themselves are being used to pose threats that military planners find intolerable. For example, electronic surveillance and scanning-radar satellites already in orbit can locate and identify ships on the high seas, and then pass this data on to ships or aircraft poised for attack. For the future, satellite designers speak of orbiting systems capable of detecting and tracking aircraft as well, over land or sea. Coupled with orbiting multi-megawatt lasers or particle-beam weapons, such systems could in principle track and destroy strategic ballistic missiles and bombers. Such systems in an ABM role, of course, are now banned by the 1972 ABM Treaty, but that has not deterred weapon designers from exuberant speculation about what they might be able to build if left unconstrained.

Threats such as these from space have already played a key role in provoking the ASAT programmes of the USA and the USSR. In the United States, the threat from Soviet ocean surveillance satellites is cited by the Defense Department as the major justification for developing a US direct-ascent ASAT interceptor. Evidently comparable threats were cited by military officials in the Soviet Union, because the operational Soviet co-orbital ASAT interceptor system had been placed under the authority of the PKO ('anti-cosmic defence') division of the PVO-Strany strategic defence forces, assigned the mission of "destroying the enemy's cosmic means of fighting".[1]

In summary, the incentives to acquire ASAT weapons are of two sorts. One stems from a competitive and reactive concern—to make certain that one can retaliate in kind for attacks on satellites and thereby prevent an imbalance in military capabilities. The other stems from a concern with the growing use of satellites for military purposes. If each side were only reacting to the other's ASAT programme (real or anticipated), then a verifiable agreement banning ASAT systems altogether would presumably break up the reactive cycle and avert arms competition in outer space. However, the problem is not so straightforward. The provocation for ASAT programmes caused by the increasing militarization of outer space must also be considered.

## *IV. 'Demilitarizing' outer space*

It is tempting to propose coping with both the ASAT problem and its major cause by expanding the scope of the Outer Space Treaty. The Treaty's terms might be redrafted, for instance, along the lines of the 1961 Antarctic Treaty. Article I of the latter agreement stipulates that: "Antarctica shall be used for peaceful purposes only. There shall be prohibited, *inter alia*, any measures of a military nature, such as the establishment of military bases and fortifications, the carrying out of military maneuvers, as well as the testing of any types of weapons." As a verification measure, Article VII (3) stipulates that: "All areas of Antarctica, including all stations, installations, and . . . points of discharging or embarking cargoes or personnel in Antarctica, shall be open at all times to inspection by any [appropriately designated]

[1] For descriptions of the Soviet and US ASAT systems, and details regarding their development and testing programmes, see reference [1].

observers . . . " Such provisions would, in one sweep, ban ASAT tests or deployments in orbit and remove the justification for acquiring ASAT systems by prohibiting the military exploitation of space.

Regrettably, there seems no prospect whatsoever of such an agreement being adopted. The Antarctic Treaty is a most extraordinary achievement. The animosity and suspicion with which the USA and the USSR view each other are of long duration and spring from profound ideological disagreements about the nature of political and social organizations. It would be gratifying, but also astonishing, were the USA and the USSR to refrain from extending their arms competition into any arena their technology allows them to reach. Antarctica became a sanctuary from military competition precisely because it was too remote and inhospitable a site. A weapon or surveillance post stationed in Antarctica is 10 000 km from the borders of either the USA or the USSR; a satellite in outer space, however, can orbit within 150 km of any point on the globe. The great powers will not readily forego such useful military vantage points.

Moreover, to make such a treaty work it would be necessary to ensure that satellites would not and could not be used for military purposes. But that cannot be ensured, given the versatility of satellites. A meteorological satellite can be used by a farmer or a tank commander, a navigation satellite by an oil tanker or an aircraft carrier, and photographs from a surveillance satellite by an arms controller or a bomber pilot. Hamstringing all satellites so that they could not perform military tasks would cripple the peaceful uses of outer space.

What we have, therefore, is a formidable obstacle to an ASAT accord. It is impossible to 'demilitarize' outer space fully, so the USA and the USSR must be persuaded that it is better to tolerate satellites carrying out military acts, or to cope with such satellites in some other way, rather than to engage in an unrestricted competition in ASAT weaponry.

## The costs of ASAT competition

A good portion of the persuading will probably be done by the costs of ASAT competition—costs both of ASAT weapons and of countermeasures to protect or replace satellite systems. The USA is prepared to spend over $600 million to carry its ASAT interceptor programme up through several space tests. The Soviet Union has certainly spent as much on interceptors and targets launched over the past 15 years. An effective, fast-reaction operational ASAT system would require a large inventory of interceptors (depending upon the interceptor's technical reliability and upon how many of the adversary's satellites were to be attacked), and several widely dispersed launch sites. Many satellite targets of interest (e.g., photographic, electronic and radar surveillance satellites) are in orbits lower than 1 000 km and can be reached with interceptors launched on common boosters. But a significant number of other satellites (e.g., early warning communications and navigation satellites) are in much higher orbits, up to 36 000 km and more. To reach geosynchronous orbits, for instance, even the US ASAT interceptor (which is to weigh only 16 kg) would require a Minuteman ICBM as a booster. The current Soviet interceptor would require a booster as large as that used for the Soyuz or Progress space launches. It is possible, therefore, that the cost of attacking each satellite target with these ASAT systems could be $10–30 million or more.

A more exotic ASAT system would use an orbiting laser or particle-beam weapon rather than Earth-launched interceptors. A laser satellite with a power output of 5–6 MW could disable or destroy target satellites hundreds or thousands of kilometres away but again the costs are daunting —perhaps $1 000–2 000 million to develop and launch a single such weapon, and a hundred times that amount to launch and maintain a fleet of such weapons capable of attacking great numbers of satellites in a short period of time. Clearly such expenditures would be justified only for very important tasks, and this explains why those who propose laser satellite systems intend to use them primarily against strategic missiles and bombers. It is not clear that such costs would be acceptable even for those purposes, since countermeasures are possible. In any case, such weapons will probably not be feasible until the 1990s or even later.

If a vigorous ASAT competition develops, the USA and the USSR will also have to devise defensive measures to protect their own satellites. This could be done by equipping satellites to evade or withstand attack, or by launching so many satellites that the other side stands little chance of attacking them all. Either way, the USA and the USSR will find they must duplicate or replace most of their satellite networks, and will have to do this at a much faster rate than now, in order to keep up with the adversary's advancing ASAT technology. Again, the expenditures involved could readily mount to thousands of millions of dollars.

The probable and ironic outcome of such an ASAT competition is that the USA and the USSR would find that satellites could not readily be defended, and they would shift away from dependence upon satellites for vital military functions. If both were to reach that conclusion now, of course, much of the incentive for engaging in an ASAT competition would disappear.

Moreover, military planners should realize that shooting down satellites with ASAT weapons is not an entirely satisfactory response to the threats from outer space. Consider the case of radar and electronic ocean surveillance satellites. How swiftly these satellite systems could be disabled by ASAT weapons would depend upon how many satellites they contained, where the satellites were in their orbits when conflict began, how rapidly they could be re-launched, and how many ASAT weapons and launch sites the defender has. It is likely that 12 to 78 hours or more would pass before most of the satellites could be successfully attacked. Obviously neither side could afford to leave its naval forces unprotected for such a period during conflict, so each will have to devise other methods for confusing surveillance satellites (e.g., radio silence, camouflage, decoys, etc.). If the surveillance satellites gather necessary targeting information before conflict breaks out, of course, shooting them down afterwards may not help at all. ASAT weapons, rather than being an indispensable response to the militarization of outer space, could well be just redundant and costly.

### The hazards of ASAT competition

Perhaps a more persuasive reason than cost, however, is the increased hazard of war that unrestrained ASAT competition may generate. Many of the surveillance satellites that the USA and the USSR would regard as prime ASAT targets in the event of war actually serve a very desirable stabilizing function in crises, ameliorating the pressures for pre-emptive strikes and reducing the opportunities for third parties to provoke a US– Soviet conflict through some mischief. A satellite which can warn that an attack is under way can also confirm that an attack is not under way or is not coming from the other great power.

In a world of many ASAT weapons, however, each side must fear that a sudden malfunction in a vital satellite network is a prelude to an attack. If the progressive loss of its satellites would degrade the effectiveness of its own military forces, the side would be under great pressure to do imprudent things. Consider the case of strategic bombers, for instance, which may be dependent upon vulnerable satellites for command and control when in flight and on alert during crises. To compensate for this dependence, the sides may be tempted to loosen or abandon certain fail-safe procedures or, still worse, to modify these procedures in a disorderly manner during the tense hours of a crisis, perhaps after losing some satellite capabilities. Under a general ban on ASAT weapons, on the other hand, the sudden malfunctioning of a vital satellite during a crisis might cause anxiety, but there would be more reason to assume that the problem was a technical malfunction and less pressure to act precipitately. Even military planners with ample budgets and strong competitive spirits can recognize that some forms of competition are just plain dangerous and foolish.

### Unpersuasive reasons

One argument that will not be decisive in producing an ASAT accord is that both the USA and the USSR ought to reconcile themselves to no better than parity in their military relationship with each other, and ought to accept an ASAT agreement as a mutual affirmation that parity is the most desirable condition. Such an argument may have appeal for devoted arms controllers, but it has little influence within the decision-making circles of these nations. Competition between the USA and the USSR is too broad and deeply engrained to be changed so radically—nor does such a change seem indispensable to progress in arms control. The two nations have different security concerns and military forces. The USA might fear that conflict was most likely in one geographical area, where loss of its satellites would be crucial; the Soviet Union might fear conflict in a different region, where it would be disadvantaged in losing satellites. Because of differences in fears and capabilities, each could emerge from ASAT negotiations believing (correctly) that it had enhanced its own security and secured an important advantage where it matters. Indeed, it is very much because of the competitive concerns of the USA and the USSR over maintaining security, within finite military budgets and at reduced risks, that the arguments noted here in favour of ASAT arms control will be persuasive to military planners.

**Issues to be resolved in an ASAT agreement**

When the USA and the USSR return to the negotiating table and begin drafting an ASAT accord, they will have to grapple with a number of important issues. In some measure, we can anticipate the points of dispute that will emerge and the solutions that might best serve ASAT arms control.

*Formulating a ban*

The first issue is: how should a ban on anti-satellite weapons be formulated? There are a number of devices that could conceivably be used to damage or destroy satellites, including weapons designed solely for that purpose as well as systems originally designed for peaceful purposes. It would be very difficult to negotiate a treaty definition of ASAT weapons which was neither too broad, and thus interfered with peaceful uses of space, nor too narrow, and thus overlooked some unanticipated ASAT weapon. The best approach here would be to prohibit an activity rather than itemized devices—an approach taken in the ABM Treaty of 1972. The signatories would undertake not to use, not to deploy, and not to test any weapon or device for damaging or destroying satellites. As in the ABM Treaty, questions regarding whether one side had or had not tested, used or deployed some device in an ASAT role (and what it should do to restore the *status quo ante*) would be resolved on a case-by-case basis, in a standing consultative commission.

This approach does leave room for mischief, which in turn may reduce the prospects of an ASAT accord being reached. Some peaceful activities in outer space (such as the rendezvous and docking of two space vehicles) will bear a general resemblance to actions appropriate for an attack on satellites, and some weapons designed and deployed for other missions (such as ABM interceptors) may have marginal ASAT capabilities. So, on the one hand, both the USA and the USSR may fear that an ASAT accord would provide the other with a pretext for constant harassment of peaceful space programmes or other military projects. On the other hand, each may fear that the other side will use other space or weapon programmes to test ASAT devices covertly. Unless each side agrees in advance that it will make extra effort to avoid provocative or ambiguous actions in its own programmes, and will forbear harassing the other side's projects, there is little prospect of an ASAT accord being reached.

The USA and the USSR are making an encouraging start on this issue. The brightest jewel in the US space programme is the space shuttle. Even before its first flight, Soviet commentators often characterized the shuttle as a military and an ASAT system. They may only have been reacting to effusive comments in the US press, including one assertion that a major impetus for the shuttle was its potential for removing Soviet 'killer satellites' from orbit. Nonetheless, these efforts to label the shuttle as an ASAT system were viewed by some in the USA as merely an attempt to retard the US space programme and to tarnish its reputation. Now the Soviet government, as part of its draft treaty regarding weapons in orbit, has proposed simply that "reusable manned space vehicles" be prohibited from placing or carrying weapons in orbit. There are crucial points yet to be clarified, including how broadly the term 'weapon' is to be construed. Still, the Soviet Union has come closer to the US position, which has been that the shuttle was not an ASAT system and would not be tested, deployed or used in any manner prohibited by an ASAT accord.

The general problem of technology in other fields touching the border of prohibited activity has been confronted by the USA and the USSR before. When the ABM Treaty was drafted, the two sides anticipated that, for instance, anti-aircraft systems and early warning radars might soon possess marginal ABM capabilities. Article VI of that Treaty addressed the problem by prohibiting tests of non-ABM devices in an ABM mode. Controversies did subsequently arise over certain equipment and testing practices. These have been resolved in the Standing Consultative Commission, though only after prolonged negotiations. Adapting a treaty to changing technology in this way, on a case-by-case basis, is a sensible approach. But it does demand patience and co-operativeness—virtues not always evident on both sides.

*Satellite interference*

A second issue to be considered is: should other forms of interference with satellites, not resulting in damage or destruction, be prohibited? Such interference would include, *inter alia*, blinding photographic surveillance satellites with lasers, jamming satellites' radio or radar receivers, giving radio commands which cause satellites to malfunction, or moving satellites from their orbits. In many of these examples it would not be feasible to ban the devices which could be used against satellites in these ways. Radio transmitters for communicating with satellites, or radars for tracking them, could both be used for radio jamming, for example, yet such transmitters obviously cannot be prohibited. Nonetheless, there is value in prohibiting such provocative acts of interference, and thus averting the alarms and suspicions they might cause. Other actions which might frustrate the missions of satellites, but which are not directed at the satellites themselves, need not be addressed in an ASAT accord. Thus, jamming the radio receivers of people on the ground who were attempting to receive satellite signals would presumably not be regarded as interference with the satellite.

*Hostile use*

A third issue of importance is: should the USA and the USSR also agree not to use satellites for 'hostile acts'? The appeal of such a provision is that it would remove one justification for ASAT weapons and thus enhance the status of outer space as a peaceful sanctuary. Since there are satellite missions which the USA or the USSR have already labelled as especially objectionable—such as, the direct broadcast of television programmes from satellites to home receivers in the other nation, or radar ocean surveillance—it is probable that the issue of 'hostile acts' will arise in ASAT negotiations. If it does, the sides risk becoming bogged down in a fruitless controversy that could put an end to ASAT talks.

There is no generic definition of 'hostile acts' that can readily be applied to satellites. Many of their functions might be regarded as hostile in times of conflict (e.g., giving meteorological, navigational or communications assistance to attacking forces), but could hardly be labelled as such or banned in peacetime. Other functions might be regarded as generally provocative or hostile (e.g., photographic, electronic or radar surveillance), but may be protected by separate agreements, such as, the SALT obligations not to interfere with national technical means of verification. Nor can the problem be solved by banning all activities not sanctioned by international law. Arguably all of the functions of current satellites, including those which either the USA or the USSR finds objectionable, are sanctioned by international law—as interpreted by one or other of these two nations. The sides could explore arrangements in which each would agree to refrain from certain specific objectionable activities directed at the other. But such arrangements would not be efficacious unless all satellites capable of the objectionable acts were deactivated or banned (including satellites of third parties). The sides would balk at this, because such satellites have legitimate peaceful uses (e.g., direct-broadcast satellites can be used for communication with remote domestic populations).

If a prohibition on 'hostile acts' by satellites is unacceptable or impractical then a ban on ASAT weaponry in effect will ask the USA and the USSR to give up using some of their technological potential against each other in the event of conflict. This is admittedly asking for a great deal of restraint, but there are precedents. A variety of prohibitions already exist, barring belligerents from using certain weapons against each other—the 1975 Biological Weapons Convention and the 1978 Environmental Modification Convention are recent examples. Such conventions are buttressed by humanitarian concerns, and this would be true also of an ASAT accord. More specifically, such conventions are buttressed by the fact that they do not ask nations to disarm themselves entirely. Again, the same would be true of an ASAT ban. It would not ask the sides to leave themselves defenceless against the actions of satellites; it would only ask that they find some means of defence other than turning outer space into a cosmic no man's land.

*Third parties*

A fourth issue of importance is: should other parties be involved in negotiating an ASAT agreement, and if they are not, would their satellites enjoy the protection of such an agreement? At the moment, it appears that only the USA and the USSR have ASAT weapon programmes. But a handful of nations might be able to build ASAT systems by the end of the decade, and if they did so, it certainly would jeopardize any ASAT agreement between the USA and the USSR. Moreover, all nations who are party to the Outer Space Treaty could claim to have interest in regulations on weapon competition in outer space. For these reasons, it is desirable that any ASAT agreement be opened for adoption by all interested nations.

This is not an argument, however, for making ASAT negotiations multilateral from the beginning. The USA and the USSR will undoubtedly pursue negotiations more vigorously if they know that other nations are following their progress with concern, but the issues involved in ASAT limitations touch on the most sensitive security concerns of these two nations, and they will not allow themselves to be pushed into hasty and ill-considered agreement under the pressure of other nations. The negotiation of possible ASAT limits is thus likely to be more frank and business-like if conducted bilaterally.

Necessarily, the satellites of third parties will enjoy some protection under an ASAT agreement between the USA and the USSR. An effective ASAT agreement must ban all ASAT devices; an accord would be worthless if it left the USA and the USSR free to test and deploy ASAT weapons on the pretext that they are directed only against third parties. On the other hand, there are some devices that can be used in non-destructive ways to interfere with satellites. It is desirable, as noted above, that the USA and the USSR agree to refrain from such interference with each other's satellites. However, if third parties fail to follow the same restraints, or use their satellites against the USA or the USSR in ways not sanctioned by international law, an ASAT agreement presumably would leave the nation concerned free to interfere with the satellites or to use any other permissible technique to frustrate their missions.

*Verification*

Finally, the most troublesome issue the great powers will have to grapple with is verification. On this issue, the obstacles to an accord are likely to be as formidable within governments as between them. It will be argued that because satellites are so few in number and so crucial for military capabilities, even a limited ASAT capability, covertly acquired by an adversary, could be decisive. It will also be argued that there are too many devices with a marginal ASAT capability, that it would be very hard to verify whether they are being prepared or tested for ASAT use and that some future ASAT systems, and the surveillance systems needed to monitor them, are as yet only hypothetical. It would therefore be impossible to judge confidently whether hypothetical verification systems will indeed be able to cope with hypothetical ASAT cheating. In sum, it will be argued that the standards for verifying an ASAT accord must be even more exacting than for other arms control agreements, so exacting as to assure each side that it will not lose even a handful of vital satellites to covertly acquired ASAT weapons in the event of conflict.

*Conclusions*

These arguments have some merit. It should not be assumed that they are offered disingenuously by people opposed in principle to arms control. However, the arguments do suffer a common flaw—they assume that the task of protecting satellites or maintaining symmetry in military capabilities would be easier if there were no restraints on ASAT competition. That is hard to believe. There are differences in the configurations of military forces and the security interests of the USA and the USSR, and this may make one nation more vulnerable to ASAT cheating and to the loss of its satellites in specific cases. The appropriate response is to make vital satellites more survivable, or to reduce dependence upon them. No ASAT interceptor will be able, for example, to protect a satellite from being blinded by a ground-based laser or from being attacked by an ABM missile. At best, ASAT weapons would enable one side to retaliate against the other's satellites. Yet if one side is genuinely more harmed by the loss of its satellites, retaliation will be of little help.

It will take time to devise defensive measures and install them on all satellites. For a while, the space powers would remain exposed to residual or covertly acquired ASAT threats, but with resources diverted from an ASAT competition into satellite defence and verification systems, it should be possible to improve verification and to build up defensive hedges against abrogation or evasion of an ASAT ban.

## Are the prospects good?

This listing of important issues is not exhaustive. Before an accord is signed, for instance, the issue of dismantling procedures for existing ASAT weapons must be resolved as well. Nevertheless, the issues listed here are extensive and troublesome enough, and in the light of them, one might well ask why the prospects for ASAT arms control should be judged good rather than poor. The main reason for being optimistic is that the United States and the Soviet Union have thus far approached ASAT negotiations in a serious and constructive manner. The recent Soviet proposal in the UN General Assembly, for a treaty banning all weapons from orbit, is a notable sign that the hiatus in ASAT negotiations has not dimmed interest in the issues. It is encouraging to hear the Soviet Union's affirmation that "it believes that outer space should always remain unsullied and free from any weapons and should not become a new arena for the arms race or a source of strained relations between states" [2]. In important respects, the draft text of the

Soviet proposal closely parallels the ASAT provisions suggested above in this paper, and that too enhances the prospects. Article 3 of the Soviet text, for instance, proposes that: "Each State Party undertakes not to destroy, damage, disturb the normal functioning or change the flight trajectory of space objects of other States Parties, if such objects were placed in orbit in strict accordance with Article 1, Paragraph 1 of this treaty". The effect of this provision would be to prohibit both the use of ASAT weapons on, and certain acts of interference with, the satellites of other nations.

The Soviet text has its flaws. Because its provisions only ban weapons from orbit, it appears not to cover ASAT weapons that would strike their targets directly from the ground without going into orbit. Moreover, the text says nothing about dismantling ASAT systems which have already been acquired and deployed. Taken together, these two omissions could be construed to permit both the United States and the Soviet Union to continue with their current ASAT interceptor programmes. Indeed, the last clause of Article 3, above, implies that ASAT systems may be acquired and used against parties who violate the accord. There may be other defects in addition to these, which will emerge only when negotiators begin clarifying key definitions and crucial aspects of verification procedures.

Nevertheless, it is worth noting that in their UN proposal, the Soviet Union appears to have laid down a challenge to the United States to match them in their earnestness on ASAT arms control. The Soviet proposal to ban all weapons from orbit can be understood as a Soviet pledge not to conduct any further tests of the orbiting Soviet ASAT interceptors, so long as serious negotiations on ASAT arms control are under way. The challenge to the United States is to show similar restraint in its ASAT programme and to rejoin the Soviet Union at the negotiating table. In these gestures, we will have signal indicators of whether the prospects for ASAT arms control remain good.

## References

1. Hafner, D.L. 'Arms control measures for anti-satellite weapons', *International Security*, Vol. 5, No. 3, Winter 1980/81, pp. 41–60.
2. Letter from Soviet Foreign Minister Andrei Gromyko to UN Secretary-General Waldheim, 10 August 1981.

# Paper 15. Banning all weapons in outer space

**O.V. Bogdanov***

*Institute of State and Law, Academy of Sciences of the USSR, 10 Frunze Street, Moscow 19, USSR*

## *I. Introduction*

The conquest of outer space can be approached in a number of different ways since it is a multifaceted international problem which has political, technological and legal sides. The papers prepared for this symposium have placed much emphasis on the scientific and technological aspects of the problem. Since this paper is a contribution from one of the two international lawyers present at the symposium, it addresses the topic from a rather different angle. It would make good sense to repeat certain important legal aspects of the outer space debate, even if they are not very new.

Generally speaking, the field of international law has seen rapid progress in directions aimed to cover some of the new aspects of possible extensions of the arms race. Outer space is undoubtedly one of the spheres in which such evolution has come under careful observation and international law has made significant progress in recent years. The extent of this progress is not always fully realized. 1982 will be the 25th anniversary of the beginning of the peaceful conquest of outer space. Before the 1950s, international law made absolutely no provision for outer space. Significant changes have since taken place. A special branch of international law has been created, called space law. It seems to be making comparatively rapid progress although it is not yet fully developed and, of course, its development could not be compared with the rate of technical innovation.

As an illustration it should be noted that during the last 15 years a number of treaties on problems related to outer space have been introduced to the international community and are becoming important instruments guiding human behaviour in this field. It is a peculiar feature of this development

* The opinions expressed in this paper are those of the author and do not necessarily reflect the official policy of his professional affiliation.

that in some cases progress has been even greater than in the field of international law in general. One example is the prohibition of weapons of mass destruction. In accordance with the 1967 Treaty these weapons are prohibited in outer space, but not fully on Earth, where this problem is a bit more complex. Only two of the existing means of mass destruction have been prohibited up to now—bacteriological and chemical weapons (originally by the Geneva Protocol of 1925).

More progress has since been made in this field. The Treaty of 1975 placing a full prohibition on biological weapons is now in force and playing a very useful role in restraining military activities in the world. A major drawback of the existing legal system, however, is that there is no treaty prohibiting nuclear weapons. Clearly, this is one of the problems which international law must attempt to solve in the forthcoming years.

New means of mass destruction bring with them an urgent need to develop international law and to draw up new treaties for their prohibition before these weapons materialize *en masse* as part of the present day military build-up.

There is sometimes a tendency to neglect or minimize progress already made in regulating the problems of outer space. For many reasons this inclination seems to be unjustifiable. Existing international treaties are, of course, not exhaustive, but at the same time their value and their potential role in reducing the threat of war should not be underestimated.

## *II. Existing international treaties*

Close study of existing international treaties on outer space is essential to an understanding of what needs to be done to make the list of international documents more complete and more effective.

In 1963 the Moscow Treaty banned nuclear tests in the atmosphere, in outer space and under water. In 1967 the Treaty on Principles Governing the Activities of States in the Exploration and Use of Outer Space, including the Moon and other Celestial Bodies, was signed. This was an important development and still constitutes the main component of international space law. This law is still being developed.

In 1979 an Agreement on Governing the Activities of States on the Moon and Other Celestial Bodies was made, and there is a continuous programme of preparations to extend outer space law. Although space law is one of the most effectively developing branches of international law at present, the rate at which development takes place varies. For a number of years there have been no new documents dealing specifically with outer space. We are now entering a new phase in which a concrete proposal for the next contribution has been drawn up. Its presentation coincided almost exactly with the time of this symposium and the document concerned is presently under discussion at the UN General Assembly.

## *III. A new treaty proposal*

Whereas it is probably inappropriate to go into the details of the proposal at this time, the initiative taken might be viewed as an indication of a possible trend in future legal developments concerning outer space. International lawyers in the Soviet Union regard this document as one which could set the main directions of legal developments in this field for some time. The Soviet proposal is a draft treaty to ban deployment of all types of weapon in outer space. It is directed mainly at strengthening the guarantee that space will be a sphere of peaceful competition between states and that it will be excluded from the arms race.

The present prohibition applies only to means of mass destruction and not to conventional weapons. Conventional weapons are not banned from space by international law and the significance of this new Soviet initiative lies in the attempt to extend the prohibition to this group of weapons. The essence of this new proposal is expressed in its first article, the first paragraph of which reads: "1. States Parties undertake not to place in orbit around the earth objects carrying weapons of any kind, install such weapons on celestial bodies, or station such weapons in outer space in any other manner, including on reusable manned space vehicles of an existing type or of other types which States Parties may develop in the future" [1]. Clearly the proposal embraces many of the kinds of activity apparent at present. There is, for example, a direct reference to prohibiting the placement of any kinds of weapon on reusable manned space vehicles, a problem which is of more than academic interest. A whole string of scientific publications exists to show the potential possibility for converting such reusable space vehicles into carriers of different kinds of weapon. This is a dangerous trend in the arms race and merits specific mention in the draft treaty. It is important to note that this reference does not only cover existing vehicles of this kind.

Article 2 of the Soviet draft underlines the necessity for strict observance of existing international legal principles regarding space in the interests of maintaining international peace and security and promoting international co-operation and mutual understanding. It is in the spirit of this article that the activities of states in space should proceed, to achieve the peaceful conquest of space.

### Verification

In Article 4 the Soviet proposal deals with verification problems. The proposal is that only national technical means be used for purposes of control. The provisions of the treaty could well be verified by these means. In most existing treaties on general problems of disarmament, the control provisions are limited to the use of national means of verification. There are, of course, some exceptions, but these are few and could not be taken as establishing any rules. The draft is therefore based on the premise that it is the national means of verification that are prevalent in the disarmament treaties. Moreover, it is quite reasonable to suppose that future years will bring many technical innovations in this field and that more and more countries will possess the technical means of monitoring and controlling activities in outer space. Thus, this problem seems likely to be more easily solved in the future.

At the time when the 1963 Moscow Test Ban Treaty was being prepared, there was much discussion on control measures. The idea arose that the means of technical verification of underground tests were being developed so quickly that practically no international machinery would be required to detect explosions, including those taking place underground throughout the world. Certainly this prediction has come true. If the situation in 1981 is compared to that of 1963, it becomes clear that tremendous technical progress has been made, enabling the detection of almost any kind of underground nuclear explosion anywhere in the world. This indicates that there is no need to create complex international infrastructures for control purposes with regard to the prohibition of stationing weapons in space.

### The legal context

The Soviet proposal must also be examined from an international legal point of view. In international law the process of demilitarization is connected with the banning of weapons in one area or another. So far the provisions of the 1967 Treaty effect a partial demilitarization of outer space—partial insofar as the prohibition covers only weapons of mass destruction. The Soviet Union is now proposing that this prohibition be widened to include all other weapons. In legal terms this would mean the next stage of demilitarization, that is, the full demilitarization of outer space. This would be a significant and satisfactory result, one which could bring about greater safety in the world and establish further restrictions on the development of the arms race.

## *IV. Conclusion*

It is a truism that it is sometimes much easier to prevent something from happening than to correct the situation after the event. As yet there are no weapons in outer space, but preparations for an arms race in space are going on. If such developments are not prohibited, outer space will become a sphere of competition in building up armaments. Once the arms race has moved into outer space, reversing the trend will be much more difficult.

The main conclusion of this paper is that action should be taken to ban all kinds of weapon in this sphere of human activity. A draft treaty such as that proposed by the Soviet Union could be instrumental in achieving this result. The proposal is now undergoing preliminary consideration by the UN General Assembly, and will surely attract the attention of those seriously interested in arms control measures and in restraining the arms race in the world.

## Reference


1. 'Draft treaty on the prohibition of the stationing of weapons of any kind in outer space', proposed to the 36th session of the UN General Assembly, 20 August 1981, A/36/192.

# Appendix 1A

## *Photographic reconnaissance satellites launched between 1977 and 1981*

| Country, satellite name and designation | Launch date and time (GMT) | Orbital inclination (deg) and period (min) | Perigee and apogee heights (km) | Comments[a] |
|---|---|---|---|---|
| **USA** | | | | |
| *1977* | | | | |
| USAF (1977-19A) | 13 Mar 1843 | 96 89 | 124 348 | Low resolution search and find; lifetime 74 days |
| USAF (1977-56A) | 27 Jun 1829 | 97 89 | 155 239 | Big Bird satellite; lifetime 179 days |
| USAF (1977-94A) | 23 Sep 1843 | 97 89 | 125 352 | Low resolution search and find; lifetime 76 days |
| *1978* | | | | |
| USAF (1978-29A) | 16 Mar 1843 | 96 89 | 160 240 | Big Bird satellite; carries several capsules which are periodically ejected from the satellite; lifetime 179 days |
| T-3D (1978-60A) | 14 Jun 1829 | 97 92 | 223 509 | New generation KH-11; lifetime 1166 days |
| *1979* | | | | |
| USAF (1979-25A) | 16 Mar 1829 | 96 89 | 170 250 | Big Bird satellite decayed after 190 days; last one, launched on 16 Mar 1978, decayed after 179 days |
| USAF (1979-44A) | 28 May 1814 | 96 89 | 131 285 | Area surveillance satellite; images transmitted to Earth; decayed after 90 days |
| *1980* | | | | |
| T-3D (1980-10A) | 7 Feb 2107 | 97 93 | 309 501 | A KH-11 very long-lived digital imaging satellite; in orbit as of 31 Jan 1982 |
| USAF (1980-52A) | 18 Jun 1829 | 96 89 | 169 265 | Big Bird satellite; lifetime 261 days |

| Country, satellite name and designation | Launch date and time (GMT) | Orbital inclination (deg) and period (min) | Perigee and apogee heights (km) | Comments[a] |
|---|---|---|---|---|
| *1981* | | | | |
| USAF (1981-19A) | 28 Feb 1912 | 96 89 | 138 336 | Close-look satellite; lifetime 112 days |
| T-3D (1981-85A) | 3 Sep 1829 | 97 92 | 244 526 | Probably a KH-11; in orbit as of 31 Jan 1982 |
| **USSR** | | | | |
| *1977* | | | | |
| Cosmos 888 (1977-01A) | 6 Jan 0950 | 65 90 | 170 325 | Manoeuvrable satellite; two-tone; lifetime 13 days |
| Cosmos 889 (1977-03A) | 20 Jan 0838 | 71 90 | 202 329 | About three such satellites are launched each year; area surveillance satellite; PDM[b]; lifetime 12 days |
| Cosmos 892 (1977-09A) | 9 Feb 1131 | 73 90 | 171 343 | Manoeuvrable satellite; two-tone; lifetime 13 days |
| Cosmos 896 (1977-16A) | 3 Mar 1033 | 73 90 | 177 343 | Manoeuvrable satellite; two-tone; lifetime 13 days |
| Cosmos 897 (1977-17A) | 10 Mar 1102 | 73 90 | 171 340 | Capsule ejected; manoeuvrable satellite; two-tone; lifetime 13 days |
| Cosmos 898 (1977-20A) | 17 Mar 0838 | 81 89 | 216 230 | Capsule ejected; PDM; also scientific mission; lifetime 13 days |
| Cosmos 902 (1977-26A) | 7 Apr 0907 | 81 89 | 168 279 | Launched in spring and autumn each year to survey Arctic ice conditions; lifetime 13 days |
| Cosmos 904 (1977-28A) | 20 Apr 0907 | 71 90 | 203 328 | About three such satellites are launched each year; area surveillance satellite; PDM; lifetime 14 days |
| Cosmos 905 (1977-30A) | 26 Apr 1453 | 67 90 | 171 339 | Fourth generation long-lived satellite; lifetime 30 days |
| Cosmos 907 (1977-33A) | 5 May 1410 | 63 90 | 181 364 | Manoeuvrable satellite; two-tone; lifetime 11 days |
| Cosmos 908 (1977-35A) | 17 May 1019 | 52 89 | 174 288 | Manoeuvrable satellite; two-tone; lifetime 14 days |
| Cosmos 912 (1977-40A) | 26 May 0712 | 81 89 | 217 231 | First unmanned satellite characterized by USSR as Earth resources satellite; lifetime 13 days |
| Cosmos 914 (1977-43A) | 31 May 0735 | 65 90 | 203 306 | PDM; also scientific mission; lifetime 13 days |
| Cosmos 915 (1977-45A) | 8 Jun 1410 | 63 89 | 173 289 | Manoeuvrable satellite; two-tone; lifetime 13 days |
| Cosmos 916 (1977-46A) | 10 Jun 0810 | 63 90 | 246 246 | Manoeuvrable satellite; two-tone; lifetime 12 days |
| Cosmos 920 (1977-52A) | 22 Jun 0810 | 65 90 | 173 342 | Manoeuvrable satellite; two-tone; lifetime 13 days |
| Cosmos 922 (1977-58A) | 30 Jun 1410 | 63 90 | 205 299 | Area surveillance satellite; PDM; lifetime 13 days |
| Cosmos 927 (1977-63A) | 12 Jul 0907 | 73 90 | 153 361 | Manoeuvrable satellite; two-tone; lifetime 13 days |

| Country, satellite name and designation | Launch date and time (GMT) | Orbital inclination (deg) and period (min) | Perigee and apogee heights (km) | Comments[a] |
|---|---|---|---|---|
| Cosmos 932 (1977-69A) | 20 Jul 0735 | 65 89 | 149 311 | Manoeuvrable satellite; two-tone; lifetime 13 days |
| Cosmos 934 (1977-62A) | 27 Jul 1814 | 63 89 | 231 255 | Manoeuvrable satellite; two-tone; lifetime 13 days |
| Cosmos 935 (1977-73A) | 29 Jul 0813 | 81 89 | 217 251 | Area surveillance satellite; PDM; lifetime 13 days |
| Cosmos 938 (1977-78A) | 24 Aug 1438 | 63 89 | 156 332 | Manoeuvrable satellite; two-tone; lifetime 13 days |
| Cosmos 947 (1977-81A) | 27 Aug 1019 | 73 90 | 203 321 | Manoeuvrable satellite; two-tone; lifetime 13 days |
| Cosmos 948 (1977-83A) | 2 Sep 0907 | 81 89 | 217 235 | Possible Earth resources satellite; capsule ejected; lifetime 13 days |
| Cosmos 949 (1977-85A) | 7 Sep 1731 | 63 90 | 177 364 | Fourth generation long-lived satellite; all such satellites have previously been launched in 67° orbit; lifetime 30 days |
| Cosmos 950 (1977-86A) | 13 Sep 1522 | 63 89 | 205 282 | Manoeuvrable satellite; two-tone; lifetime 14 days |
| Cosmos 953 (1977-89A) | 16 Sep 1438 | 63 89 | 151 300 | Manoeuvrable satellite; two-tone; lifetime 13 days |
| Cosmos 957 (1977-98A) | 30 Sep 0950 | 65 90 | 150 350 | Manoeuvrable satellite; two-tone; lifetime 13 days |
| Cosmos 958 (1977-100A) | 11 Oct 1522 | 63 91 | 257 351 | Interlaced ground track to get complete coverage; similar to Cosmos 867; lifetime 13 days |
| Cosmos 964 (1977-110A) | 4 Dec | 73 90 | 180 391 | High resolution, manoeuvrable satellite; two-tone; lifetime 13 days |
| Cosmos 966 (1977-115A) | 12 Dec | 65 90 | 210 316 | Area surveillance satellite; PDM; also scientific mission; lifetime 12 days |
| Cosmos 969 (1977-120A) | 20 Dec | 63 90 | 188 340 | High resolution, manoeuvrable satellite; two-tone; lifetime 13 days |
| Cosmos 973 (1977-124A) | 27 Dec | 71 90 | 210 348 | Area surveillance satellite; PDM; also scientific mission; lifetime 13 days |
| *1978* | | | | |
| Cosmos 974 (1978-01A) | 6 Jan 1550 | 63 90 | 178 334 | High resolution, manoeuvrable satellite; lifetime 13 days |
| Cosmos 984 (1978-06A) | 13 Jan 1522 | 63 90 | 206 291 | Low resolution, PDM; lifetime 13 days |
| Cosmos 986 (1978-10A) | 24 Jan 0950 | 65 89 | 172 318 | High resolution, manoeuvrable satellite; lifetime 14 days |
| Cosmos 987 (1978-13A) | 31 Jan 1453 | 63 89 | 175 321 | High resolution, manoeuvrable satellite; lifetime 14 days |
| Cosmos 988 (1978-15A) | 8 Feb 1214 | 73 90 | 201 335 | Low resolution, special subset; lifetime 12 days |
| Cosmos 989 (1978-17A) | 14 Feb 0936 | 65 89 | 169 318 | High resolution; lifetime 14 days |
| Cosmos 992 (1978-25A) | 4 Mar 0735 | 71 90 | 203 323 | Low resolution, PDM; lifetime 13 days |

| Country, satellite name and designation | Launch date and time (GMT) | Orbital inclination (deg) and period (min) | Perigee and apogee heights (km) | Comments[a] |
|---|---|---|---|---|
| Cosmos 993 (1978-27A) | 10 Mar 1048 | 73 90 | 171 340 | High resolution, manoeuvrable satellite; lifetime 13 days |
| Cosmos 995 (1978-30A) | 17 Mar 1048 | 81 90 | 217 235 | Low resolution, PDM; lifetime 13 days |
| Cosmos 999 (1978-33A) | 30 Mar 0755 | 71 90 | 174 352 | High resolution, manoeuvrable satellite; lifetime 13 days |
| Cosmos 1002 (1978-37A) | 6 Apr 0922 | 65 89 | 205 283 | Low resolution; lifetime 13 days |
| Cosmos 1003 (1978-40A) | 20 Apr 1536 | 63 90 | 178 328 | High resolution, manoeuvrable satellite; lifetime 14 days |
| Cosmos 1004 (1978-43A) | 5 May 1526 | 63 89 | 205 290 | Low resolution, PDM; scientific pickaback; the last of the first generation extended-duration satellites; the satellites transmitted PDM signals on 19.994 MHz; lifetime 13 days |
| Cosmos 1007 (1978-48A) | 16 May 1048 | 73 90 | 168 350 | High resolution, manoeuvrable satellite; lifetime 13 days |
| Cosmos 1010 (1978-52A) | 23 May 0726 | 81 89 | 215 229 | Probably also performed Earth resources mission; low resolution; satellite launched to aid shipping; lifetime 13 days |
| Cosmos 1012 (1978-54A) | 25 May 1438 | 63 89 | 202 265 | First of the new third generation non-manoeuvrable low resolution satellites; they transmit two-tone short-wave beacon signals without telemetry on 19.994 MHz (19.989 MHz for high resolution types); lifetime 13 days |
| Cosmos 1021 (1978-57A) | 10 Jun 0838 | 65 89 | 173 313 | Manoeuvrable; high resolution; lifetime 13 days |
| Cosmos 1022 (1978-59A) | 12 Jun 1033 | 73 90 | 171 344 | Manoeuvrable; high resolution; lifetime 13 days |
| Cosmos 1026 (1978-69A) | 2 Jul 0936 | 52 89 | 207 247 | Low resolution, third generation; lifetime 4 days |
| Cosmos 1028 (1978-76A) | 5 Aug 1453 | 67 89 | 170 248 | Manoeuvrable fourth generation long-lived satellite, high resolution; capsule returned; lifetime 30 days |
| Cosmos 1029 (1978-82A) | 29 Aug 1507 | 63 90 | 179 330 | High resolution; a 20 kg 0.6 m piece of the spacecraft's engine landed near Garnat-sur-Engièvre (Allier), France; lifetime 10 days |
| Cosmos 1031 (1978-85A) | 9 Sep 1507 | 63 90 | 182 329 | High resolution, manoeuvrable satellite; lifetime 13 days |
| Cosmos 1032 (1978-88A) | 19 Sep 0810 | 81 89 | 215 215 | Low resolution, third generation; ice survey satellite; lifetime 13 days |
| Cosmos 1033 (1978-89A) | 3 Oct 1102 | 81 89 | 212 231 | Low resolution; second ice survey satellite in two weeks; Earth resources observations also; lifetime 13 days |
| Cosmos 1042 (1978-92A) | 8 Oct 1536 | 63 89 | 179 299 | High resolution, manoeuvrable satellite; lifetime 13 days |

| Country, satellite name and designation | Launch date and time (GMT) | Orbital inclination (deg) and period (min) | Perigee and apogee heights (km) | Comments[a] |
|---|---|---|---|---|
| Cosmos 1044<br>(1978-97A) | 17 Oct<br>1507 | 63<br>90 | 203<br>295 | Low resolution; third generation satellite; lifetime 13 days |
| Cosmos 1046<br>(1978-102A) | 1 Nov<br>1200 | 73<br>90 | 202<br>324 | Low resolution; special subset; lifetime 12 days |
| Cosmos 1047<br>(1978-104A) | 15 Nov<br>1146 | 73<br>90 | 171<br>354 | High resolution, manoeuvrable satellite; lifetime 13 days |
| Cosmos 1049<br>(1978-107A) | 21 Nov<br>1200 | 73<br>90 | 169<br>338 | High resolution; long delay in locating return capsule; lifetime 13 days |
| Cosmos 1050<br>(1978-108A) | 28 Nov<br>1619 | 63<br>90 | 254<br>278 | High resolution, manoeuvrable satellite; lifetime 14 days |
| Cosmos 1059<br>(1978-110A) | 7 Dec<br>1536 | 63<br>90 | 180<br>338 | High resolution, manoeuvrable satellite; lifetime 13 days |
| Cosmos 1060<br>(1978-111A) | 8 Dec<br>0936 | 65<br>90 | 206<br>292 | Low resolution, third generation; lifetime 13 days |
| Cosmos 1061<br>(1978-114A) | 14 Dec<br>1522 | 63<br>90 | 203<br>310 | Low resolution, third generation; lifetime 13 days |
| Cosmos 1068<br>(1978-123A) | 26 Dec<br>1536 | 63<br>90 | 177<br>391 | High resolution, manoeuvrable satellite; lifetime 13 days |
| Cosmos 1069<br>(1978-124A) | 28 Dec<br>1634 | 63<br>90 | 241<br>285 | Special subset, low resolution; lifetime 13 days |
| *1979* | | | | |
| Cosmos 1070<br>(1979-01A) | 11 Jan<br>1507 | 63<br>89 | 205<br>293 | Lifetime 9 days; low resolution |
| Cosmos 1071<br>(1979-02A) | 13 Jan<br>1336 | 63<br>90 | 179<br>159 | Lifetime 13 days, high resolution |
| Cosmos 1073<br>(1979-06A) | 30 Jan<br>1522 | 63<br>90 | 182<br>328 | Lifetime 13 days; high resolution; manoeuvrable |
| Cosmos 1078<br>(1979-16A) | 22 Feb<br>1214 | 73<br>89 | 168<br>250 | Lifetime 8 days; high resolution; manoeuvrable |
| Cosmos 1079<br>(1979-19A) | 27 Feb<br>1507 | 67<br>90 | 174<br>337 | Lifetime 12 days; fourth generation; high resolution; satellite may have malfunctioned |
| Cosmos 1080<br>(1979-23A) | 14 Mar<br>1048 | 73<br>89 | 169<br>294 | Lifetime 14 days; high resolution; manoeuvrable; TF recovery beacon |
| Cosmos 1090<br>(1979-27A) | 31 Mar<br>1048 | 73<br>90 | 202<br>326 | Lifetime 13 days; high resolution; TG recovery beacon |
| Cosmos 1095<br>(1979-34A) | 20 Apr<br>1131 | 73<br>90 | 199<br>379 | Lifetime 14 days; manoeuvrable; TF recovery beacon |
| Cosmos 1097<br>(1979-37A) | 23 Apr<br>1717 | 63<br>89 | 173<br>331 | Lifetime 30 days; fourth generation |
| Cosmos 1098<br>(1979-40A) | 15 May<br>1146 | 73<br>90 | 170<br>354 | Lifetime 13 days; high resolution; manoeuvrable; TF recovery beacon |
| Cosmos 1099<br>(1979-41A) | 17 May<br>0712 | 81<br>89 | 215<br>247 | Lifetime 13 days; high resolution; manoeuvrable; also Earth resources mission |
| Cosmos 1102<br>(1979-43A) | 25 May<br>0658 | 81<br>89 | 212<br>260 | Lifetime 13 days; high resolution; also Earth resources mission |

| Country, satellite name and designation | Launch date and time (GMT) | Orbital inclination (deg) and period (min) | Perigee and apogee heights (km) | Comments[a] |
|---|---|---|---|---|
| Cosmos 1103 (1979-45A) | 31 May 1634 | 63 91 | 257 375 | Lifetime 14 days; medium resolution; TF recovery beacon |
| Cosmos 1105 (1979-52A) | 8 Jun 0712 | 81 89 | 212 254 | Lifetime 13 days; high resolution; manoeuvrable; TK recovery beacon; also Earth resources mission |
| Cosmos 1106 (1979-54A) | 12 Jun 0658 | 81 89 | 216 237 | Lifetime 13 days; low resolution; TG recovery beacon; also Earth resources mission |
| Cosmos 1107 (1979-55A) | 15 Jun 1102 | 73 90 | 198 301 | Lifetime 14 days; high resolution; manoeuvrable; TF recovery beacon |
| Cosmos 1108 (1979-56A) | 22 Jun 0658 | 81 89 | 214 245 | Lifetime 13 days; high resolution; TK recovery beacon; series of Earth resources satellites; data transmitted to Priroda State Research and Production Centre |
| Cosmos 1111 (1979-61A) | 29 Jun 1605 | 63 91 | 255 328 | Lifetime 15 days; medium resolution; TF recovery beacon |
| Cosmos 1113 (1979-64A) | 10 Jul 0907 | 65 90 | 173 330 | Lifetime 13 days; high resolution; manoeuvrable; TF recovery beacon |
| Cosmos 1115 (1979-66A) | 13 Jul 0824 | 81 89 | 217 235 | Lifetime 13 days; high resolution; manoeuvrable; 2000th satellite; TK recovery beacon; Earth resources mission |
| Cosmos 1117 (1979-68A) | 25 Jul 1522 | 63 90 | 179 325 | Lifetime 13 days; high resolution; manoeuvrable; TF recovery beacon |
| Cosmos 1118 (1979-69A) | 27 Jul 0712 | 81 89 | 217 243 | Lifetime 13 days; low resolution |
| Cosmos 1119 (1979-71A) | 3 Aug 1048 | 81 89 | 216 242 | Lifetime 12 days; low resolution; TL recovery beacon; special subset |
| Cosmos 1120 (1979-73A) | 11 Aug 0922 | 71 90 | 170 362 | Lifetime 13 days; high resolution; fourth generation; manoeuvrable; TF recovery beacon |
| Cosmos 1121 (1979-74A) | 14 Aug 1536 | 67 90 | 171 348 | Lifetime 30 days; high resolution; manoeuvrable; fourth generation |
| Cosmos 1122 (1979-75A) | 17 Aug 0741 | 81 89 | 214 239 | Lifetime 13 days; low resolution; Earth resources mission |
| Cosmos 1123 (1979-76A) | 21 Aug 1117 | 81 89 | 212 234 | Lifetime 13 days; high resolution; TK recovery beacon; also Earth resources mission |
| Cosmos 1126 (1979-79A) | 31 Aug 1117 | 73 90 | 197 395 | Lifetime 14 days; medium resolution; TF recovery beacon |
| Cosmos 1127 (1979-80A) | 5 Sep 1019 | 81 89 | 215 272 | Lifetime 13 days; High resolution; manoeuvrable; TF recovery beacon; Earth resources mission also |
| Cosmos 1128 (1979-81A) | 28 Sep 1536 | 73 89 | 199 328 | Lifetime 14 days; medium resolution; TF recovery beacon |
| Cosmos 1139 (1979-88A) | 5 Oct 1131 | 73 90 | 202 327 | Lifetime 13 days; low resolution; TL recovery beacon; special subset |

| Country, satellite name and designation | Launch date and time (GMT) | Orbital inclination (deg) and period (min) | Perigee and apogee heights (km) | Comments[a] |
|---|---|---|---|---|
| Cosmos 1142 (1979-92A) | 22 Oct 1243 | 73 90 | 198 382 | Lifetime 13 days; medium resolution; TF recovery beacon |
| Cosmos 1144 (1979-97A) | 2 Nov 1605 | 67 89 | 158 337 | Lifetime 32 days; high resolution; manoeuvrable; fourth generation; TF recovery beacon |
| Cosmos 1147 (1979-102A) | 12 Dec 1229 | 73 90 | 196 382 | Lifetime 14 days; medium resolution |
| Cosmos 1148 (1979-106A) | 29 Dec 0419 | 67 90 | 173 343 | Lifetime 13 days; high resolution; manoeuvrable |
| *1980* | | | | |
| Cosmos 1149 (1980-01A) | 9 Jan 1214 | 73 90 | 118 392 | Lifetime 14 days; third generation; high resolution; manoeuvrable; TF recovery beacon |
| Cosmos 1152 (1980-06A) | 24 Jan 1550 | 67 90 | 173 345 | Lifetime 13 days; fourth generation; high resolution |
| Cosmos 1155 (1980-09A) | 7 Feb 1102 | 73 90 | 195 397 | Lifetime 14 days; third generation; high resolution; manoeuvrable |
| Cosmos 1165 (1980-17A) | 21 Feb 1200 | 73 90 | 170 350 | Lifetime 13 days; third generation; TF recovery beacon |
| Cosmos 1166 (1980-20A) | 4 Mar 1034 | 73 90 | 198 382 | Lifetime 14 days; third generation; high resolution; manoeuvrable; TF recovery beacon |
| Cosmos 1170 (1980-25A) | 1 Apr 0810 | 70 90 | 174 366 | Lifetime 11 days; third generation |
| Cosmos 1173 (1980-29A) | 17 Apr 0838 | 70 90 | 166 354 | Lifetime 11 days; third generation; high resolution; manoeuvrable; TF recovery beacon |
| Cosmos 1177 (1980-35A) | 29 Apr 1312 | 67 90 | 174 346 | Lifetime 44 days; fourth generation; high resolution; manoeuvrable |
| Cosmos 1178 (1980-36A) | 7 May 1258 | 73 90 | 200 386 | Lifetime 15 days; third generation; high resolution; manoeuvrable |
| Cosmos 1180 (1980-38A) | 15 May 0546 | 63 90 | 238 291 | Lifetime 12 days; third generation; TF recovery beacon |
| Cosmos 1182 (1980-40A) | 23 May 0712 | 82 89 | 211 251 | Lifetime 13 days; third generation |
| Cosmos 1183 (1980-42A) | 28 May 1200 | 73 90 | 201 389 | Lifetime 14 days; third generation; high resolution; manoeuvrable; TF recovery beacon |
| Cosmos 1185 (1980-46A) | 6 Jun 0658 | 82 90 | 214 282 | Lifetime 14 days; third generation; high resolution; manoeuvrable; TF recovery beacon |
| Cosmos 1187 (1980-48A) | 12 Jun 1229 | 73 90 | 199 307 | Lifetime 14 days; third generation; high resolution; manoeuvrable |
| Cosmos 1189 (1980-54A) | 26 Jun 1229 | 73 90 | 198 305 | Lifetime 14 days; third generation; TF recovery beacon |
| Cosmos 1200 (1980-59A) | 9 Jul 1243 | 73 90 | 198 307 | Lifetime 14 days; third generation; high resolution; manoeuvrable; TF recovery beacon |

| Country, satellite name and designation | Launch date and time (GMT) | Orbital inclination (deg) and period (min) | Perigee and apogee heights (km) | Comments[a] |
| --- | --- | --- | --- | --- |
| Cosmos 1201 (1980-61A) | 15 Jul 0735 | 82 89 | 213 247 | Lifetime 13 days; third generation; TK recovery beacon; also Earth resources mission |
| Cosmos 1202 (1980-65A) | 24 Jul 1243 | 73 90 | 198 307 | Lifetime 14 days; third generation; TF recovery beacon |
| Cosmos 1203 (1980-66A) | 31 Jul 0755 | 82 90 | 213 270 | Lifetime 14 days; third generation; high resolution; manoeuvrable; TF recovery beacon |
| Cosmos 1205 (1980-68A) | 12 Aug 1200 | 73 90 | 199 306 | Lifetime 14 days; third generation; high resolution; manoeuvrable; TF recovery beacon |
| Cosmos 1207 (1980-70A) | 22 Aug 1005 | 82 89 | 211 256 | Lifetime 13 days; third generation; TK recovery beacon |
| Cosmos 1208 (1980-71A) | 26 Aug 1536 | 67 90 | 173 339 | Lifetime 29 days; fourth generation; high resolution |
| Cosmos 1209 (1980-72A) | 3 Sep 1019 | 82 89 | 211 280 | Lifetime 14 days; third generation; high resolution; manoeuvrable; TF recovery beacon |
| Cosmos 1210 (1980-76A) | 19 Sep 1019 | 82 89 | 180 244 | Lifetime 14 days; third generation; high resolution; manoeuvrable; TF recovery beacon |
| Cosmos 1211 (1980-77A) | 23 Sep 1033 | 82 89 | 216 242 | Lifetime 11 days; third generation; TL recovery beacon |
| Cosmos 1212 (1980-78A) | 26 Sep 1019 | 82 89 | 209 249 | Lifetime 13 days; third generation; TK recovery beacon; also Earth resources mission |
| Cosmos 1213 (1980-80A) | 3 Oct 1200 | 73 90 | 229 289 | Lifetime 14 days; third generation; high resolution; manoeuvrable; TF recovery beacon |
| Cosmos 1214 (1980-82A) | 10 Oct 1312 | 67 90 | 174 345 | Lifetime 13 days; third generation; |
| Cosmos 1216 (1980-84A) | 16 Oct 1229 | 73 90 | 198 379 | Lifetime 14 days; third generation; high resolution; manoeuvrable; TF recovery beacon |
| Cosmos 1218 (1980-86A) | 30 Oct 1005 | 65 90 | 171 353 | Lifetime 43 days; fourth generation; manoeuvrable; high resolution |
| Cosmos 1219 (1980-88A) | 31 Oct 1200 | 73 90 | 228 291 | Lifetime 13 days; third generation; TF recovery beacon |
| Cosmos 1221 (1980-90A) | 12 Nov 1229 | 73 91 | 196 399 | Lifetime 14 days; third generation; high resolution; manoeuvrable; TF recovery beacon |
| Cosmos 1224 (1980-96A) | 1 Dec 1214 | 73 90 | 198 378 | Lifetime 14 days; third generation; high resolution; manoeuvrable; TF recovery beacon |
| Cosmos 1227 (1980-101A) | 16 Dec 1214 | 73 90 | 109 300 | Lifetime 12 days; third generation; TF recovery beacon |
| Cosmos 1236 (1980-105A) | 26 Dec 1912 | 67 90 | 169 363 | Lifetime 26 days; third generation; |

| Country, satellite name and designation | Launch date and time (GMT) | Orbital inclination (deg) and period (min) | Perigee and apogee heights (km) | Comments[a] |
|---|---|---|---|---|
| *1981* | | | | |
| Cosmos 1237<br>(1981-01A) | 6 Jan<br>1214 | 73<br>90 | 195<br>387 | Lifetime 14 days; subsequently orbited at high perigee |
| Cosmos 1239<br>(1981-04A) | 16 Jan<br>1200 | 82<br>89 | 216<br>234 | Lifetime 12 days; geodetic and mapping; non-manoeuvrable |
| Cosmos 1240<br>(1981-05A) | 20 Jan<br>1102 | 65<br>90 | 171<br>357 | Lifetime 28 days; fourth generation; high resolution |
| Cosmos 1245<br>(1981-14A) | 13 Feb<br>1117 | 73<br>92 | 356<br>416 | Lifetime 14 days; high perigee |
| Cosmos 1246<br>(1981-15A) | 18 Feb<br>0907 | 65<br>90 | 198<br>272 | Lifetime 23 days; fourth generation; high resolution |
| Cosmos 1248<br>(1981-20A) | 5 Mar<br>1507 | 67<br>90 | 173<br>345 | Lifetime 30 days; fourth generation; high resolution |
| Cosmos 1259<br>(1981-26A) | 17 Mar<br>0836 | 70<br>90 | 208<br>383 | Lifetime 14 days; high perigee; TF recovery beacon |
| Cosmos 1262<br>(1981-32A) | 7 Apr<br>1048 | 73<br>90 | 197<br>393 | Lifetime 14 days; high resolution; manoeuvrable |
| Cosmos 1264<br>(1981-35A) | 15 Apr<br>1033 | 70<br>91 | 208<br>388 | Lifetime 14 days; subsequently orbited at high perigee; TF recovery beacon |
| Cosmos 1265<br>(1981-36A) | 16 Apr<br>1131 | 73<br>90 | 226<br>288 | Lifetime 12 days; high resolution; manoeuvrable; TF recovery beacon |
| Cosmos 1268<br>(1981-40A) | 28 Apr<br>0907 | 70<br>90 | 210<br>358 | Lifetime 14 days; high resolution; manoeuvrable; TF recovery beacon |
| Cosmos 1270<br>(1981-45A) | 18 May<br>1200 | 65<br>90 | 173<br>349 | Lifetime 30 days; fourth generation; high resolution |
| Cosmos 1272<br>(1981-47A) | 21 May<br>0922 | 70<br>92 | 362<br>417 | Lifetime 14 days; high perigee |
| Cosmos 1273<br>(1981-48A) | 22 May<br>0712 | 82<br>89 | 210<br>264 | Lifetime 13 days; high resolution; Earth resources; data received by Priroda Nature Station |
| Cosmos 1274<br>(1981-52A) | 3 Jun<br>1410 | 67<br>90 | 172<br>355 | Lifetime 30 days; fourth generation; high resolution |
| Cosmos 1276<br>(1981-55A) | 16 Jun<br>0658 | 82<br>89 | 216<br>239 | Lifetime 13 days; Earth resources; TF recovery beacon |
| Cosmos 1277<br>(1981-56A) | 17 Jun<br>0936 | 70<br>90 | 208<br>379 | Lifetime 14 days; subsequently orbited at high perigee |
| Cosmos 1279<br>(1981-62A) | 1 Jul<br>0936 | 70<br>90 | 212<br>363 | Lifetime 14 days; high resolution; manoeuvrable; TF recovery beacon |
| Cosmos 1280<br>(1981-63A) | 2 Jul<br>0712 | 82<br>90 | 211<br>286 | Lifetime 14 days; high resolution; Earth resources; data received by Priroda Nature Station |
| Cosmos 1281<br>(1981-64A) | 7 Jul<br>1229 | 73<br>90 | 197<br>394 | Lifetime 14 days; subsequently orbited at high perigee; TF recovery beacon |
| Cosmos 1282<br>(1981-66A) | 15 Jul<br>1312 | 65<br>90 | 173<br>337 | Lifetime 30 days; fourth generation; high resolution |
| Cosmos 1283<br>(1981-67A) | 17 Jul<br>0810 | 82<br>89 | 182<br>250 | Lifetime 14 days; subsequently orbited at high perigee; Earth resources; TF recovery beacon |

| Country, satellite name and designation | Launch date and time (GMT) | Orbital inclination (deg) and period (min) | Perigee and apogee heights (km) | Comments[a] |
|---|---|---|---|---|
| Cosmos 1284 (1981-68A) | 29 Jul 1326 | 82 89 | 183 241 | Lifetime 14 days; subsequently orbited at high perigee; earth resources; TF recovery beacon |
| Cosmos 1296 (1981-78A) | 13 Aug 1919 | 67 90 | 172 354 | Lifetime 31 days; fourth generation; high resolution |
| Cosmos 1297 (1981-79A) | 18 Aug 0936 | 73 90 | 199 364 | Lifetime 12 days; high resolution; manoeuvrable |
| Cosmos 1298 (1981-80A) | 21 Aug 1019 | 65 90 | 174 331 | Lifetime 42 days; fourth generation; high resolution |
| Cosmos 1301 (1981-83A) | 27 Aug 1033 | 82 89 | 213 272 | Lifetime 14 days; high resolution; Earth resources; TF recovery beacon |
| Cosmos 1303 (1981-86A) | 4 Sep 0810 | 70 90 | 208 376 | Lifetime 14 days; subsequently orbited at high perigee; TF recovery beacon |
| Cosmos 1307 (1981-90A) | 15 Sep 1131 | 73 90 | 198 394 | Lifetime 14 days; subsequently orbited at high perigee; TF recovery beacon |
| Cosmos 1309 (1981-92A) | 18 Sep 0936 | 82 89 | 212 257 | Lifetime 13 days; geodetic and mapping; non-manoeuvrable; TL recovery beacon |
| Cosmos 1313 (1981-99A) | 1 Oct 0907 | 70 91 | 206 291 | Lifetime 14 days; high resolution; manoeuvrable; TF recovery beacon |
| Cosmos 1314 (1981-101A) | 9 Oct 1048 | 82 89 | 214 237 | Lifetime 13 days; high resolution; Earth resources; data received by Priroda Nature Station; TK recovery beacon |
| Cosmos 1316 (1981-104A) | 15 Oct 0922 | 70 91 | 209 385 | Lifetime 14 days; high resolution; TF recovery beacon |
| Cosmos 1318 (1981-109A) | 3 Nov 1312 | 67 90 | 172 353 | Lifetime 31 days; fourth generation; high resolution; |
| Cosmos 1319 (1981-112A) | 13 Nov 0936 | 70 90 | 209 377 | Lifetime 14 days; high perigee; TF recovery beacon |
| Cosmos 1329 (1981-118A) | 4 Dec 0950 | 65 90 | 232 264 | Lifetime 14 days; high resolution; TF recovery beacon |
| Cosmos 1330 (1981-121A) | 19 Dec 1200 | 70 90 | 168 379 | Lifetime 31 days; fourth generation; |
| **People's Republic of China** | | | | |
| *1978* | | | | |
| China 8 (1978-11A) | 25 Jan 0502 | 57.03 91 | 161 479 | Lifetime 12 days; capsule was recovered |

[a] Recovery beacon data supplied by the Kettering Group
[b] PDM = Pulse Duration Moderation

# Appendix 1B

## *Electronic reconnaissance satellites launched between 1977 and 1981*

| Country, satellite name and designation | Launch date and time (GMT) | Orbital inclination (deg) and period (min) | Perigee and apogee heights (km) | Comments |
|---|---|---|---|---|
| **USA** | | | | |
| *1978* | | | | |
| USAF<br>(1978-29B) | 16 Mar<br>1843 | 96<br>98 | 639<br>645 | Launched from Big Bird satellite (1978-29A); lifetime 60 years |
| *1979* | | | | |
| USAF<br>(1979-25B) | 16 Mar<br>1829 | 96<br>97 | 621<br>628 | Launched from Big Bird satellite (1979-25A); lifetime 60 years |
| *1980* | | | | |
| USAF<br>(1980-52C) | 18 Jun<br>1829 | 97<br>112 | 1 331<br>1 333 | Launched from Big Bird satellite (1980-52A); lifetime 5000 years |
| **USSR** | | | | |
| *1977* | | | | |
| Cosmos 895<br>(1977-15A) | 26 Feb<br>2122 | 81<br>97 | 611<br>635 | Lifetime 60 years |
| Cosmos 899<br>(1977-22A) | 24 Mar<br>2219 | 74<br>95 | 503<br>547 | Lifetime 104 years |
| Cosmos 901<br>(1977-25A) | 5 Apr<br>1033 | 71<br>96 | 269<br>820 | Lifetime 17 months |
| Cosmos 919<br>(1977-51A) | 18 Jun<br>1033 | 71<br>96 | 269<br>822 | Lifetime 7 months |
| Cosmos 924<br>(1977-60A) | 4 Jul<br>2219 | 74<br>95 | 513<br>550 | Lifetime 10 years |
| Cosmos 925<br>(1977-61A) | 7 Jul<br>0726 | 81<br>97 | 609<br>934 | Lifetime 60 years |

| Country, satellite name and designation | Launch date and time (GMT) | Orbital inclination (deg) and period (min) | Perigee and apogee heights (km) | Comments |
|---|---|---|---|---|
| Cosmos 955 (1977-91A) | 20 Sep 0112 | 81 98 | 630 641 | Lifetime 60 years |
| Cosmos 960 (1977-103A) | 25 Oct 0531 | 74 95 | 502 546 | Lifetime 10 years |
| *1978* | | | | |
| Cosmos 975 (1978-04A) | 10 Jun 1326 | 81 98 | 634 653 | Lifetime 60 years |
| Cosmos 1005 (1978-45A) | 12 May 0405 | 81 98 | 627 653 | Lifetime 60 years |
| Cosmos 1008 (1978-49A) | 17 May 1438 | 74 95 | 499 594 | Lifetime 10 years |
| Cosmos 1043 (1978-94A) | 10 Oct 1938 | 81 97 | 622 635 | Lifetime 60 years |
| Cosmos 1062 (1978-115A) | 15 Dec 1326 | 74 95 | 504 550 | Lifetime 9 years |
| Cosmos 1063 (1978-117A) | 19 Dec 0141 | 81 97 | 631 634 | Lifetime 60 years |
| *1979* | | | | |
| Cosmos 1077 (1979-12A) | 13 Feb 2150 | 81 97 | 625 631 | Lifetime 60 years |
| Cosmos 1093 (1979-32A) | 14 Apr 0531 | 81 97 | 621 635 | Lifetime 60 years |
| Cosmos 1114 (1979-65A) | 11 Jul 1550 | 74 95 | 506 552 | Lifetime 9 years |
| Cosmos 1116 (1979-67A) | 20 Jul 1200 | 81 97 | 590 643 | Lifetime 60 years |
| Cosmos 1143 (1979-93A) | 26 Oct 1814 | 81 97 | 624 646 | Lifetime 60 years |
| Cosmos 1145 (1979-99A) | 27 Nov 1005 | 81 97 | 624 635 | Lifetime 60 years |
| *1980* | | | | |
| Cosmos 1154 (1980-08A) | 30 Jun 1258 | 81 97 | 630 644 | Lifetime 60 years |
| Cosmos 1184 (1980-44A) | 4 Jun 0735 | 81 97 | 623 647 | Lifetime 60 years |
| Cosmos 1186 (1980-47A) | 6 Jun 1102 | 74 95 | 473 519 | Lifetime 18 months; lower orbital period than usual; does not fit the orbital plane spacing of the main group |
| Cosmos 1206 (1980-69A) | 15 Aug 0531 | 81 97 | 630 634 | Lifetime 60 years |
| Cosmos 1215 (1980-83A) | 14 Oct 2038 | 74 95 | 498 550 | Lifetime 10 years |
| Cosmos 1222 (1980-93A) | 21 Nov 1200 | 81 97 | 631 633 | Lifetime 60 years |

| Country, satellite name and designation | Launch date and time (GMT) | Orbital inclination (deg) and period (min) | Perigee and apogee heights (km) | Comments |
|---|---|---|---|---|
| *1981* | | | | |
| Cosmos 1242 (1981-08A) | 27 Jan 1507 | 81 98 | 626 658 | Lifetime 60 years |
| Cosmos 1271 (1981-46A) | 19 May 0350 | 81 98 | 628 650 | Lifetime 60 years |
| Cosmos 1311 (1981-97A) | 28 Sep 2234 | 83 95 | 463 519 | Lifetime 3 years; may be the start of a new system |
| Cosmos 1315 (1981-103A) | 13 Oct 2324 | 81 98 | 627 667 | Lifetime 60 years |

# Appendix 1C

## *Ocean surveillance and oceanographic satellites launched between 1977 and 1981*

| Country, satellite name and designation | Launch date and time (GMT) | Orbital inclination (deg) and period (min) | Perigee and apogee heights (km) | Comments |
|---|---|---|---|---|
| **USA** | | | | |
| *1977* | | | | |
| NOSS-2<br>(1977-112A) | 8 Dec<br>1746 | 63<br>108 | 1 054<br>1 169 | Navy ocean-surveillance satellites; quadruple launch; lifetimes 1600 years |
| NOSS-2<br>(1977/112C) | 8 Dec<br>1746 | 63<br>108 | 1 054<br>1 169 | |
| NOSS-2<br>(1977-112D) | 8 Dec<br>1746 | 63<br>108 | 1 054<br>1 169 | |
| NOSS-2<br>(1977-112E) | 8 Dec<br>1746 | 63<br>108 | 1 055<br>1 168 | |
| *1978* | | | | |
| NASA/Seasat 1<br>(1978-64A) | 27 Jun<br>0112 | 108<br>101 | 776<br>800 | Lifetime 200 years |
| *1980* | | | | |
| NOSS-3<br>(1980-19A) | 3 Mar<br>1117 | 63<br>107 | 1 035<br>1 150 | Navy ocean-surveillance satellites; quadruple launch; sub-satellites are SSU satellites; lifetimes 1600 years |
| NOSS-3<br>(1980-19C) | 3 Mar<br>1117 | 63<br>107 | 1 048<br>1 166 | |
| NOSS-3<br>(1980-19D) | 3 Mar<br>1117 | 63<br>107 | 1 048<br>1 166 | |
| NOSS-3<br>(1980-19E) | 3 Mar<br>1117 | 63<br>107 | 1 048<br>1 166 | |

| Country, satellite name and designation | Launch date and time (GMT) | Orbital inclination (deg) and period (min) | Perigee and apogee heights (km) | Comments |
|---|---|---|---|---|
| **USSR** | | | | |
| *1977* | | | | |
| Cosmos 937<br>(1977-77A) | 24 Aug<br>0712 | 65<br>92 | 149<br>597 | Lifetime 421 days |
| Cosmos 952<br>(1977-88A) | 16 Sep<br>1424 | 65<br>90 | 251<br>265 | Lifetime 600 years after satellite moved into higher orbit |
| Cosmos 954<br>(1977-90A) | 18 Sep<br>1355 | 65<br>90 | 251<br>265 | Lifetime 128 days |
| *1979* | | | | |
| Cosmos 1076<br>(1979-11A) | 12 Feb<br>1258 | 83<br>98 | 637<br>666 | Oceanographic satellite; lifetime 60 years; carries two telescopes |
| Cosmos 1094<br>(1979-33A) | 18 Apr<br>1200 | 65<br>93 | 426<br>442 | Orbit 200 km higher than that of Cosmos 954 which carried a nuclear reactor and fell over Canada; lifetime 2 years |
| Cosmos 1096<br>(1979-36A) | 25 Apr<br>1005 | 65<br>93 | 428<br>442 | EORSAT satellite placed into same orbit as Cosmos 1094 but 23 minutes ahead of Cosmos 1094; three of the seven EORSAT exploded in orbit; lifetime 2 years |
| *1980* | | | | |
| Cosmos 1151<br>(1980-05A) | 23 Jan<br>0658 | 83<br>98 | 637<br>666 | Oceanographic satellite; lifetime 60 years |
| Cosmos 1167<br>(1980-21A) | 14 Mar<br>1048 | 65<br>93 | 426<br>442 | Equipped with ion-thrusters to maintain a near circular orbit; thought performing naval electronic reconnaissance |
| Cosmos 1176<br>(1980-34A) | 29 Apr<br>1005 | 65<br>90 | 250<br>266 | Jettisoned rocket and platform; manoeuvred into circular orbit with a perigee of 870 km |
| Cosmos 1220<br>(1980-89A) | 4 Nov<br>1507 | 65<br>93 | 427<br>440 | Similar satellite to Cosmos 1167 above; these two are working as a pair |
| *1981* | | | | |
| Cosmos 1249<br>(1981-21A) | 5 Mar<br>1814 | 65<br>90 | 252<br>265 | Nuclear-powered radar; manoeuvred into higher orbit on 1 Jul (65.0°, 103.89 minutes, 898 km × 985 km) |
| Cosmos 1260<br>(1981-28A) | 20 Mar<br>2346? | 65<br>93 | 425<br>444 | Passive satellite with ion thruster |
| Cosmos 1266<br>(1981-37A) | 21 Apr<br>0350 | 65<br>90 | 249<br>268 | Nuclear-powered radar; manoeuvred into higher orbit on 1 May (64.76°, 103.65 minutes, 891 km × 965 km) |
| Cosmos 1286<br>(1981-72A) | 4 Aug<br>0824 | 65<br>93 | 432<br>445 | Passive satellite with ion thruster |
| Cosmos 1299<br>(1981-81A) | 24 Aug<br>1634 | 65<br>90 | 248<br>267 | Nuclear-powered radar; manoeuvred into higher orbit on 5 Sep (65.12°, 104.00 minutes, 910 km × 984 km) |
| Cosmos 1300<br>(1981-82A) | 24 Aug<br>2136 | 83<br>98 | 638<br>666 | Presumed oceanographic satellite; similar to Cosmos 1076 and 1151 |
| Cosmos 1306<br>(1981-89A) | 14 Sep<br>2038 | 65<br>93 | 409<br>462 | Passive satellite with ion thruster |
| Cosmos 1328<br>(1981-117A) | 3 Dec<br>1146 | 83<br>98 | 637<br>665 | Presumed to be oceanographic; similar to Cosmos 1076 and 1151 |

# Appendix 1D

## *Early warning satellites launched between 1977 and 1981*

| Country, satellite name and designation | Launch date and time (GMT) | Orbital inclination (deg) and period (min) | Perigee and apogee heights (km) | Comments |
|---|---|---|---|---|
| **USA** | | | | |
| *1977* | | | | |
| T-3C<br>(1977-07A) | 6 Feb | 0.5<br>1 433 | 35 620<br>35 862 | Satellite equipped with infra-red detector and system to detect interference from other satellites |
| RH-2<br>(1977-38A) | 23 May<br>1424 | 0.2<br>1 435 | 35 679<br>35 855 | RH-1 (1973-13A) was launched on 6 March 1973 |
| RH-3<br>(1977-114A) | 11 Dec<br>1258 | 0.2<br>1 435 | 35 679<br>35 855 | |
| *1978* | | | | |
| RH-4<br>(1978-38A) | 8 Apr<br>0043 | 28<br>616 | 150<br>35 033 | Probably eventually orbited in synchronous orbit similar to RH-3; lifetime > $10^6$ years |
| IMEWS-8<br>(1978-58A) | 10 Jun<br>1912 | 0.5<br>1 433 | 35 620<br>35 860 | Lifetime > $10^6$ |
| *1979* | | | | |
| IMEWS-9<br>(1979-53A) | 10 Jun<br>1341 | 2<br>1 446 | 35 801<br>36 261 | Satellite to monitor Soviet and Chinese missile tests as well as space launches; lifetime > $10^6$ years |
| T-3C<br>(1979-86A) | 10 Oct<br>1131 | 8<br>1 446 | 30 443<br>41 497 | Lifetime > $10^6$ years |
| *1981* | | | | |
| IMEWS-11<br>(1981-25A) | 16 Mar<br>2107 | 2<br>1 421 | 35 463<br>35 527 | Placed in geostationary orbit at longitude 71° E |
| IMEWS-12<br>(1981-107A) | 31 Oct<br>0936 | | | Orbit similar to 1981-25A |

| Country, satellite name and designation | Launch date and time (GMT) | Orbital inclination (deg) and period (min) | Perigee and apogee heights (km) | Comments |
|---|---|---|---|---|
| **USSR** | | | | |
| *1977* | | | | |
| Cosmos 903 (1977-27A) | 11 Apr 0141 | 63 726 | 603 40 153 | Lifetime 10 years |
| Cosmos 917 (1977-47A) | 16 Jun 0210 | 63 719 | 586 39 818 | Lifetime 10 years |
| Cosmos 931 (1977-68A) | 20 Jul 0448 | 63 724 | 604 40 065 | Lifetime 12 years |
| *1978* | | | | |
| Cosmos 1024 (1978-66A) | 28 Jun 0307 | 63 725 | 605 40 094 | Lifetime 12 years |
| Cosmos 1030 (1978-83A) | 6 Sep 0307 | 63 726 | 613 40 129 | Lifetime 12 years |
| *1979* | | | | |
| Cosmos 1109 (1979-58A) | 27 Jun 1814 | 63 724 | 613 40 060 | Lifetime 12 years |
| Cosmos 1124 (1979-77A) | 28 Aug 0000 | 63 727 | 561 40 267 | Lifetime 12 years |
| *1980* | | | | |
| Cosmos 1172 (1980-28A) | 12 Apr 2024 | 63 726 | 608 40 155 | Lifetime 16 years |
| Cosmos 1188 (1980-50A) | 14 Jun 2053 | 63 726 | 609 40 129 | Lifetime 100 years |
| Cosmos 1191 (1980-57A) | 2 Jul 0058 | 63 725 | 605 40 127 | Lifetime 100 years |
| Cosmos 1217 (1980-85A) | 24 Oct 1229 | 63 725 | 596 40 131 | Lifetime 100 years |
| Cosmos 1223 (1980-95A) | 27 Nov 2317 | 63 718 | 605 39 749 | Lifetime 100 years |
| *1981* | | | | |
| Cosmos 1247 (1981-16A) | 19 Feb 1131 | 63 707 | 688 39 232 | Lifetime 25 years |
| Cosmos 1261 (1981-31A) | 31 Mar 0950 | 63 710 | 589 39 406 | Lifetime 25 years |
| Cosmos 1278 (1981-58A) | 19 Jun 1938 | 63 727 | 623 40 213 | Lifetime 25 years |
| Cosmos 1285 (1981-71A) | 4 Aug 0014 | 63 728 | 594 40 250 | Ground tracks never stabilized; did not separate from escape stage? lifetime 25 years |
| Cosmos 1317 (1981-108A) | 31 Oct 2324 | 63 726 | 584 40 163 | Lifetime 25 years |

# Appendix 1E

## *Communications satellites launched between 1977 and 1981*

| Country, satellite name and designation | Launch date and time (GMT) | Orbital inclination (deg) and period (min) | Perigee and apogee heights (km) | Comments |
|---|---|---|---|---|
| **USA** | | | | |
| *1977* | | | | |
| DSCS-7 (1977-34A) | 12 May 1438 | 2 1 427 | 35 438 35 762 | Defense Satellite Communications System; two satellites, equipped with anti-jamming systems; one placed over Atlantic and the other over Pacific; lifetimes > $10^6$ years |
| DSCS-8 (1977-34B) | 12 May 1438 | 2 1 437 | 35 781 35 792 | |
| SDS-2 (1977-34A) | 23 May 0126 | 0.2 1 435 | 35 681 35 855 | Satellite Data System; lifetime > $10^6$ years |
| *1978* | | | | |
| FLTSATCOM-1 (1978-16A) | 9 Feb 2122 | 3 1426 | 35 522 35 666 | Provides communications for shipborne, airborne and ground tactical users, and relays ocean surveillance data; lifetime > $10^6$ years; first of a series of five |
| SDS-3 (1978-21A) | 25 Feb 0448 | 63 704 | 311 39 377 | Approximate orbit; lifetime 10 years |
| DSCS-9 and 10 DSCS 11 | 25 Mar | | | Two satellites failed to orbit; malfunction in T-3C booster rocket; the launcher exploded from ground eight minutes after launching |
| SDS-4 (1978-75A) | 5 Aug 0448 | 63 704 | 380 39 315 | Orbit and launch time unconfirmed; lifetime 10 years |

| Country, satellite name and designation | Launch date and time (GMT) | Orbital inclination (deg) and period (min) | Perigee and apogee heights (km) | Comments |
|---|---|---|---|---|
| DSCS-11 (1978-113A) | 14 Dec 0043 | 3 1 452 | 35 796 36 412 | DSCS-11 is placed over the Eastern Pacific and DSCS-12 over the Western Pacific; the orbits are unconfirmed; lifetimes > $10^6$ years |
| DSCS-12 (1978-113B) | 14 Dec 0043 | 3 1 464 | 36 261 36 413 | |
| *1979* | | | | |
| FLTSATCOM-2 (1979-38A) | 4 May 1858 | 26 635 | 166 35 986 | Provides communications for USAF and USN between aircraft, ships, submarines, ground stations, SAC and NCA for the Atlantic, European and African areas; entered geosynchronous orbit on 6 May 1979; lifetime > $10^6$ years |
| DSCS-13 (1979-98A) | 21 Nov 0210 | 2 1 413 | 35 594 35 789 | Lifetimes > $10^6$ years; stationed at 135°W and 175°E |
| DSCS-14 (1979-98B) | 21 Nov 0210 | 2 1 451 | 35 792 36 357 | |
| *1980* | | | | |
| FLTSATCOM-3 (1980-04A) | 18 Jan 0126 | 2 1 423 | 35 405 35 661 | Third of a series designed to provide communication links between small ground stations in 200–400 MHz band; located at 172°E; lifetime > $10^6$ years |
| FLTSATCOM-4 (1980-87A) | 31 Oct 0350 | 2 1 419 | 34 903 35 991 | Location 172°E; lifetime > $10^6$ years |
| SDS-6 (1980-100A) | 13 Dec 1605 | 64 697 | 250 39 130 | Lifetime 100 years |
| *1981* | | | | |
| SDS-7 (1981-38A) | 24 Apr | | | SDS mission providing data relay functions and supports communications for USAF strategic bombers flying in polar regions; orbit similar to 1980-100A; lifetime 10 years |
| FLTSATCOM-5 (1981-73A) | 6 Aug 0810 | 6 1 558 | 35 102 41 185 | Last in the series, intended as an in-orbit spare for the other four now in operation; lifetime > $10^6$ years |
| **USSR** | | | | |
| *1977* | | | | |
| Molniya 2-17 (1977-10A) | 11 Feb 1507 | 63 735 | 464 40 756 | Lifetime 10 years |
| Molniya 1-36 (1977-21A) | 24 Mar 0014 | 64 718 | 465 39 879 | Lifetime 10 years |

| Country, satellite name and designation | Launch date and time (GMT) | Orbital inclination (deg) and period (min) | Perigee and apogee heights (km) | Comments | |
|---|---|---|---|---|---|
| Molniya 3-7<br>(1977-32A) | 28 Apr<br>0922 | 63<br>736 | 436<br>40 817 | Lifetime 12 years | |
| Molniya 1-37<br>(1977-54A) | 24 Jun<br>0546 | 63<br>700 | 447<br>39 011 | Lifetime 12 years | |
| Raduga 3<br>(1977-61A) | 23 Jul<br>2122 | 0.2<br>1 436 | 35 730<br>42 170 | Lifetime $10^6$ years; in Statsionar 2 position | |
| Cosmos 939<br>(1977-79A) | 24 Aug<br>1814 | 74<br>115 | 1 435<br>1 464 | Lifetime 8 000 years | Octuple launch |
| Cosmos 940<br>(1977-79B) | 24 Aug<br>1814 | 74<br>115 | 1 397<br>1 464 | Lifetime 6 000 years | |
| Cosmos 941<br>(1977-79C) | 24 Aug<br>1814 | 74<br>115 | 1 416<br>1 464 | Lifetime 7 000 years | |
| Cosmos 942<br>(1977-79D) | 24 Aug<br>1814 | 74<br>116 | 1 464<br>1 535 | Lifetime 10 000 years | |
| Cosmos 943<br>(1977-79E) | 24 Aug<br>1814 | 74<br>115 | 1 453<br>1 464 | Lifetime 9 000 years | |
| Cosmos 944<br>(1977-79F) | 24 Aug<br>1814 | 74<br>115 | 1 464<br>1 473 | Lifetime 9 000 years | |
| Cosmos 945<br>(1977-79G) | 24 Aug<br>1814 | 74<br>116 | 1 464<br>1 493 | Lifetime 10 000 years | |
| Cosmos 946<br>(1977-79H) | 24 Aug<br>1814 | 74<br>116 | 1 464<br>1 512 | Lifetime 10 000 years | |
| Molniya 1-38<br>(1977-82A) | 30 Aug<br>1814 | 63<br>736 | 445<br>40 785 | Lifetime 10 years | |
| Ekran 2<br>(1977-92A) | 20 Sep<br>1731 | 0.04<br>1 427 | 35 580<br>35 622 | Lifetime $10^6$ years; in Statsionar T position | |
| Molniya 3-8<br>(1977-105A) | 28 Oct<br>0141 | 63<br>736 | 420<br>40 842 | Lifetime 12 years | |
| *1978* | | | | | |
| Cosmos 976<br>(1978-05A) | 10 Jan<br>2053 | 74<br>115 | 1 457<br>1 465 | Lifetime 9 000 years | Octuple launch |
| Cosmos 977<br>(1978-05B) | 10 Jan<br>2053 | 74<br>115 | 1 403<br>1 465 | Lifetime 7 000 years | |
| Cosmos 978<br>(1978-05C) | 10 Jan<br>2053 | 74<br>115 | 1 421<br>1 465 | Lifetime 8 000 years | |
| Cosmos 979<br>(1978-05D) | 10 Jan<br>2053 | 74<br>115 | 1 440<br>1 465 | Lifetime 9 000 years | |
| Cosmos 980<br>(1978-05E) | 10 Jan<br>2053 | 74<br>115 | 1 465<br>1 478 | Lifetime 10 000 years | |
| Cosmos 981<br>(1978-05F) | 10 Jan<br>2053 | 74<br>116 | 1 465<br>1 478 | Lifetime 10 000 years | |
| Cosmos 982<br>(1978-05G) | 10 Jan<br>2053 | 74<br>116 | 1 465<br>1 518 | Lifetime 1 000 years | |
| Cosmos 983<br>(1978-05H) | 10 Jan<br>2053 | 74<br>116 | 1 465<br>1 540 | Lifetime 10 000 years | |
| Molniya 3-9<br>(1978-09A) | 24 Jan<br>0658 | 63<br>736 | 646<br>40 618 | Lifetime 12 years; replaced Molniya 3-3 | |

| Country, satellite name and designation | Launch date and time (GMT) | Orbital inclination (deg) and period (min) | Perigee and apogee heights (km) | Comments | |
|---|---|---|---|---|---|
| Cosmos 990<br>(1978-19A) | 17 Feb<br>1634 | 74<br>101 | 783<br>809 | Store-dump communications satellite; lifetime 120 years | |
| Molniya 1-39<br>(1978-24A) | 2 Mar<br>2248 | 63<br>738 | 617<br>40 739 | Replaces Molniya 1-30; lifetime 12 years | |
| Molniya 1-40<br>(1978-55A) | 2 Jun<br>1214 | 63<br>736 | 422<br>40 842 | Replaces Molniya 1-34; lifetime 12 years | |
| Cosmos 1013<br>(1978-56A) | 7 Jun<br>2150 | 74<br>116 | 1 480<br>1 557 | Lifetime 10 000 years | Octuple launch |
| Cosmos 1014<br>(1978-56B) | 7 Jun<br>2150 | 74<br>116 | 1 480<br>1 534 | Lifetime 10 000 years | |
| Cosmos 1015<br>(1978-56C) | 7 Jun<br>2150 | 74<br>116 | 1 475<br>1 519 | Lifetime 10 000 years | |
| Cosmos 1016<br>(1978-56D) | 7 Jun<br>2150 | 74<br>116 | 1 473<br>1 501 | Lifetime 10 000 years | |
| Cosmos 1017<br>(1978-56E) | 7 Jun<br>2150 | 74<br>116 | 1 460<br>1 495 | Lifetime 9 000 years | |
| Cosmos 1018<br>(1978-56F) | 7 Jun<br>2150 | 74<br>115 | 1 444<br>1 491 | Lifetime 9 000 years | |
| Cosmos 1019<br>(1978-56G) | 7 Jun<br>2150 | 74<br>115 | 1 425<br>1 491 | Lifetime 8 000 years | |
| Cosmos 1020<br>(1978-56H) | 7 Jun<br>2150 | 74<br>115 | 1 410<br>1 487 | Lifetime 8 000 years | |
| Cosmos 1023<br>(1978-63A) | 21 Jun<br>0936 | 74<br>101 | 783<br>805 | Store-dump communications satellite; lifetime 120 years | |
| Molniya 1-41<br>(1978-72A) | 14 Jul<br>1507 | 63<br>736 | 607<br>40 666 | Replaces Molniya 1-35; lifetime 12 years | |
| Raduga 4<br>(1978-73A) | 18 Jul<br>2248 | 0.5<br>1 478 | 36 473<br>36 730 | Replaces Raduga 3 in Statsionar 2 position; lifetime > $10^6$ years | |
| Molniya 1-42<br>(1978-80A) | 22 Aug<br>2346 | 63<br>736 | 443<br>40 793 | Replaces Molniya 1-33; lifetime 12 years | |
| Cosmos 1034<br>(1978-91A) | 4 Oct<br>0350 | 74<br>115 | 1 423<br>1 484 | Lifetime 8 000 years | Octuple launch |
| Cosmos 1035<br>(1978-91B) | 4 Oct<br>0350 | 74<br>115 | 1 405<br>1 482 | Lifetime 7 000 years | |
| Cosmos 1036<br>(1978-91C) | 4 Oct<br>0350 | 74<br>115 | 1 443<br>1 484 | Lifetime 9 000 years | |
| Cosmos 1037<br>(1978-91D) | 4 Oct<br>0350 | 74<br>115 | 1 463<br>1 484 | Lifetime 9 000 years | |
| Cosmos 1038<br>(1978-91E) | 4 Oct<br>0350 | 74<br>116 | 1 480<br>1 488 | Lifetime 10 000 years | |
| Cosmos 1039<br>(1978-91F) | 4 Oct<br>0350 | 74<br>116 | 1 481<br>1 554 | Lifetime 10 000 years | |
| Cosmos 1040<br>(1978-91G) | 4 Oct<br>0350 | 74<br>116 | 1 481<br>1 529 | Lifetime 10 000 years | |
| Cosmos 1041<br>(1978-91H) | 4 Oct<br>0350 | 74<br>116 | 1 480<br>1 510 | Lifetime 10 000 years | |
| Molniya 3-10<br>(1978-95A) | 13 Oct<br>0517 | 63<br>736 | 452<br>40 829 | Probably a replacement for Molniya 3-6 (1976-127A); lifetime 12 years | |

| Country, satellite name and designation | Launch date and time (GMT) | Orbital inclination (deg) and period (min) | Perigee and apogee heights (km) | Comments |
|---|---|---|---|---|
| Molniya 3-10 (1978-95A) | 13 Oct 0517 | 63 736 | 452 40 829 | Probably a replacement for Molniya 3-6 (1976-127A); lifetime 12 years |
| Cosmos 1048 (1978-105A) | 16 Nov 2150 | 74 101 | 785 816 | Store-dump communications satellite; lifetime 120 years |
| Cosmos 1051 (1978-109A) | 5 Dec 1814 | 74 115 | 1 397 1 487 | Lifetime 7 000 years (Octuple launch) |
| Cosmos 1052 (1978-109B) | 5 Dec 1814 | 74 115 | 1 412 1 490 | Lifetime 8 000 years (Octuple launch) |
| Cosmos 1053 (1978-109C) | 5 Dec 1814 | 74 115 | 1 433 1 488 | Lifetime 9 000 years (Octuple launch) |
| Cosmos 1054 (1978-109D) | 5 Dec 1814 | 74 115 | 1 449 1 491 | Lifetime 9 000 years (Octuple launch) |
| Cosmos 1055 (1978-109E) | 5 Dec 1814 | 74 116 | 1 460 1 500 | Lifetime 10 000 years (Octuple launch) |
| Cosmos 1056 (1978-109F) | 5 Dec 1814 | 74 116 | 1 472 1 508 | Lifetime 10 000 years (Octuple launch) |
| Cosmos 1057 (1978-109G) | 5 Dec 1814 | 74 116 | 1 482 1 518 | Lifetime 9 000 years (Octuple launch) |
| Cosmos 1058 (1978-109H) | 5 Dec 1814 | 74 116 | 1 481 1 541 | Lifetime 10 000 years (Octuple launch) |
| *1979* | | | | |
| Molniya 3-11 (1979-04A) | 18 Jan 1550 | 63 736 | 433 40 817 | Third generation; lifetime 20 years |
| Cosmos 1081 (1979-24A) | 15 Mar 0253 | 74 115 | 1 406 1 467 | Lifetime 7 000 years (Octuple launch; probable operational life 5–6 months) |
| Cosmos 1082 (1979-24B) | 15 Mar 0253 | 74 115 | 1 424 1 466 | Lifetime 8 000 years (Octuple launch; probable operational life 5–6 months) |
| Cosmos 1083 (1979-24C) | 15 Mar 0253 | 74 115 | 1 443 1 465 | Lifetime 9 000 years (Octuple launch; probable operational life 5–6 months) |
| Cosmos 1084 (1979-24D) | 15 Mar 0253 | 74 115 | 1 463 1 463 | Lifetime 9 000 years (Octuple launch; probable operational life 5–6 months) |
| Cosmos 1085 (1979-24E) | 15 Mar 0253 | 74 116 | 1 467 1 507 | Lifetime 10 000 years (Octuple launch; probable operational life 5–6 months) |
| Cosmos 1086 (1979-24F) | 15 Mar 0253 | 74 116 | 1 468 1 484 | Lifetime 10 000 years (Octuple launch; probable operational life 5–6 months) |
| Cosmos 1087 (1979-24G) | 15 Mar 0253 | 74 116 | 1 468 1 526 | Lifetime 10 000 years (Octuple launch; probable operational life 5–6 months) |
| Cosmos 1088 (1979-24H) | 15 Mar 0253 | 74 116 | 1 466 1 548 | Lifetime 10 000 years (Octuple launch; probable operational life 5–6 months) |
| Molniya 1-43 (1979-31A) | 12 Apr 0029 | 63 735 | 623 630 | Replaces Molniya 1-32; lifetime 11 years |
| Raduga 5 (1979-35A) | 25 Apr 0350 | 0.4 1 436 | 35 789 35 789 | Statsionar series satellite located over Indian Ocean at Statsionar 1 position; lifetime $> 10^6$ years |
| Molniya 3-12 (1979-48A) | 5 Jun 2331 | 63 735 | 439 449 | Lifetime 12 years |
| Cosmos 1110 (1979-60A) | 28 Jun 2010 | 74 101 | 791 814 | Store-dump communications satellite; lifetime 120 years |

| Country, satellite name and designation | Launch date and time (GMT) | Orbital inclination (deg) and period (min) | Perigee and apogee heights (km) | Comments |
|---|---|---|---|---|
| Gorizont 2 (1979-62A) | 5 Jul 2331 | 0.8 1 477 | 36 550 36 550 | Lifetime > $10^6$ years |
| Molniya 1-44 (1979-70A) | 31 Jul 0405 | 63 718 | 452 39 902 | Lifetime 20 years |
| Cosmos 1125 (1979-78A) | 28 Aug 0000 | 74 101 | 784 812 | Store-dump communications satellite; lifetime 120 years |
| Cosmos 1130 (1979-84A) | 25 Sep 2053 | 74 115 | 1 400 1 482 | Lifetime 7 000 years (Octuple launch) |
| Cosmos 1131 (1979-84B) | 25 Sep 2053 | 74 115 | 1 410 1 486 | Lifetime 8 000 years (Octuple launch) |
| Cosmos 1132 (1979-84C) | 25 Sep 2053 | 74 115 | 1 429 1 483 | Lifetime 9 000 years (Octuple launch) |
| Cosmos 1133 (1979-84D) | 25 Sep 2053 | 74 115 | 1 441 1 485 | Lifetime 10 000 years (Octuple launch) |
| Cosmos 1134 (1979-84E) | 25 Sep 2053 | 74 115 | 1 455 1 486 | Lifetime 10 000 years (Octuple launch) |
| Cosmos 1135 (1979-84F) | 25 Sep 2053 | 74 116 | 1 465 1 493 | Lifetime 10 000 years (Octuple launch) |
| Cosmos 1136 (1979-84G) | 25 Sep 2053 | 74 116 | 1 472 1 501 | Lifetime 10 000 years (Octuple launch) |
| Cosmos 1137 (1979-84H) | 25 Sep 2053 | 74 116 | 1 472 1 519 | Lifetime 10 000 years (Octuple launch) |
| Cosmos 1140 (1979-89A) | 11 Oct 1634 | 74 101 | 780 805 | Store-dump communications satellite; lifetime 120 years |
| Molniya 1-45 (1979-91A) | 20 Oct 1200 | 63 736 | 818 40 627 | Lifetime 15 years |
| Gorizont 3 (1979-105A) | 28 Dec | 0.6 1 436 | 35 689 35 883 | Synchronous orbit similar to (1979-62A); lifetime > $10^6$ years; near Statsionar 5 location at 58°E |
| *1980* | | | | |
| Molniya 1-46 (1980-02A) | 11 Jan 1229 | 63 718 | 442 39 920 | Lifetime 17 years |
| Cosmos 1156 (1980-12A) | 11 Feb 2331 | 74 115 | 1 400 1 475 | Lifetime 8 000 years (Octuple launch) |
| Cosmos 1157 (1980-12B) | 11 Feb 2331 | 74 115 | 1 417 1 477 | Lifetime 9 000 years (Octuple launch) |
| Cosmos 1158 (1980-12C) | 11 Feb 2331 | 74 115 | 1 435 1 478 | Lifetime 9 000 years (Octuple launch) |
| Cosmos 1159 (1980-12D) | 11 Feb 2331 | 74 115 | 1 453 1 481 | Lifetime 9 000 years (Octuple launch) |
| Cosmos 1160 (1980-12E) | 11 Feb 2331 | 74 115 | 1 467 1 486 | Lifetime 9 000 years (Octuple launch) |
| Cosmos 1161 (1980-12F) | 11 Feb 2331 | 74 116 | 1 469 1 505 | Lifetime 10 000 years (Octuple launch) |
| Cosmos 1162 (1980-12G) | 11 Feb 2331 | 74 116 | 1 472 1 523 | Lifetime 10 000 years (Octuple launch) |
| Cosmos 1163 (1980-12H) | 11 Feb 2331 | 74 116 | 1 472 1 545 | Lifetime 10 000 years (Octuple launch) |

| Country, satellite name and designation | Launch date and time (GMT) | Orbital inclination (deg) and period (min) | Perigee and apogee heights (km) | Comments |
|---|---|---|---|---|
| Cosmos 1164 1980-13A) | 12 Feb 0058 | 63 764 | 435 40 858 | Failed to replace Molniya 1-43; lifetime < 1 day |
| Raduga 6 (1980-16A) | 20 Feb 0810 | 0.6 1 486 | 36 087 37 438 | Lifetime > $10^6$ years; in Statsionar 2 position |
| Cosmos 1175 1980-31C) | 18 Apr 1731 | 63 92 | 313 463 | Failed to replace Molniya 3-9; lifetime 164 days |
| Gorizont 4 (1980-49A) | 14 Jun 0058 | 0.8 1 436 | 35 744 35 828 | Lifetime > $10^6$ years; in Statsionar 4 position |
| Molniya 1-47 (1980-53A) | 21 Jun 2010 | 63 738 | 631 40 703 | Lifetime 12 years |
| Cosmos 1190 (1980-56A) | 1 Jul 0712 | 74 101 | 792 806 | Store-dump communications satellite? lifetime 120 years |
| Cosmos 1192 (1980-58A) | 9 Jul 0043 | 74 115 | 1 398 1 476 | Lifetime 7 000 years (Octuple launch) |
| Cosmos 1193 (1980-58B) | 9 Jul 0043 | 74 115 | 1 452 1 477 | Lifetime 8 000 years (Octuple launch) |
| Cosmos 1194 (1980-58C) | 9 Jul 0043 | 74 115 | 1 414 1 479 | Lifetime 9 000 years (Octuple launch) |
| Cosmos 1195 (1980-58D) | 9 Jul 0043 | 74 115 | 1 470 1 477 | Lifetime 10 000 years (Octuple launch) |
| Cosmos 1196 (1980-58E) | 9 Jul 0043 | 74 116 | 1 473 1 494 | Lifetime 10 000 years (Octuple launch) |
| Cosmos 1197 (1980-58F) | 9 Jul 0043 | 74 116 | 1 475 1 510 | Lifetime 10 000 years (Octuple launch) |
| Cosmos 1198 (1980-58G) | 9 Jul 0043 | 74 116 | 1 475 1 533 | Lifetime 10 000 years (Octuple launch) |
| Cosmos 1199 (1980-58H) | 9 Jul 0043 | 74 115 | 1 433 1 478 | Lifetime 9 000 years (Octuple launch) |
| Ekran 5 (1980-60A) | 15 Jul 0307 | 0.4 ~1 420 | ~35 474 ~35 474 | Statsionar T; lifetime > $10^6$ years |
| Molniya 3-13 (1980-63A) | 18 Jul 1048 | 63 736 | 457 40 818 | Lifetime 100 years |
| Raduga 7 (1980-81A) | 5 Oct 1717 | 0.3 1 436 | 35 730 35 840 | Lifetime > $10^6$ years; in Statsionar 3 position |
| Molniya 1-48 (1980-92A) | 16 Nov 0434 | 63 736 | 601 40 662 | Lifetime 16 years |
| Cosmos 1228 (1980-102A) | 24 Dec 2248 | 74 114 | 1 391 1 464 | Lifetime 7 000 years (Octuple launch) |
| Cosmos 1229 (1980-102B) | 24 Dec 2248 | 74 115 | 1 416 1 498 | Lifetime 9 000 years (Octuple launch) |
| Cosmos 1230 (1980-102C) | 24 Dec 2248 | 74 115 | 1 412 1 452 | Lifetime 8 000 years (Octuple launch) |
| Cosmos 1231 (1980-102D) | 24 Dec 2248 | 74 115 | 1 410 1 461 | Lifetime 8 000 years (Octuple launch) |
| Cosmos 1232 (1980-102E) | 24 Dec 2248 | 74 115 | 1 414 1 458 | Lifetime 8 000 years (Octuple launch) |

| Country, satellite name and designation | Launch date and time (GMT) | Orbital inclination (deg) and period (min) | Perigee and apogee heights (km) | Comments |
|---|---|---|---|---|
| Cosmos 1233 (1980-102F) | 24 Dec 2248 | 74 114 | 1 372 1 452 | Lifetime 6 000 years |
| Cosmos 1234 (1980-102G) | 24 Dec 2248 | 74 114 | 1 404 1 454 | Lifetime 8 000 years; Octuple launch (Cosmos 1233–1235) |
| Cosmos 1235 (1980-102H) | 24 Dec 2248 | 74 114 | 1 392 1 455 | Lifetime 7 000 years |
| Ekran 6 (1980-104A) | 26 Dec 1326 | 0.07 1 440 | 35 859 35 859 | Statsionar T; lifetime > $10^6$ years |
| *1981* | | | | |
| Molniya 3-14 (1981-02A) | 9 Jan 1507 | 63 736 | 439 40 800 | Replaces Molniya 3-10; lifetime 12 years |
| Molniya 1-49 (1981-09A) | 30 Jan 1800 | 63 736 | 430 40 805 | Probably serving military communications in Siberia and northern USSR; to replace Molniya 1-41; lifetime 12 years |
| Cosmos 1250 (1981-22A) | 6 Mar 1131 | 74 115 | 1 399 1 467 | Lifetime 7 000 years |
| Cosmos 1251 (1981-22B) | 6 Mar 1131 | 74 115 | 1 406 1 474 | Lifetime 7 000 years |
| Cosmos 1252 (1981-22C) | 6 Mar 1131 | 74 115 | 1 420 1 474 | Lifetime 8 000 years |
| Cosmos 1253 (1981-22D) | 6 Mar 1131 | 74 115 | 1 442 1 485 | Lifetime 9 000 years; Octuple launch (Cosmos 1250–1257) |
| Cosmos 1254 (1981-22E) | 6 Mar 1131 | 74 115 | 1 434 1 474 | Lifetime 9 000 years |
| Cosmos 1255 (1981-22F) | 6 Mar 1131 | 74 115 | 1 448 1 474 | Lifetime 9 000 years |
| Cosmos 1256 (1981-22G) | 6 Mar 1131 | 74 115 | 1 459 1 479 | Lifetime 10 000 years |
| Cosmos 1257 (1981-22H) | 6 Mar 1131 | 74 116 | 1 470 1 482 | Lifetime 10 000 years |
| Raduga 8 (1981-27A) | 18 Mar 0448 | 0.8 1 475 | 36 551 36 551 | Replaces Raduga 6 at Statsionar 2 position at longitude 35°E; lifetime > $10^6$ years |
| Molniya 3-15 (1981-30A) | 24 Mar 0336 | 63 736 | 609 40 643 | Replaces Molniya 3-11; lifetime 12 years |
| Cosmos 1269 (1981-41A) | 7 May 1326 | 74 102 | 796 810 | Possibly store-dump communications satellite; lifetime 120 years |
| Molniya 3-16 (1981-54A) | 9 Jun 0336 | 63 737 | 434 40 844 | Replaces Molniya 3-14; lifetime 12 years |
| Molniya 1-50 (1981-60A) | 24 Jun 1926 | 63 736 | 617 40 641 | Replaces Molniya 1-42; lifetime 12 years |
| Ekran 7 (1981-61A) | 26 Jun 2400 | 0.1 1 427 | 35 599 35 600 | At Statsionar T; initial Tass announcement specified Statsionar 1 but all previous Ekrans have been at Statsionar T; lifetime > $10^6$ years |

| Country, satellite name and designation | Launch date and time (GMT) | Orbital inclination (deg) and period (min) | Perigee and apogee heights (km) | Comments | |
|---|---|---|---|---|---|
| Raduga 9 (1981-69A) | 30 Jul 2136 | 0.4 1 477 | 36 582 36 583 | Replaces Raduga 8 at Statsionar 2; lifetime > $10^6$ years | |
| Cosmos 1287 (1981-74A) | 6 Aug 1146 | 74 116 | 1 466 1 515 | Lifetime 10 000 years | Octuple launch; Navy tactical communication |
| Cosmos 1288 (1981-74B) | 6 Aug 1146 | 74 116 | 1 468 1 494 | Lifetime 10 000 years | |
| Cosmos 1289 (1981-74C) | 6 Aug 1146 | 74 115 | 1 462 1 481 | Lifetime 10 000 years | |
| Cosmos 1290 (1981-74D) | 6 Aug 1146 | 74 115 | 1 460 1 466 | Lifetime ? years | |
| Cosmos 1291 (1981-74E) | 6 Aug 1146 | 74 115 | 1 460 1 466 | Lifetime ? years | |
| Cosmos 1292 (1981-74F) | 6 Aug 1146 | 74 115 | 1 428 1 466 | Lifetime ? years | |
| Cosmos 1293 (1981-74G) | 6 Aug 1146 | 74 115 | 1 411 1 467 | Lifetime ? years | |
| Cosmos 1294 (1981-74H) | 6 Aug 1146 | 74 115 | 1 395 1 466 | Lifetime ? years | |
| Cosmos 1302 (1981-84A) | 28 Aug 1919 | 74 101 | 783 812 | Possibly store-dump communications satellite; lifetime 120 years | |
| Cosmos 1305 (1981-88A) | 11 Sep 0853 | 63 264 | 626 13 865 | Failure to replace Molniya 3-12 due to under-burn of e-stage; lifetime 10 years | |
| Raduga 10 (1981-102A) | 9 Oct 1702 | 0.1 1 444 | 35 932 35 932 | Replaces Raduga 7 at Statsionar 3; lifetime > $10^6$ years | |
| Molniya 3-17 (1981-105A) | 17 Oct 0600 | 63 736 | 619 40 648 | Replaces Molniya 3-12; see Cosmos 1305; lifetime 12 years | |
| Molniya 1-51 (1981-113A) | 17 Nov 1536 | 63 702 | 441 39 136 | Replaces Molniya 1-39; lifetime 12 years | |
| Cosmos 1320 (1981-116A) | 28 Nov 1800 | 74 117 | 1 482 1 638 | Lifetime 10 000 years | Octuple launch; each satellite has a period of 117 minutes which is higher than normal for such launches |
| Cosmos 1321 (1981-116B) | 28 Nov 1800 | 74 117 | 1 482 1 635 | Lifetime 10 000 years | |
| Cosmos 1322 (1981-116C) | 28 Nov 1800 | 74 117 | 1 483 1 631 | Lifetime 10 000 years | |
| Cosmos 1323 (1981-116D) | 28 Nov 1800 | 74 117 | 1 483 1 627 | Lifetime 10 000 years | |
| Cosmos 1324 (1981-116E) | 28 Nov 1800 | 74 117 | 1 482 1 623 | Lifetime 10 000 years | |
| Cosmos 1325 (1981-116F) | 28 Nov 1800 | 74 117 | 1 483 1 619 | Lifetime 10 000 years | |
| Cosmos 1326 (1981-116G) | 28 Nov 1800 | 74 117 | 1 485 1 617 | Lifetime 10 000 years | |
| Cosmos 1327 (1981-116H) | 28 Nov 1800 | 74 117 | 1 486 1 609 | Lifetime 10 000 years | |

| Country, satellite name and designation | Launch date and time (GMT) | Orbital inclination (deg) and period (min) | Perigee and apogee heights (km) | Comments |
|---|---|---|---|---|
| Molniya 1-52 (1981-123A) | 23 Dec 1326 | 63 699 | 484 13 960 | Replaces Molniya 1-45; lifetime 12 years |
| **NATO** | | | | |
| *1977* | | | | |
| NATO 3B (1977-05A) | 28 Jan 0058 | 2.6 1 436 | 35 777 35 797 | Lifetime > $10^6$ years |
| *1978* | | | | |
| NATO 3C (1977-106A) | 16 Nov 0058 | 4.4 1 429 | 35 516 35 768 | Lifetime > $10^6$ years |

# Appendix 1F

## *Navigation satellites launched between 1977 and 1981*

| Country, satellite name and designation | Launch date and time (GMT) | Orbital inclination (deg) and period (min) | Perigee and apogee heights (km) | Comments[a] |
|---|---|---|---|---|
| **USA** | | | | |
| *1977* | | | | |
| NTS2 (1977-53A) | 23 Jun 0058 | 63 352 | 160 20 106 | Used to test highly accurate caesium-type atomic clock for NAVSTAR Global Positioning System; lifetime $10^6$ years |
| *1978* | | | | |
| NDS 1 (1978-20A) | 20 Feb 2346 | 63 719 | 20 095 20 309 | Navigation Development Satellites; first four of six NAVSTAR Global Positioning System (GPS); GPS was originally planned to consist of 24 satellites in 12-h circular orbits; the satellites transmit the positioning signals and provide continuous global coverage; the main control station in the USA monitors and updates the satellite navigation subsystems and is able to determine the navigator's position to within 10m accuracy in three dimensions, and the velocity to within a few cm/s; the GPS is designed to replace the Transit navigation satellite system; lifetime $10^6$ years |
| NDS 2 (1978-47A) | 13 May 1033 | 63 711 | 19 952 20 084 | |
| NDS 3 (1978-93A) | 7 Oct 0029 | 63 723 | 20 285 20 312 | |
| NDS 4 (1978-112A) | 11 Dec 0419 | 63 722 | 20 267 20 316 | |
| *1980* | | | | |
| NAVSTAR 5 (1980-11A) | 9 Feb 2317 | 64 717 | 20 083 20 147 | Fifth in a network of 18 satellites; lifetime $10^6$ years |
| NAVSTAR 6 (1980-32A) | 26 Apr 2150 | 63 708 | 19 628 30 232 | Sixth in a network of 18 satellites; lifetime $10^6$ years |

| Country, satellite name and designation | Launch date and time (GMT) | Orbital inclination (deg) and period (min) | Perigee and apogee heights (km) | Comments[a] |
|---|---|---|---|---|
| *1981* | | | | |
| Nova-1 (1981-44A) | 15 May 0600 | 90 109 | 1 170 1 187 | Improved Transit; lifetime 3 000 years |
| **USSR** | | | | |
| *1977* | | | | |
| Cosmos 890 (1977-04A) | 20 Jan 2010 | 83 105 | 983 1 020 | Lifetime 1 200 years |
| Cosmos 894 (1977-13A) | 21 Feb 1717 | 83 105 | 972 1 014 | Lifetime 1 200 years |
| Cosmos 911 (1977-39A) | 25 May 1102 | 83 105 | 970 1 004 | Lifetime 1 200 years |
| Cosmos 926 (1977-62A) | 8 Jul 1731 | 83 105 | 976 1 022 | Lifetime 1 200 years |
| Cosmos 928 (1977-64A) | 13 Jul 0502 | 83 105 | 956 1 011 | Lifetime 1 200 years |
| Cosmos 951 (1977-87A) | 14 Sep 1955 | 83 105 | 968 1 017 | Lifetime 1 200 years |
| Cosmos 962 (1977-107A) | 28 Oct 1550 | 83 105 | 968 1 012 | Lifetime 1 000 years |
| Cosmos 971 (1977-122A) | 23 Dec 1919 | 83 105 | 993 1 021 | Lifetime 1 200 years |
| *1978* | | | | |
| Cosmos 985 (1978-07A) | 17 Jan 0322 | 83 105 | 945 1 022 | Number 4 replaces Cosmos 894 (number 8); lifetime 1 200 years |
| Cosmos 991 (1978-22A) | 28 Feb 0643 | 83 105 | 963 1 009 | Number 5 replaces Cosmos 887 (number 7); lifetime 1 200 years |
| Cosmos 994 (1978-28A) | 15 Mar 1550 | 83 105 | 980 1 011 | Lifetime 1 200 years |
| Cosmos 996 (1978-31A) | 28 Mar 0126 | 83 105 | 957 1 010 | Number 7 replaces Cosmos 804 (number 6); lifetime 1 200 years |
| Cosmos 1000 (1978-34A) | 31 Mar 1355 | 83 105 | 965 1 012 | The first satellite to be described officially by the USSR as part of a global navigation system; number 13; lifetime 1 200 years |
| Cosmos 1011 (1978-53A) | 28 May 1648 | 83 105 | 960 1 014 | Number 6 replaces Cosmos 928 (number 1); lifetime 1 200 years |
| Cosmos 1027 (1978-74A) | 27 Jul 0448 | 83 105 | 166 1 004 | Lifetime 1 200 years |
| Cosmos 1064 (1978-119A) | 19 Dec 1229 | 83 99 | 424 965 | Number 1 replaces Cosmos 991 (number 5): failed to circularize orbit; subsequently augmented by Cosmos 1072 (number 8); lifetime 8 years |

| Country, satellite name and designation | Launch date and time (GMT) | Orbital inclination (deg) and period (min) | Perigee and apogee heights (km) | Comments[a] |
|---|---|---|---|---|
| *1979* | | | | |
| Cosmos 1072 (1979-03A) | 16 Jan 1731 | 83 105 | 965 1 020 | Number 5 to replace Cosmos 991; lifetime 1 200 years |
| Cosmos 1089 (1979-26A) | 21 Mar 0405 | 83 105 | 973 1 005 | Number 2; it has been referred to as maritime support navigation system; lifetime 1 200 years |
| Cosmos 1091 (1979-28A) | 7 Apr 0614 | 83 105 | 969 1 012 | Number 4; lifetime 1 200 years |
| Cosmos 1092 (1979-30A) | 11 Apr 2150 | 83 105 | 969 1 009 | Number 14; lifetime 1 200 years |
| Cosmos 1104 (1979-46A) | 31 May 1800 | 83 105 | 962 1 010 | Number 1; lifetime 1 200 years |
| Cosmos 1141 (1979-90A) | 16 Oct 1214 | 83 105 | 961 1 003 | Number 6; lifetime 1 200 years |
| *1980* | | | | |
| Cosmos 1150 (1980-03A) | 14 Jan 1938 | 83 105 | 971 1 017 | Number 3 to replace Cosmos 1011 which was renumbered 8; lifetime 1 200 years |
| Cosmos 1153 (1980-07A) | 25 Jan 2219 | 83 105 | 967 1 020 | Number 4 to replace Cosmos 1091 which was renumbered 7; lifetime 1 200 years |
| Cosmos 1168 (1980-22A) | 17 Mar 2136 | 83 105 | 964 1 015 | Number 11 to replace Cosmos 883; lifetime 1 200 years |
| Cosmos 1181 (1980-39A) | 30 May 0922 | 83 105 | 976 1 008 | Number 5 to replace Cosmos 1072 which was renumbered 8; Cosmos 1011 became obsolete; lifetime 1 200 years |
| Cosmos 1225 (1980-97A) | 5 Dec 0419 | 83 105 | 950 1 031 | Number 2 to replace Cosmos 1089 which was renumbered 7; Cosmos 1091 became obsolete; lifetime 1 200 years |
| Cosmos 1226 (1980-99A) | 10 Dec 2234 | 83 105 | 966 1 014 | Number 13 to replace Cosmos 1000 which stopped functioning; lifetime 1 200 years |
| *1981* | | | | |
| Cosmos 1244 (1981-13A) | 12 Feb 1814 | 83 105 | 963 1 014 | Replaces Cosmos 1104; lifetime 1 200 years |
| Cosmos 1275 (1981-53A) | 4 Jun 1536 | 83 105 | 964 1 014 | Replaces Cosmos 1141; lifetime 1 200 years |
| Cosmos 1295 (1981-77A) | 12 Aug 0546 | 83 105 | 952 1 015 | Replaces Cosmos 1181; lifetime 1 200 years |
| Cosmos 1304 (1981-87A) | 4 Sep 1102 | 83 104 | 912 980 | Replaces Cosmos 926; orbit not quite nominal; lifetime 1 200 years |
| Cosmos 1308 (1981-91A) | 18 Sep 0336 | 83 105 | 970 1 004 | Replaces Cosmos 1275; lifetime 1 200 years |

[a] Numbers are the Soviet identity numbers for each satellite. These numbers allow positive identification of the satellite without the need to use predictions or orbital calculations; numbers have been changed since about August 1978 (G.E. Perry, private communication).

# Appendix 1G

## *Meteorological satellites launched between 1977 and 1981*

| Country, satellite name and designation | Launch date and time (GMT) | Orbital inclination (deg) and period (min) | Perigee and apogee heights (km) | Comments |
|---|---|---|---|---|
| **USA** | | | | |
| *1977* | | | | |
| AMS 2 (1977-44A) | 5 Jun 0307 | 99 102 | 811 869 | Advanced Meteorological Satellite; lifetime 80 years |
| GOES 2 (1977-48A) | 16 Jun 1048 | 0.9 1 436 | 35 266 36 304 | Geostationary Operational Environmental Satellite; lifetime $10^6$ years |
| *1978* | | | | |
| AMS 3 (1978-42A) | 1 May 0307 | 99 102 | 820 835 | Lifetime 80 years |
| GOES 3 (1978-62A) | 16 Jun 1102 | 2 1 447 | 35 473 36 521 | Satellite, operated by the US National Oceanographic and Atmospheric Administration, is placed at about 135°W; this type of satellite is also used for obtaining ocean data; it replaces GOES 1 (1975-100A) which was placed at 60°E; lifetime > $10^6$ years |
| Tiros 11 (1978-96A) | 13 Oct 1117 | 99 102 | 850 866 | Monitors ocean currents and upwellings; designed to operate for two years; lifetime 500 years |
| Nimbus 7 (1978-98A) | 24 Oct 0810 | 99 104 | 943 953 | Nimbus second stage carried CAMEO-Chemically Active Materials Ejected from Orbit (Barium released on 29 Oct 1978); lifetime 1 000 years |

| Country, satellite name and designation | Launch date and time (GMT) | Orbital inclination (deg) and period (min) | Perigee and apogee heights (km) | Comments |
|---|---|---|---|---|
| *1979* | | | | |
| AMS 4<br>(1979-50A) | 6 Jun<br>1829 | 99<br>102 | 819<br>838 | Incorporates a block-5D bus; besides USAF, weather pictures available to US weather service; lifetime 80 years |
| NOAA 6<br>(1979-57A) | 27 Jun<br>1550 | 99<br>101 | 812<br>828 | Second in a series of eight spacecraft; first was Tiros N; lifetime 120 years |
| *1980* | | | | |
| NOAA B<br>(1980-43A) | 29 May<br>1048 | 92<br>102 | 264<br>1 445 | Intended orbit not achieved; lifetime one year |
| GOES 4<br>(1980-74A) | 9 Sep<br>2234 | 0.2<br>1 436 | 35 776<br>35 800 | Planned initial location at longitude 90°W; lifetime > $10^6$ years |
| *1981* | | | | |
| GOES 5<br>(1981-49A) | 22 May<br>2234 | 0.5<br>1 430 | 35 453<br>35 887 | Placed above longitude 85°W in a geosynchronous orbit; lifetime > $10^6$ years |
| NOAA 7<br>(1981-59A) | 23 Jun<br>1048 | 99<br>102 | 845<br>863 | Replaces Tiros 11 launched in Oct 1978; lifetime 350 years |
| **USSR** | | | | |
| *1977* | | | | |
| Meteor 2-02<br>(1977-02A) | 6 Jan<br>2317 | 81<br>103 | 890<br>906 | Lifetime 500 years |
| Meteor 27<br>(1977-24A) | 5 Apr<br>0210 | 81<br>103 | 854<br>897 | Lifetime 500 years |
| Meteor 28<br>(1977-57A) | 29 Jun<br>1843 | 98<br>98 | 601<br>670 | First Soviet satellite to use Sun-synchronous retrograde orbit; it has been reported that satellite also carried multi-spectral sensor for Earth resources applications; lifetime 60 years |
| Meteor 2-03<br>(1977-117A) | 14 Dec<br>0936 | 81<br>103 | 864<br>904 | Lifetime 500 years |
| *1979* | | | | |
| Meteor 29<br>(1979-05A) | 25 Jan<br>0546 | 98<br>97 | 622<br>654 | Lifetime 60 years |
| Meteor 2-04<br>(1979-21A) | 1 Mar<br>1843 | 81<br>102 | 839<br>897 | Second generation; lifetime 500 years |
| Meteor 2-05<br>(1979-95A) | 31 Oct<br>0936 | 81<br>103 | 873<br>890 | Lifetime 500 years |
| *1980* | | | | |
| Meteor 30<br>(1980-51A) | 18 Jun<br>0614 | 98<br>97 | 584<br>667 | Lifetime 60 years |
| Meteor 2-06<br>(1980-73A) | 9 Sep<br>1102 | 81<br>102 | 848<br>894 | Lifetime 500 years |

| Country, satellite name and designation | Launch date and time (GMT) | Orbital inclination (deg) and period (min) | Perigee and apogee heights (km) | Comments |
|---|---|---|---|---|
| *1981* | | | | |
| Meteor 2-07<br>(1981-43A) | 14 May<br>2150 | 81<br>102 | 855<br>893 | Lifetime 500 years |
| Meteor 31<br>(1981-65A) | 10 Jul<br>0517 | 98<br>98 | 610<br>671 | Lifetime 60 years |

# Appendix 1H

## *Interceptor/destructor satellites launched between 1977 and 1981*

| Country, satellite name and designation | Launch date and time (GMT) | Orbital inclination (deg) and period (min) | Perigee and apogee heights (km) | Comments |
|---|---|---|---|---|
| **USSR** | | | | |
| *1977* | | | | |
| Cosmos 909 (1977-36A) | 19 May 1634 | 66 117 | 990 2 109 | Target satellite; lifetime 4 000 years |
| Cosmos 910 (1977-37A) | 23 May 1229 | 66 100 | 300 1 774 | Interceptor/destructor satellite probably intended to pass close to Cosmos 909; the satellite entered atmosphere before completing first orbit; lifetime 0.05 day |
| Cosmos 918 (1977-50A) | 17 Jun 0726 | 65 88 | 128 243 | Interceptor/destructor satellite which may have temporarily entered orbit passing close to Cosmos 909 in new "pop-up" technique; lifetime 1 day |
| Cosmos 959 (1977-101A) | 21 Oct 1005 | 66 95 | 146 850 | Target satellite; lifetime 42 years |
| Cosmos 961 (1977-104A) | 26 Oct 0517 | 66 89 66 102 | 125 302 269 1 421 | Interceptor/destructor satellite—"pop-up" type—which fired propulsion system to climb to the higher orbit of Cosmos 959; orbits are provisional; lifetime 0.78 day |
| Cosmos 967 (1977-116A) | 13 Dec 1550 | 66 105 | 973 1 013 | Target satellite; lifetime 1 200 years |
| Cosmos 970 (1977-121A) | 21 Dec 1033 | 65 95 | 144 854 | Interceptor/destructor satellite; approached Cosmos 967 in co-orbital type interception and exploded |

| Country, satellite name and designation | Launch date and time (GMT) | Orbital inclination (deg) and period (min) | Perigee and apogee heights (km) | Comments |
|---|---|---|---|---|
| *1978* | | | | |
| Cosmos 1009 (1978-50A) | 19 May 0014 | 66 109 | 966 1 364 | Interceptor/destructor satellite passed close to Cosmos 967 (1977-116A) target satellite launched on 13 December 1977; test made just before the Helsinki meeting between the USA and the USSR to ban the ASAT system; lifetime 0.17 day |
| *1979* | | | | |
| Cosmos 1075 (1979-10A) | 8 Feb 1005 | 66 95 | 473 516 | Orbital characteristics similar to target satellites; lifetime 3 years |
| Cosmos 1146 (1979-100A) | 5 Dec 1033 | 66 94 | 444 494 | Calibration mission; orbital characteristics similar to target satellites; lifetime 1.5 years |
| *1980* | | | | |
| Cosmos 1169 (1980-23A) | 27 Mar 0735 | 66 95 | 477 515 | Probably served as a calibration target for ground radars; lifetime 3 years |
| Cosmos 1171 (1980-26A) | 3 Apr 0735 | 66 105 | 947 1 033 | Target satellite like Cosmos 400 and 967; orbital heights similar to Soviet navigation satellites; lifetime 1 200 years |
| Cosmos 1174 (1980-30A) | 18 Apr 0058 | 66 99 | 362 1 025 | Interceptor |
| *1981* | | | | |
| Cosmos 1241 (1981-06A) | 21 Jan 0824 | 66 105 | 977 1 011 | ASAT target; orbital height similar to that of Soviet navigation satellites; lifetime 1 200 years |
| Cosmos 1243 (1981-10A) | 2 Feb 0014 | 66 98 | 297 1 017 | Probably passed closer than 8 km to Cosmos 1241 on 2 Feb then de-orbited; first successful R&D test of a new optical thermal guidance system; lifetime < 1 day |
| Cosmos 1258 (1981-24A) | 14 Mar 1702 | 66 98 | 303 1 026 | Probably passed closer than 8 km to Cosmos 1241, then de-orbited; lifetime < 1 day |

# Appendix 2

## *Six treaties that contain provisions aimed at some form of arms control in space*

Treaty Banning Nuclear Weapons Tests in the Atmosphere, in Outer Space and Under Water (1963)

Treaty on Principles Governing the Activities of States in the Exploration and Use of Outer Space, including the Moon and Other Celestial Bodies (1967)

Treaty Between the United States of America and the Union of Soviet Socialist Republics on the Limitation of Anti-Ballistic Missile Systems (1972)

Interim Agreement Between the United States of America and the Union of Soviet Socialist Republics on Certain Measures With Respect to the Limitation of Strategic Offensive Arms (1972)

Agreement Governing the Activities of States on the Moon and Other Celestial Bodies (1979)

Proposed Soviet draft treaty on the prohibition of the stationing of weapons of any kind in outer space (1981)

# Treaty Banning Nuclear Weapon Tests in the Atmosphere, in Outer Space and Under Water

ENTERED INTO FORCE: 10 October 1963

---

The Governments of the United States of America, the United Kingdom of Great Britain and Northern Ireland, and the Union of Soviet Socialist Republics, hereinafter referred to as the "Original Parties",

Proclaiming as their principal aim the speediest possible achievement of an agreement on general and complete disarmament under strict international control in accordance with the objectives of the United Nations which would put an end to the armaments race and eliminate the incentive to the production and testing of all kinds of weapons, including nuclear weapons,

Seeking to achieve the discontinuance of all test explosions of nuclear weapons for all time, determined to continue negotiations to this end, and desiring to put an end to the contamination of man's environment by radioactive substances,

Have agreed as follows:

## Article I

1. Each of the Parties to this Treaty undertakes to prohibit, to prevent, and not to carry out any nuclear weapon test explosion, or any other nuclear explosion, at any place under its jurisdiction or control:

(*a*) in the atmosphere; beyond its limits, including outer space; or under water, including territorial waters or high seas; or

(*b*) in any other environment if such explosion causes radioactive debris to be present outside the territorial limits of the State under whose jurisdiction or control such explosion is conducted. It is understood in this connection that the provisions of this subparagraph are without prejudice to the conclusion of a treaty resulting in the permanent banning of all nuclear test explosions, including all such explosions underground, the conclusion of which, as the Parties have stated in the Preamble to this Treaty, they seek to achieve.

2. Each of the Parties to this Treaty undertakes furthermore to refrain from causing, encouraging, or in any way participating in, the carrying out of any nuclear weapon test explosion, or any other nuclear explosion, anywhere which would take place in any of the environments described, or have the effect referred to, in paragraph 1 of this Article.

## Article II

1. Any Party may propose amendments to this Treaty. The text of any proposed amendment shall be submitted to the Depositary Governments which shall circulate it to all Parties to this Treaty. Thereafter, if requested to do so by one-third or more of the Parties, the Depositary Governments shall convene a conference, to which they shall invite all the Parties, to consider such amendment.

2. Any amendment to this Treaty must be approved by a majority of the votes of all the Parties to this Treaty, including the votes of all of the Original Parties. The amendment shall enter into force for all Parties upon the deposit of instruments of ratification by a majority of all the Parties, including the instruments of ratification of all of the Original Parties.

## Article III

1. This Treaty shall be open to all States for signature. Any State which does not sign this Treaty before its entry into force in accordance with paragraph 3 of this Article may accede to it at any time.

2. This Treaty shall be subject to ratification by signatory States. Instruments of ratification and instruments of accession shall be deposited with the Governments of the Original Parties—the United States of America, the United Kingdom of Great Britain and Northern Ireland, and the Union of Soviet Socialist Republics—which are hereby designated the Depositary Governments.

3. This Treaty shall enter into force after its ratification by all the Original Parties and the deposit of their instruments of ratification.

4. For States whose instruments of ratification or accession are deposited subsequent to the entry into force of this Treaty, it shall enter into force on the date of the deposit of their instruments of ratification or accession.

5. The Depositary Governments shall promptly inform all signatory and acceding States of the date of each signature, the date of deposit of each instrument of ratification of and accession to this Treaty, the date of its entry into force, and the date of receipt of any requests for conferences or other notices.

6. This Treaty shall be registered by the Depositary Governments pursuant to Article 102 of the Charter of the United Nations.

## Article IV

This Treaty shall be of unlimited duration.

Each Party shall in exercising its national sovereignty have the right to withdraw from the Treaty if it decides that extraordinary events, related to the subject matter of this Treaty, have jeopardized the supreme interests of its country. It shall give notice of such withdrawal to all other Parties to the Treaty three months in advance.

## Article V

This Treaty, of which the English and Russian texts are equally authentic, shall be deposited in the archives of the Depositary Governments. Duly certified copies of this Treaty shall be transmitted by the Depositary Governments to the Governments of the signatory and acceding States.

IN WITNESS WHEREOF the undersigned, duly authorized, have signed this Treaty.

DONE in triplicate at the city of Moscow the fifth day of August, one thousand nine hundred and sixty-three.

*For the Government of the United States of America:*

Dean Rusk

*For the Government of the United Kingdom of Great Britain and Northern Ireland:*

Home

*For the Government of the Union of Soviet Socialist Republics:*

A. Gromyko

# Treaty on Principles Governing the Activities of States in the Exploration and Use of Outer Space, including the Moon and Other Celestial Bodies

ENTERED INTO FORCE: 10 October 1967

---

The States Parties to this Treaty,

Inspired by the great prospects opening up before mankind as a result of man's entry into outer space,

Recognizing the common interest of all mankind in the progress of the exploration and use of outer space for peaceful purposes,

Believing that the exploration and use of outer space should be carried on for the benefit of all peoples irrespective of the degree of their economic or scientific development,

Desiring to contribute to broad international co-operation in the scientific as well as the legal aspects of the exploration and use of outer space for peaceful purposes,

Believing that such co-operation will contribute to the development of mutual understanding and to the strengthening of friendly relations between States and peoples,

Recalling resolution 1962 (XVIII), entitled "Declaration of Legal Principles Governing the Activities of States in the Exploration and Use of Outer Space", which was adopted unanimously by the United Nations General Assembly on 13 December 1963,

Recalling resolution 1884 (XVIII), calling upon States to refrain from placing in orbit around the earth any objects carrying nuclear weapons or any other kinds of weapons of mass destruction or from installing such weapons on celestial bodies, which was adopted unanimously by the United Nations General Assembly on 17 October 1963,

Taking account of United Nations General Assembly resolution 110 (II) of 3 November 1947, which condemned propaganda designed or likely to provoke or encourage any threat to the peace, breach of the peace or act of aggression, and considering that the aforementioned resolution is applicable to outer space,

Convinced that a Treaty on Principles Governing the Activities of States in the Exploration and Use of Outer Space, including the Moon and Other Celestial Bodies, will further the Purposes and Principles of the Charter of the United Nations,

Have agreed on the following:

## Article I

The exploration and use of outer space, including the moon and other celestial bodies, shall be carried out for the benefit and in the interests of all countries, irrespective of their degree of economic or scientific development, and shall be the province of all mankind.

Outer space, including the moon and other celestial bodies, shall be free for exploration and use by all States without discrimination of any kind, on a basis of equality and in accordance with international law, and there shall be free access to all areas of celestial bodies.

There shall be freedom of scientific investigation in outer space, including the moon and other celestial bodies, and States shall facilitate and encourage international co-operation in such investigation.

Article II

Outer space, including the moon and other celestial bodies, is not subject to national appropriation by claim of sovereignty, by means of use or occupation, or by any other means.

Article III

States Parties to the Treaty shall carry on activities in the exploration and use of outer space, including the moon and other celestial bodies, in accordance with international law, including the Charter of the United Nations, in the interest of maintaining international peace and security and promoting international co-operation and understanding.

Article IV

States Parties to the Treaty undertake not to place in orbit around the earth any objects carrying nuclear weapons or any other kinds of weapons of mass destruction, instal such weapons on celestial bodies, or station such weapons in outer space in any other manner.

The moon and other celestial bodies shall be used by all States Parties to the Treaty exclusively for peaceful purposes. The establishment of military bases, installations and fortifications, the testing of any type of weapons and the conduct of military manœuvres on celestial bodies shall be forbidden. The use of military personnel for scientific research or for any other peaceful purposes shall not be prohibited. The use of any equipment or facility necessary for peaceful exploration of the moon and other celestial bodies shall also not be prohibited.

Article V

States Parties to the Treaty shall regard astronauts as envoys of mankind in outer space and shall render to them all possible assistance in the event of accident, distress, or emergency landing on the territory of another State Party or on the high seas. When astronauts make such a landing, they shall be safely and promptly returned to the State of registry of their space vehicle.

In carrying on activities in outer space and on celestial bodies, the astronauts of one State Party shall render all possible assistance to the astronauts of other States Parties.

States Parties to the Treaty shall immediately inform the other States Parties to the Treaty or the Secretary-General of the United Nations of any phenomena they discover in outer space, including the moon and other celestial bodies, which could constitute a danger to the life or health of astronauts.

## Article VI

States Parties to the Treaty shall bear international responsibility for national activities in outer space, including the moon and other celestial bodies, whether such activities are carried on by governmental agencies or non-governmental entities, and for assuring that national activities are carried out in conformity with the provisions set forth in the present Treaty. The activities of non-governmental entities in outer space, including the moon and other celestial bodies, shall require authorization and continuing supervision by the appropriate State Party to the Treaty. When activities are carried on in outer space, including moon and other celestial bodies, by an international organization, responsibility for compliance with this Treaty shall be borne both by the international organization and by the States Parties to the Treaty participating in such organization.

## Article VII

Each State Party to the Treaty that launches or procures the launching of an object into outer space, including the moon and other celestial bodies, and each State Party from whose territory or facility an object is launched, is internationally liable for damage to another State Party to the Treaty or to its natural or juridical persons by such object or its component parts on the Earth, in air space or in outer space, including the moon and other celestial bodies.

## Article VIII

A State Party to the Treaty on whose registry an object launched into outer space is carried shall retain jurisdiction and control over such object, and over any personnel thereof, while in outer space or on a celestial body. Ownership of objects launched into outer space, including objects landed or constructed on a celestial body, and of their component parts, is not affected by their presence in outer space or on a celestial body or by their return to the Earth. Such objects or component parts found beyond the limits of the State Party to the Treaty on whose registry they are carried shall be returned to that State Party, which shall, upon request, furnish identifying data prior to this return.

## Article IX

In the exploration and use of outer space, including the moon and other celestial bodies, States Parties to the Treaty shall be guided by the principle of co-operation and mutual assistance and shall conduct all their activities in outer space, including the moon and other celestial bodies, with due regard to the corresponding interests of all other States Parties to the Treaty. States Parties to the Treaty shall pursue studies of outer space, including the moon and other celestial bodies, and conduct exploration of them so as to avoid their harmful contamination and also adverse changes in the environment of the Earth resulting from the introduction of extraterrestrial matter and, where necessary, shall adopt appropriate measures for this purpose. If a State Party to the Treaty has reason to believe that an activity or experiment planned by it or its nationals in outer space, including the moon and other celestial bodies, would cause potentially harmful interference with activities of other States Parties in the peaceful exploration and use of outer space, including the moon and other celestial bodies, it shall undertake appropriate international consultations before proceeding with any such activity or experiment. A State Party to the Treaty which has reason to believe that an activity or experiment planned by another State Party in outer space, including the moon and other celestial bodies, would cause potentially harmful interference with activities in the peaceful exploration and use of outer space, including the moon and other celestial bodies, may request consultation concerning the activity or experiment.

## Article X

In order to promote international co-operation in the exploration and use of outer space, including the moon and other celestial bodies, in conformity with the purposes of this Treaty, the States Parties to the Treaty shall consider on a basis of equality any requests by other States Parties to the Treaty to be afforded an opportunity to observe the flight of space objects launched by those States.

The nature of such an opportunity for observation and the conditions under which it could be afforded shall be determined by agreement between the States concerned.

## Article XI

In order to promote international co-operation in the peaceful exploration and use of outer space, States Parties to the Treaty conducting activities in outer space, including the moon and other celestial bodies, agree to inform the Secretary-General of the United Nations as well as the public and the international scientific community, to the greatest extent feasible and practicable, of the nature, conduct, locations and results of such activities. On receiving the said information, the Secretary-General of the United Nations should be prepared to disseminate it immediately and effectively.

## Article XII

All stations, installations, equipment and space vehicles on the moon and other celestial bodies shall be open to representatives of other States Parties to the Treaty on a basis of reciprocity. Such representatives shall give reasonable advance notice of a projected visit, in order that appropriate consultations may be held and that maximum precautions may be taken to assure safety and to avoid interference with normal operations in the facility to be visited.

## Article XIII

The provisions of this Treaty shall apply to the activities of States Parties to the Treaty in the exploration and use of outer space, including the moon and other celestial bodies, whether such activities are carried on by a single State Party to the Treaty or jointly with other States, including cases where they are carried on within the framework of international inter-governmental organizations.

Any practical questions arising in connexion with activities carried on by international inter-governmental organizations in the exploration and use of outer space, including the moon and other celestial bodies, shall be resolved by the States Parties to the Treaty either with the appropriate international organization or with one or more States members of that international organization, which are Parties to this Treaty.

## Article XIV

1. This Treaty shall be open to all States for signature. Any State which does not sign this Treaty before its entry into force in accordance with paragraph 3 of this Article may accede to it at any time.

2. This Treaty shall be subject to ratification by signatory States. Instruments of ratification and instruments of accession shall be deposited with the Governments of the United Kingdom of Great Britain and Northern Ireland, the Union of Soviet Socialist Republics and the United States of America, which are hereby designated the Depositary Governments.

3. This Treaty shall enter into force upon the deposit of instruments of ratification by five Governments including the Governments designated as Depositary Governments under this Treaty.

4. For States whose instruments of ratification or accession are deposited subsequent to the entry into force of this Treaty, it shall enter into force on the date of the deposit of their instruments of ratification or accession.

5. The Depositary Governments shall promptly inform all signatory and acceding States of the date of each signature, the date of deposit of each instrument of ratification of and accession to this Treaty, the date of its entry into force and other notices.

6. This Treaty shall be registered by the Depositary Governments pursuant to Article 102 of the Charter of the United Nations.

## Article XV

Any State Party to the Treaty may propose amendments to this Treaty. Amendments shall enter into force for each State Party to the Treaty accepting the amendments upon their acceptance by a majority of the States Parties to the Treaty and thereafter for each remaining State Party to the Treaty on the date of acceptance by it.

## Article XVI

Any State Party to the Treaty may give notice of its withdrawal from the Treaty one year after its entry into force by written notification to the Depositary Governments. Such withdrawal shall take effect one year from the date of receipt of this notification.

## Article XVII

This Treaty, of which the English, Russian, French, Spanish and Chinese texts are equally authentic, shall be deposited in the archives of the Depositary Governments. Duly certified copies of this Treaty shall be transmitted by the Depositary Governments to the Governments of the signatory and acceding States.

In witness whereof the undersigned, duly authorized, have signed this Treaty.

Done in triplicate, at the cities of London, Moscow and Washington, the twenty-seventh day of January, one thousand nine hundred and sixty-seven.

## Treaty Between the United States of America and the Union of Soviet Socialist Republics on the Limitation of Anti-Ballistic Missile Systems

ENTERED INTO FORCE: 3 October 1972

---

The United States of America and the Union of Soviet Socialist Republics, hereinafter referred to as the Parties,

Proceeding from the premise that nuclear war would have devastating consequences for all mankind,

Considering that effective measures to limit anti-ballistic missile systems would be a substantial factor in curbing the race in strategic offensive arms and would lead to a decrease in the risk of outbreak of war involving nuclear weapons,

Proceeding from the premise that the limitation of anti-ballistic missile systems, as well as certain agreed measures with respect to the limitation of strategic offensive arms, would contribute to the creation of more favorable conditions for further negotiations on limiting strategic arms,

Mindful of their obligations under Article VI of the Treaty on the Non-Proliferation of Nuclear Weapons,

Declaring their intention to achieve at the earliest possible date the cessation of the nuclear arms race and to take effective measures towards reductions in strategic arms, nuclear disarmament, and general and complete disarmament,

Desiring to contribute to the relaxation of international tension and the strengthening of trust between States,

Have agreed as follows:

### Article I

1. Each party undertakes to limit anti-ballistic missile (ABM) systems and to adopt other measures in accordance with the provisions of this Treaty.
2. Each Party undertakes not to deploy ABM systems for a defense of the territory of its country and not to provide a base for such a defense, and not to deploy ABM systems for defense of an individual region except as provided for in Article III of this Treaty.

### Article II

1. For the purpose of this Treaty an ABM system is a system to counter strategic ballistic missiles or their elements in flight trajectory, currently consisting of:

(*a*) ABM interceptor missiles, which are interceptor missiles constructed and deployed for an ABM role, or of a type tested in an ABM mode;

(*b*) ABM launchers, which are launchers constructed and deployed for launching ABM interceptor missiles; and (*c*) ABM radars, which are radars constructed and deployed for an ABM role, or of a type tested in an ABM mode.

2. The ABM system components listed in paragraph 1 of this Article include those which are:

(*a*) operational;
(*b*) under construction;
(*c*) undergoing testing;
(*d*) undergoing overhaul, repair or conversion; or
(*e*) mothballed.

## Article III

Each Party undertakes not to deploy ABM systems or their components except that:

(*a*) within one ABM system deployment area having a radius of one hundred and fifty kilometers and centered on the Party's national capital, a Party may deploy: (1) no more than one hundred ABM launchers and no more than one hundred ABM interceptor missiles at launch sites, and (2) ABM radars within no more than six ABM radar complexes, the area of each complex being circular and having a diameter of no more than three kilometers; and

(*b*) within one ABM system deployment area having a radius of one hundred and fifty kilometers and containing ICBM silo launchers, a Party may deploy: (1) no more than one hundred ABM launchers and no more than one hundred ABM interceptor missiles at launch sites, (2) two large phased-array ABM radars comparable in potential to corresponding ABM radars operational or under construction on the date of signature of the Treaty in an ABM system deployment area containing ICBM silo launchers, and (3) no more than eighteen ABM radars each having a potential less than the potential of the smaller of the above-mentioned two large phased-array ABM radars.

## Article IV

The limitations provided for in Article III shall not apply to ABM systems or their components used for development or testing, and located within current or additionally agreed test ranges. Each Party may have no more than a total of fifteen ABM launchers at test ranges.

## Article V

1. Each Party undertakes not to develop, test, or deploy ABM systems or components which are sea-based, air-based, space-based, or mobile land-based.

2. Each Party undertakes not to develop, test, or deploy ABM launchers for launching more than one ABM interceptor missile at a time from each launcher, not to modify deployed launchers to provide them with such a capability, not to develop, test, or deploy automatic or semi-automatic or other similar systems for rapid reload of ABM launchers.

### Article VI

To enhance assurance of the effectiveness of the limitations on ABM systems and their components provided by the Treaty, each Party undertakes:

(*a*) not to give missiles, launchers, or radars, other than ABM interceptor missiles, ABM launchers, or ABM radars, capabilities to counter strategic ballistic missiles or their elements in flight trajectory, and not to test them in an ABM mode; and

(*b*) not to deploy in the future radars for early warning of strategic ballistic missile attack except at locations along the periphery of its national territory and oriented outward.

### Article VII

Subject to the provisions of this Treaty, modernization and replacement of ABM systems or their components may be carried out.

### Article VIII

ABM systems or their components in excess of the numbers or outside the areas specified in this Treaty, as well as ABM systems or their components prohibited by this Treaty, shall be destroyed or dismantled under agreed procedures within the shortest possible agreed period of time.

### Article IX

To assure the viability and effectiveness of this Treaty, each Party undertakes not to transfer to other States, and not to deploy outside its national territory, ABM systems or their components limited by this Treaty.

### Article X

Each Party undertakes not to assume any international obligations which would conflict with this Treaty.

### Article XI

The Parties undertake to continue active negotiations for limitations on strategic offensive arms.

### Article XII

1. For the purpose of providing assurance of compliance with the provisions of this Treaty, each Party shall use national technical means of verification at its disposal in a manner consistent with generally recognized principles of international law.
2. Each Party undertakes not to interfere with the national technical means of verification of the other Party operating in accordance with paragraph 1 of this Article.

3. Each Party undertakes not to use deliberate concealment measures which impede verification by national technical means of compliance with the provisions of this Treaty. This obligation shall not require changes in current construction, assembly, conversion, or overhaul practices.

## Article XIII

1. To promote the objectives and implementation of the provisions of this Treaty, the Parties shall establish promptly a Standing Consultative Commission, within the framework of which they will:

(*a*) consider questions concerning compliance with the obligations assumed and related situations which may be considered ambiguous;

(*b*) provide on a voluntary basis such information as either Party considers necessary to assure confidence in compliance with the obligations assumed;

(*c*) consider questions involving unintended interference with national technical means of verification;

(*d*) consider possible changes in the strategic situation which have a bearing on the provisions of this Treaty;

(*e*) agree upon procedures and dates for destruction or dismantling of ABM systems or their components in cases provided for by the provisions of this Treaty;

(*f*) consider, as appropriate, possible proposals for further increasing the viability of this Treaty; including proposals for amendments in accordance with the provisions of this Treaty;

(*g*) consider, as appropriate, proposals for further measures aimed at limiting strategic arms.

2. The Parties through consultation shall establish, and may amend as appropriate, Regulations for the Standing Consultative Commission governing procedures, composition and other relevant matters.

## Article XIV

1. Each Party may propose amendments to this Treaty. Agreed amendments shall enter into force in accordance with the procedures governing the entry into force of this Treaty.

2. Five years after entry into force of this Treaty, and at five-year intervals thereafter, the Parties shall together conduct a review of this Treaty.

## Article XV

1. This Treaty shall be of unlimited duration.

2. Each Party shall, in exercising its national sovereignty, have the right to withdraw from this Treaty if it decides that extraordinary events related to the subject matter of this Treaty have jeopardized its supreme interests. It shall give notice of its decision to the other Party six months prior to withdrawal from the Treaty. Such notice shall include a statement of the extraordinary events the notifying Party regards as having jeopardized its supreme interests.

## Article XVI

1. This Treaty shall be subject to ratification in accordance with the constitutional procedures of each Party. The Treaty shall enter into force on the day of the exchange of instruments of ratification.

2. This Treaty shall be registered pursuant to Article 102 of the Charter of the United Nations.

DONE at Moscow on May 26, 1972, in two copies, each in the English and Russian languages, both texts being equally authentic.

| | |
|---|---|
| *For the United States of America* | *For the Union of Soviet Socialist Republics* |
| President of the United States of America | General Secretary of the Central Committee of the CPSU |

---

# Agreed Statements, Common Understandings, and Unilateral Statements Regarding the Treaty Between the United States of America and the Union of Soviet Socialist Republics on the Limitation of Anti-Ballistic Missiles

## 1. Agreed Statements

The document set forth below was agreed upon and initialed by the Heads of the Delegations on May 26, 1972 (letter designations added);

### AGREED STATEMENTS REGARDING THE TREATY BETWEEN THE UNITED STATES OF AMERICA AND THE UNION OF SOVIET SOCIALIST REPUBLICS ON THE LIMITATION OF ANTI-BALLISTIC MISSILE SYSTEMS

[A]

The Parties understand that, in addition to the ABM radars which may be deployed in accordance with subparagraph (a) of Article III of the Treaty, those non-phased-array ABM radars operational on the date of signature of the Treaty within the ABM system deployment area for defense of the national capital may be retained.

[B]

The Parties understand that the potential (the product of mean emitted power in watts and antenna area in square meters) of the smaller of the two large phased-array ABM radars referred to in subparagraph (b) of Article III of the Treaty is considered for purposes of the Treaty to be three million.

[C]

The Parties understand that the centre of the ABM system deployment area centered on the national capital and the centre of the ABM system deployment area containing ICBM silo launchers for each Party shall be separated by no less than thirteen hundred kilometers.

[D]

In order to insure fulfillment of the obligation not to deploy ABM systems and their components except as provided in Article III of the Treaty, the Parties agree that in the event ABM systems based on other physical principles and including components capable of substituting for ABM interceptor missiles, ABM launchers, or ABM radars are created in the future, specific limitations on such systems and their components would be subject to discussion in accordance with Article XIII and agreement in accordance with Article XIV of the Treaty.

[E]

The Parties understand that Article V of the Treaty includes obligations not to develop, test or deploy ABM interceptor missiles for the delivery by each ABM interceptor missile of more than one independently guided warhead.

[F]

The Parties agree not to deploy phased-array radars having a potential (the product of mean emitted power in watts and antenna area in square meters) exceeding three million, except as provided for in Articles III, IV and VI of the Treaty, or except for the purposes of tracking objects in outer space or for use as national technical means of verification.

[G]

The Parties understand that Article IX of the Treaty includes the obligation of the US and the USSR not to provide to other States technical descriptions or blue prints specially worked out for the construction of ABM systems and their components limited by the Treaty.

## 2. Common Understandings

Common understanding of the Parties on the following matters was reached during the negotiations:

### A. Location of ICBM Defenses

The U.S. Delegation made the following statement on May 26, 1972:

> Article III of the ABM Treaty provides for each side one ABM system deployment area centered on its national capital and one ABM system deployment area containing ICBM silo launchers. The two sides have registered

agreement on the following statement: "The Parties understand that the centre of the ABM system deployment area centered on the national capital and the center of the ABM system deployment area containing ICBM silo launchers for each Party shall be separated by no less than thirteen hundred kilometers." In this connection, the U.S. side notes that its ABM system deployment area for defense of ICBM silo launchers, located west of the Mississippi River, will be centered in the Grand Forks ICBM silo launcher deployment area. (See Agreed Statement [C].)

### B. ABM Test Ranges

The U.S. Delegation made the following statement on April 26, 1972:

> Article IV of the ABM Treaty provides that "the limitations provided for in Article III shall not apply to ABM systems or their components used for development or testing, and located within current or additionally agreed test ranges." We believe it would be useful to assure that there is no misunderstanding as to current ABM test ranges. It is our understanding that ABM test ranges encompass the area within which ABM components are located for test purposes. The current U.S. ABM test ranges are at White Sands, New Mexico, and at Kwajalein Atoll, and the current Soviet ABM test range is near Sary Shagan in Kazakhstan. We consider that non-phased array radars of types used for range safety or instrumentation purposes may be located outside of ABM test ranges. We interpret the reference in Article IV to "additionally agreed test ranges" to mean that ABM components will not be located at any other test ranges without prior agreement between our Governments that there will be such additional ABM test ranges.

On May 5, 1972, the Soviet Delegation stated that there was a common understanding on what ABM test ranges were, that the use of the types of non ABM radars for range safety or instrumentation was not limited under the Treaty, that the reference in Article IV to "additionally agreed" test ranges was sufficiently clear, and that national means permitted identifying current test ranges.

### C. Mobile ABM Systems

On January 29, 1972, the U.S. Delegation made the following statement:

> Article V(1) of the Joint Draft Text of the ABM Treaty includes an undertaking not to develop, test, or deploy mobile land-based ABM systems and their components. On May 5, 1971, the U.S. side indicated that, in its view, a prohibition on deployment of mobile ABM systems and components would rule out the deployment of ABM launchers and radars which were not permanent fixed types. At that time, we asked for the Soviet view of this interpretation. Does the Soviet side agree with the U.S. side's interpretation put forward on May 5, 1971?

On April 13, 1972, the Soviet Delegation said there is a general common understanding on this matter.

### D. Standing Consultative Commission

Ambassador Smith made the following statement on May 22, 1972:

The United States proposes that the sides agree that, with regard to initial implementation of the ABM Treaty's Article XIII on the Standing Consultative Commission (SCC) and of the consultation Articles to the Interim Agreement on offensive arms and the Accidents Agreement,[1] agreement establishing the SCC will be worked out early in the follow-on SALT negotiations; until that is completed, the following arrangements will prevail: when SALT is in session, any consultation desired by either side under these Articles can be carried out by the two SALT Delegations; when SALT is not in session, *ad hoc* arrangements for any desired consultations under these Articles may be made through diplomatic channels.

Minister Semenov replied that, on an *ad referendum* basis, he could agree that the U.S. statement corresponded to the Soviet understanding.

E. Standstill

On May 6, 1972, Minister Semenov made the following statement:

In an effort to accommodate the wishes of the U.S. side, the Soviet Delegation is prepared to proceed on the basis that the two sides will in fact observe the obligations of both the Interim Agreement and the ABM Treaty beginning from the date of signature of these two documents.

In reply, the U.S. Delegation made the following statement on May 20, 1972:

The U.S. agrees in principle with the Soviet statement made on May 6 concerning observance of obligations beginning from date of signature but we would like to make clear our understanding that this means that, pending ratification and acceptance, neither side would take any action prohibited by the agreements after they had entered into force. This understanding would continue to apply in the absence of notification by either signatory of its intention not to proceed with ratification or approval.

The Soviet Delegation indicated agreement with the U.S. statement.

## 3. Unilateral Statements

The following noteworthy unilateral statements were made during the negotiations by the United States Delegation:

A. Withdrawal from the ABM Treaty

On May 9, 1972, Ambassador Smith made the following statement:

The U.S. Delegation has stressed the importance the U.S. Government

[1] See Article 7 of Agreement to Reduce the Risk of Outbreak of Nuclear War Between the United States of America and the Union of Soviet Socialist Republics, signed Sept. 30, 1971.

attaches to achieving agreement on more complete limitations on strategic offensive arms, following agreement on an ABM Treaty and on an Interim Agreement on certain measures with respect to the limitation of strategic offensive arms. The U.S. Delegation believes that an objective of the follow-on negotiations should be to constrain and reduce on a long-term basis threats to the survivability of our respective strategic retaliatory forces. The USSR Delegation has also indicated that the objectives of SALT would remain unfulfilled without the achievement of an agreement providing for more complete limitations on strategic offensive arms. Both sides recognize that the initial agreements would be steps toward the achievement of more complete limitations on strategic arms. If an agreement providing for more complete strategic offensive arms limitations were not achieved within five years, U.S. supreme interests could be jeopardized. Should that occur, it would constitute a basis for withdrawal from the ABM Treaty. The U.S. does not wish to see such a situation occur, nor do we believe that the USSR does. It is because we wish to prevent such a situation that we emphasize the importance the U.S. Government attaches to achievement of more complete limitations on strategic offensive arms. The U.S. Executive will inform the Congress, in connection with Congressional consideration of the ABM Treaty and the Interim Agreement, of this statement of the U.S. position.

## B. Tested in ABM Mode

On April 7, 1972, the U.S. Delegation made the following statement:

Article II of the Joint Text Draft uses the term "tested in an ABM mode," in defining ABM components, and Article VI includes certain obligations concerning such testing. We believe that the sides should have a common understanding of this phrase. First, we would note that the testing provisions of the ABM Treaty are intended to apply to testing which occurs after the date of signature of the Treaty, and not to any testing which may have occurred in the past. Next, we would amplify the remarks we have made on this subject during the previous Helsinki phase by setting forth the objectives which govern the U.S. view on the subject, namely, while prohibiting testing of non-ABM components for ABM purposes: not to prevent testing of ABM components, and not to prevent testing of non-ABM components for non-ABM purposes. To clarify our interpretation of "tested in an ABM mode," we note that we would consider a launcher, missile or radar to be "tested in an ABM mode" if, for example, any of the following events occur: (1) a launcher is used to launch an ABM interceptor missile, (2) an interceptor missile is flight tested against a target vehicle which has a flight trajectory with characteristics of a strategic ballistic missile flight trajectory, or is flight tested in conjunction with the test of an ABM interceptor missile or an ABM radar at the same test range, or is flight tested to an altitude inconsistent with interception of targets against which air defenses are deployed, (3) a radar makes measurements on a cooperative target vehicle of the kind referred to in item (2) above during the reentry portion of its trajectory or makes measurements in conjunction with the test of an ABM interceptor missile or an ABM radar at the same test range. Radars used for purposes such as range safety or instrumentation would be exempt from application of these criteria.

### C. No-Transfer Article of ABM Treaty

On April 18, 1972, the U.S. Delegation made the following statement:

> In regard to this Article [IX], I have a brief and I believe self-explanatory statement to make. The U.S. side wishes to make clear that the provisions of this Article do not set a precedent for whatever provision may be considered for a Treaty on Limiting Strategic Offensive Arms. The question of transfer of strategic offensive arms is a far more complex issue, which may require a different solution.

### D. No Increase in Defense of Early Warning Radars

On July 28, 1970, the U.S. Delegation made the following statement:

> Since Hen House radars [Soviet ballistic missile early warning radars] can detect and track ballistic missile warheads at great distances, they have a significant ABM potential. Accordingly, the U.S. would regard any increase in the defenses of such radars by surface-to-air missiles as inconsistent with an agreement.

## Interim Agreement Between the United States of America and the Union of Soviet Socialist Republics on Certain Measures With Respect to the Limitation of Strategic Offensive Arms

ENTERED INTO FORCE: 3 October 1972

---

The United States of America and the Union of Soviet Socialist Republics, hereinafter referred to as the Parties,

Convinced that the Treaty on the Limitation of Anti-Ballistic Missile Systems and this Interim Agreement on Certain Measures with Respect to the Limitation of Strategic Offensive Arms will contribute to the creation of more favorable conditions for active negotiations on limiting strategic arms as well as to the relaxation of international tension and the strengthening of trust between States,

Taking into account the relationship between strategic offensive and defensive arms,

Mindful of their obligations under Article VI of the Treaty on the Non-Proliferation of Nuclear Weapons,

Have agreed as follows:

### Article I

The Parties undertake not to start construction of additional fixed land-based intercontinental ballistic missile (ICBM) launchers after July 1, 1972.

### Article II

The Parties undertake not to convert land-based launchers for light ICBMs, or for ICBMs of older types deployed prior to 1964, into land-based launchers for heavy ICBMs of types deployed after that time.

### Article III

The Parties undertake to limit submarine-launched ballistic missile (SLBM) launchers and modern ballistic missile submarines to the numbers operational and under construction on the date of signature of this Interim Agreement, and in addition to launchers and submarines constructed under procedures established by the Parties as replacements for an equal number of ICBM launchers of older types deployed prior to 1964 or for launchers on older submarines.

### Article IV

Subject to the provisions of this Interim Agreement, modernization and replacement of strategic offensive ballistic missiles and launchers covered by this Interim Agreement may be undertaken.

## Article V

1. For the purpose of providing assurance of compliance with the provisions of this Interim Agreement, each Party shall use national technical means of verification at its disposal in a manner consistent with generally recognized principles of international law.

2. Each party undertakes not to interfere with the national technical means of verification of the other Party operating in accordance with paragraph 1 of this Article.

3. Each Party undertakes not to use deliberate concealment measures which impede verification by national technical means of compliance with the provisions of this Interim Agreement. This obligation shall not require changes in current construction, assembly, conversion, or overhaul practices.

## Article VI

To promote the objectives and implementation of the provisions of this Interim Agreement, the Parties shall use the Standing Consultative Commission established under Article XIII of the Treaty on the Limitation of Anti-Ballistic Missile Systems in accordance with the provisions of that Article.

## Article VII

The Parties undertake to continue active negotiations for limitations on strategic offensive arms. The obligations provided for in this Interim Agreement shall not prejudice the scope or terms of the limitations on strategic offensive arms which may be worked out in the course of further negotiations.'

## Article VIII

1. This Interim Agreement shall enter into force upon exchange of written notices of acceptance by each Party, which exchange shall take place simultaneously with the exchange of instruments of ratification of the Treaty on the Limitation of Anti-Ballistic Missile Systems.

2. This Interim Agreement shall remain in force for a period of five years unless replaced earlier by an agreement on more complete measures limiting strategic offensive arms. It is the objective of the Parties to conduct active follow-on negotiations with the aim of concluding such an agreement as soon as possible.

3. Each Party shall, in exercising its national sovereignty, have the right to withdraw from this Interim Agreement if it decides that extraordinary events related to the subject matter of this Interim Agreement have jeopardized its supreme interests. It shall give notice of its decision to the other Party six months prior to withdrawal from this Interim Agreement. Such notice shall include a statement of the extraordinary events the notifying Party regards as having jeopardized its supreme interests.

DONE at Moscow on May 26, 1972, in two copies, each in the English and Russian languages, both texts being equally authentic.

| | |
|---|---|
| *For the United States of America* | *For the Union of Soviet Socialist Republics* |
| President of the United States of America | General Secretary of the Central Committee of the CPSU |

---

## Protocol to the Interim Agreement Between the United States of America and the Union of Soviet Socialist Republics on Certain Measures With Respect to the Limitation of Strategic Offensive Arms

The United States of America and the Union of Soviet Socialist Republics, hereinafter referred to as the Parties,

Having agreed on certain limitations relating to submarine-launched ballistic missile launchers and modern ballistic missile submarines, and to replacement procedures, in the Interim Agreement,

Have agreed as follows:

The Parties understand that, under Article III of the Interim Agreement, for the period during which that Agreement remains in force:

The U.S. may have no more than 710 ballistic missile launchers on submarines (SLBMs) and no more than 44 modern ballistic missile submarines. The Soviet Union may have no more than 950 ballistic missile launchers on submarines and no more than 62 modern ballistic missile submarines.

Additional ballistic missile launchers on submarines up to the above-mentioned levels, in the U.S.—over 656 ballistic missile launchers on nuclear-powered submarines, and in the U.S.S.R.—over 740 ballistic missile launchers on nuclear-powered submarines, operational and under construction, may become operational as replacements for equal numbers of ballistic missile launchers of older types deployed prior to 1964 or of ballistic missile launchers on older submarines.

The deployment of modern SLBMs on any submarine, regardless of type, will be counted against the total level of SLBMs permitted for the U.S. and the U.S.S.R.

This Protocol shall be considered an integral part of the Interim Agreement.

DONE at Moscow this 26th day of May, 1972

| | |
|---|---|
| *For the United States of America* | *For the Union of Soviet Socialist Republics* |
| President of the United States of America | General Secretary of the Central Committee of the CPSU |

---

## Agreed Statements, Common Understandings, and Unilateral Statements Regarding the Interim Agreement Between the United States of America and the Union of Soviet Socialist Republics on Certain Measures With Respect to the Limitation of Strategic Offensive Arms

### 1. Agreed Statements

The document set forth below was agreed upon and initialed by the Heads of the Delegations on May 26, 1972 (letter designations added):

AGREED STATEMENTS REGARDING THE INTERIM AGREEMENT BETWEEN THE UNITED STATES OF AMERICA AND THE UNION OF SOVIET SOCIALIST REPUBLICS ON CERTAIN MEASURES WITH RESPECT TO THE LIMITATION OF STRATEGIC OFFENSIVE ARMS

[A]

The Parties understand that land-based ICBM launchers referred to in the Interim Agreement are understood to be launchers for strategic ballistic missiles capable of ranges in excess of the shortest distance between the northeastern border of the continental U.S. and the northwestern border of the continental U.S.S.R.

[B]

The Parties understand that fixed land-based ICBM launchers under active construction as of the date of signature of the Interim Agreement may be completed.

[C]

The Parties understand that in the process of modernization and replacement the dimensions of land-based ICBM silo launchers will not be significantly increased.

[D]

The Parties understand that during the period of the Interim Agreement there shall be no significant increase in the number of ICBM or SLBM test and training launchers, or in the number of such launchers for modern land-based heavy ICBMs. The Parties further understand that construction or conversion of ICBM launchers at test ranges shall be undertaken only for purposes of testing and training.

[E]

The Parties understand that dismantling or destruction of ICBM launchers of older types deployed prior to 1964 and ballistic missile launchers on older submarines being replaced by new SLBM launchers on modern submarines will be initiated at the time of the beginning of sea trials of a replacement submarine, and will be completed in the shortest possible agreed period of time. Such dismantling or destruction, and timely notification thereof, will be accomplished under procedures to be agreed in the Standing Consultative Commission.

## 2. Common Understandings

Common understanding of the Parties on the following matters was reached during the negotiations:

### A. Increase in ICBM Silo Dimensions

Ambassador Smith made the following statement on May 26, 1972:

> The Parties agree that the term "significantly increased" means that an increase will not be greater than 10–15 percent of the present dimensions of land-based ICBM silo launchers.

Minister Semenov replied that this statement corresponded to the Soviet understanding.

### B. Standing Consultative Commission

Ambassador Smith made the following statement on May 22, 1972:

> The United States proposes that the sides agree that, with regard to initial implementation of the ABM Treaty's Article XIII on the Standing Consultative Commission (SCC) and of the consultation Articles to the Interim Agreement on offensive arms and the Accidents Agreement,[1] agreement establishing the SCC will be worked out early in the follow-on SALT negotiations; until that is completed, the following arrangements will prevail: when SALT is in session, any consultation desired by either side under these Articles can be carried out by the two SALT Delegations; when SALT is not in session, *ad hoc* arrangements for any desired consultations under these Articles may be made through diplomatic channels.

Minister Semenov replied that, on an *ad referendum* basis, he could agree that the U.S. statement corresponded to the Soviet understanding.

### C. Standstill

On May 6, 1972, Minister Semenov made the following statement:

> In an effort to accommodate the wishes of the U.S. side, the Soviet Delegation is prepared to proceed on the basis that the two sides will in fact observe the obligations of both the Interim Agreement and the ABM Treaty beginning from the date of signature of these two documents.

In reply, the U.S. Delegation made the following statement on May 20, 1972:

> The U.S. agrees in principle with the Soviet statement made on May 6 concerning observance of obligations beginning from date of signature but we

---

[1] See Article 7 of Agreement to Reduce the Risk of Outbreak of Nuclear War Between the United States of America and the Union of Soviet Socialist Republics, signed Sept. 30, 1971.

would like to make clear our understanding that this means that, pending ratification and acceptance, neither side would take any action prohibited by the agreements after they had entered into force. This understanding would continue to apply in the absence of notification by either signatory of its intention not to proceed with ratification or approval.

The Soviet Delegation indicated agreement with the U.S. statement.

### 3. Unilateral Statements

(*a*) The following noteworthy unilateral statements were made during the negotiations by the United States Delegation:

#### A. Withdrawal from the ABM Treaty

On May 9, 1972, Ambassador Smith made the following statement:

The U.S. Delegation has stressed the importance the U.S. Government attaches to achieving agreement on more complete limitations on strategic offensive arms, following agreement on an ABM Treaty and on an Interim Agreement on certain measures with respect to the limitation of strategic offensive arms. The U.S. Delegation believes that an objective of the follow-on negotiations should be to constrain and reduce on a long-term basis threats to the survivability of our respective strategic retaliatory forces. The USSR Delegation has also indicated that the objectives of SALT would remain unfulfilled without the achievement of an agreement providing for more complete limitations on strategic offensive arms. Both sides recognize that the initial agreements would be steps toward the achievement of more complete limitations on strategic arms. If an agreement providing for more complete strategic offensive arms limitations were not achieved within five years, U.S. supreme interests could be jeopardized. Should that occur, it would constitute a basis for withdrawal from the ABM Treaty. The U.S. does not wish to see such a situation occur, nor do we believe that the USSR does. It is because we wish to prevent such a situation that we emphasize the importance the U.S. Government attaches to achievement of more complete limitations on strategic offensive arms. The U.S. Executive will inform the Congress, in connection with Congressional consideration of the ABM Treaty and the Interim Agreement, of this statement of the U.S. position.

#### B. Land-Mobile ICBM Launchers

The U.S. Delegation made the following statement on May 20, 1972:

In connection with the important subject of land-mobile ICBM launchers, in the interest of concluding the Interim Agreement the U.S. Delegation now withdraws its proposal that Article I or an agreed statement explicitly prohibit the deployment of mobile land-based ICBM launchers. I have been instructed to inform you that, while agreeing to defer the question of limitation of operational land-mobile ICBM launchers to the subsequent negotiations on more complete limitations on strategic offensive arms, the U.S. would consider the deployment of operational land-mobile ICBM launchers during the period of the Interim Agreement as inconsistent with the objectives of that Agreement.

C. Covered Facilities

The U.S. Delegation made the following statement on May 20, 1972:

I wish to emphasize the importance that the United States attaches to the provisions of Article V, including in particular their application to fitting out or berthing submarines.

D. "Heavy" ICBM's

The U.S. Delegation made the following statement on May 26, 1972:

The U.S. Delegation regrets that the Soviet Delegation has not been willing to agree on a common definition of a heavy missile. Under these circumstances, the U.S. Delegation believes it necessary to state the following: The United States would consider any ICBM having a volume significantly greater than that of the largest light ICBM now operational on either side to be a heavy ICBM. The U.S. proceeds on the premise that the Soviet side will give due account to this consideration.

(*b*) The following noteworthy unilateral statement was made by the Delegation of the U.S.S.R. and is shown here with the U.S. reply:

On May 17, 1972, Minister Semenov made the following unilateral "Statement of the Soviet Side":

Taking into account that modern ballistic missile submarines are presently in the possession of not only the U.S., but also of its NATO allies, the Soviet Union agrees that for the period of effectiveness of the Interim 'Freeze' Agreement the U.S. and its NATO allies have up to 50 such submarines with a total of up to 800 ballistic missile launchers thereon (including 41 U.S. submarines with 656 ballistic missile launchers). However, if during the period of effectiveness of the Agreement U.S. allies in NATO should increase the number of their modern submarines to exceed the numbers of submarines they would have operational or under construction on the date of signature of the Agreement, the Soviet Union will have the right to a corresponding increase in the number of its submarines. In the opinion of the Soviet side, the solution of the question of modern ballistic missile submarines provided for in the Interim Agreement only partially compensates for the strategic imbalance in the deployment of the nuclear-powered missile submarines of the USSR and the U.S. Therefore, the Soviet side believes that this whole question, and above all the question of liquidating the American missile submarine bases outside the U.S., will be appropriately resolved in the course of follow-on negotiations.

On May 24, Ambassador Smith made the following reply to Minister Semenov:

The United States side has studied the "statement made by the Soviet side" of May 17 concerning compensation for submarine basing and SLBM submarines belonging to third countries. The United States does not accept the validity of the considerations in that statement.

On May 26 Minister Semenov repeated the unilateral statement made on May 17. Ambassador Smith also repeated the U.S. rejection on May 26.

# Agreement Governing the Activities of States on the Moon and Other Celestial Bodies

*The States Parties to this Agreement,*

*Noting* the achievements of States in the exploration and use of the moon and other celestial bodies,

*Recognizing* that the moon, as a natural satellite of the earth, has an important role to play in the exploration of outer space,

*Determined* to promote on the basis of equality the further development of co-operation among States in the exploration and use of the moon and other celestial bodies,

*Desiring* to prevent the moon from becoming an area of international conflict,

*Bearing in mind* the benefits which may be derived from the exploitation of the natural resources of the moon and other celestial bodies,

*Recalling* the Treaty on Principles Governing the Activities of States in the Exploration and Use of Outer Space, including the Moon and Other Celestial Bodies, the Agreement on the Rescue of Astronauts, the Return of Astronauts and the Return of Objects Launched into Outer Space, the Convention on International Liability for Damage Caused by Space Objects, and the Convention on Registration of Objects Launched into Outer Space,

*Taking into account* the need to define and develop the provisions of these international instruments in relation to the moon and other celestial bodies, having regard to further progress in the exploration and use of outer space,

*Have agreed* on the following:

Article 1

1. The provisions of this Agreement relating to the moon shall also apply to other celestial bodies within the solar system, other than the earth, except in so far as specific legal norms enter into force with respect to any of these celestial bodies.

2. For the purposes of this Agreement reference to the moon shall include orbits around or other trajectories to or around it.

3. This Agreement does not apply to extraterrestrial materials which reach the surface of the earth by natural means.

Article 2

All activities on the moon, including its exploration and use, shall be carried out in accordance with international law, in particular the Charter of the United Nations, and taking into account the Declaration on Principles of International Law concerning Friendly Relations and Co-operation among States in accordance with the Charter of the United Nations, adopted by the General Assembly on 24 October 1970, in the interests of maintaining international peace and security and promoting international co-operation and mutual understanding, and with due regard to the corresponding interests of all other States Parties.

## Article 3

1. The moon shall be used by all States Parties exclusively for peaceful purposes.

2. Any threat or use of force or any other hostile act or threat of hostile act on the moon is prohibited. It is likewise prohibited to use the moon in order to commit any such act or to engage in any such threat in relation to the earth, the moon, spacecraft, the personnel of spacecraft or man-made space objects.

3. States Parties shall not place in orbit around or other trajectory to or around the moon objects carrying nuclear weapons or any other kinds of weapons of mass destruction or place or use such weapons on or in the moon.

4. The establishment of military bases, installations and fortifications, the testing of any type of weapons and the conduct of military manoeuvres on the moon shall be forbidden. The use of military personnel for scientific research or for any other peaceful purposes shall not be prohibited. The use of any equipment or facility necessary for peaceful exploration and use of the moon shall also not be prohibited.

## Article 4

1. The exploration and use of the moon shall be the province of all mankind and shall be carried out for the benefit and in the interests of all countries, irrespective of their degree of economic or scientific development. Due regard shall be paid to the interests of present and future generations as well as to the need to promote higher standards of living and conditions of economic and social progress and development in accordance with the Charter of the United Nations.

2. States Parties shall be guided by the principle of co-operation and mutual assistance in all their activities concerning the exploration and use of the moon. International co-operation in pursuance of this Agreement should be as wide as possible and may take place on a multilateral basis, on a bilateral basis or through international intergovernmental organizations.

## Article 5

1. States Parties shall inform the Secretary-General of the United Nations as well as the public and the international scientific community, to the greatest extent feasible and practicable, of their activities concerned with the exploration and use of the moon. Information on the time, purposes, locations, orbital parameters and duration shall be given in respect of each mission to the moon as soon as possible after launching, while information on the results of each mission, including scientific results, shall be furnished upon completion of the mission. In the case of a mission lasting more than thirty days, information on conduct of the mission, including any scientific results, shall be given periodically at thirty days' intervals. For missions lasting more than six months, only significant additions to such information need be reported thereafter.

2. If a State Party becomes aware that another State Party plans to operate simultaneously in the same area of or in the same orbit around or trajectory to or

around the moon, it shall promptly inform the other State of the timing of and plans for its own operations.

3. In carrying out activities under this Agreement, States Parties shall promptly inform the Secretary-General, as well as the public and the international scientific community, of any phenomena they discover in outer space, including the moon, which could endanger human life or health, as well as of any indication of organic life.

## Article 6

1. There shall be freedom of scientific investigation on the moon by all States Parties without discrimination of any kind, on the basis of equality and in accordance with international law.

2. In carrying out scientific investigations and in furtherance of the provisions of this Agreement, the States Parties shall have the right to collect on and remove from the moon samples of its mineral and other substances. Such samples shall remain at the disposal of those States Parties which caused them to be collected and may be used by them for scientific purposes. States Parties shall have regard to the desirability of making a portion of such samples available to other interested States Parties and the international scientific community for scientific investigations. States Parties may in the course of scientific investigations also use mineral and other substances of the moon in quantities appropriate for the support of their missions.

3. States Parties agree on the desirability of exchanging scientific and other personnel on expeditions to or installations on the moon to the greatest extent feasible and practicable.

## Article 7

1. In exploring and using the moon, States Parties shall take measures to prevent the disruption of the existing balance of its environment whether by introducing adverse changes in that environment, by its harmful contamination through the introduction of extra-environmental matter or otherwise. States Parties shall also take measures to avoid harmfully affecting the environment of the earth through the introduction of extraterrestrial matter or otherwise.

2. States Parties shall inform the Secretary-General of the United Nations of the measures being adopted by them in accordance with paragraph 1 of this article and shall also, to the maximum extent feasible, notify him in advance of all placements by them of radio-active materials on the moon and of the purposes of such placements.

3. States Parties shall report to other States Parties and to the Secretary-General concerning areas of the moon having special scientific interest in order that, without prejudice to the rights of other States Parties, consideration may be given to the designation of such areas as international scientific preserves for which special protective arrangements are to be agreed upon in consultation with the competent bodies of the United Nations.

## Article 8

1. States Parties may pursue their activities in the exploration and use of the moon anywhere on or below its surface, subject to the provisions of this Agreement.

2. For these purposes States Parties may, in particular:

(*a*) Land their space objects on the moon and launch them from the moon;

(*b*) Place their personnel, space vehicles, equipment, facilities, stations and installations anywhere on or below the surface of the moon.

Personnel, space vehicles, equipment, facilities, stations and installations may move or be moved freely over or below the surface of the moon.

3. Activities of States Parties in accordance with paragraphs 1 and 2 of this article shall not interfere with the activities of other States Parties on the moon. Where such interference may occur, the States Parties concerned shall undertake consultations in accordance with article 15, paragraphs 2 and 3 of this Agreement.

## Article 9

1. States Parties may establish manned and unmanned stations on the moon. A State Party establishing a station shall use only that area which is required for the needs of the station and shall immediately inform the Secretary-General of the United Nations of the location and purposes of that station. Subsequently, at annual intervals that State shall likewise inform the Secretary-General whether the station continues in use and whether its purposes have changed.

2. Stations shall be installed in such a manner that they do not impede the free access to all areas of the moon by personnel, vehicles and equipment of other States Parties conducting activities on the moon in accordance with the provisions of this Agreement or of article I of the Treaty on Principles Governing the Activities of States in the Exploration and Use of Outer Space, Including the Moon and Other Celestial Bodies.

## Article 10

1. States Parties shall adopt all practicable measures to safeguard the life and health of persons on the moon. For this purpose they shall regard any person on the moon as an astronaut within the meaning of article V of the Treaty on Principles Governing the Activities of States in the Exploration and Use of Outer Space, including the Moon and Other Celestial Bodies and as part of the personnel of a spacecraft within the meaning of the Agreement on the Rescue of Astronauts, the Return of Astronauts and the Return of Objects Launched into Outer Space.

2. States Parties shall offer shelter in their stations, installations, vehicles and other facilities to persons in distress on the moon.

## Article 11

1. The moon and its natural resources are the common heritage of mankind, which finds its expression in the provisions of this Agreement and in particular in paragraph 5 of this article.

2. The moon is not subject to national appropriation by any claim of sovereignty, by means of use or occupation, or by any other means.

3. Neither the surface nor the subsurface of the moon, nor any part thereof or natural resources in place, shall become property of any State, international intergovernmental or non-governmental organization, national organization or non-governmental entity or of any natural person. The placement of personnel, space vehicles, equipment, facilities, stations and installations on or below the surface of

the moon, including structures connected with its surface or subsurface, shall not create a right of ownership over the surface or the subsurface of the moon or any areas thereof. The foregoing provisions are without prejudice to the international régime referred to in paragraph 5 of this article.

4. States Parties have the right to exploration and use of the moon without discrimination of any kind, on a basis of equality and in accordance with international law and the terms of this Agreement.

5. States Parties to this Agreement hereby undertake to establish an international régime, including appropriate procedures, to govern the exploitation of the natural resources of the moon as such exploitation is about to become feasible. This provision shall be implemented in accordance with article 18 of this Agreement.

6. In order to facilitate the establishment of the international régime referred to in paragraph 5 of this article, States Parties shall inform the Secretary-General of the United Nations as well as the public and the international scientific community, to the greatest extent feasible and practicable, of any natural resources they may discover on the moon.

7. The main purposes of the international régime to be established shall include:

(*a*) The orderly and safe development of the natural resources of the moon;

(*b*) The rational management of those resources;

(*c*) The expansion of opportunities in the use of those resources;

(*d*) An equitable sharing by all States Parties in the benefits derived from those resources, whereby the interests and needs of the developing countries, as well as the efforts of those countries which have contributed either directly or indirectly to the exploration of the moon, shall be given special consideration.

8. All the activities with respect to the natural resources of the moon shall be carried out in a manner compatible with the purposes specified in paragraph 7 of this article and the provisions of article 6, paragraph 2, of this Agreement.

## Article 12

1. States Parties shall retain jurisdiction and control over their personnel, vehicles, equipment, facilities, stations and installations on the moon. The ownership of space vehicles, equipment, facilities, stations and installations shall not be affected by their presence on the moon.

2. Vehicles, installations and equipment or their component parts found in places other than their intended location shall be dealt with in accordance with article 5 of the Agreement on Rescue of Astronauts, the Return of Astronauts and the Return of Objects Launched into Outer Space.

3. In the event of an emergency involving a threat to human life, States Parties may use the equipment, vehicles, installations, facilities or supplies of other States Parties on the moon. Prompt notification of such use shall be made to the Secretary-General of the United Nations or the State Party concerned.

## Article 13

A State Party which learns of the crash landing, forced landing or other unintended landing on the moon of a space object, or its component parts, that were not launched by it, shall promptly inform the launching State Party and the Secretary-General of the United Nations.

Article 14

1. States Parties to this Agreement shall bear international responsibility for national activities on the moon, whether such activities are carried on by governmental agencies or by non-governmental entities, and for assuring that national activities are carried out in conformity with the provisions set forth in this Agreement. States Parties shall ensure that non-governmental entities under their jurisdiction shall engage in activities on the moon only under the authority and continuing supervision of the appropriate State Party.

2. States Parties recognize that detailed arrangements concerning liability for damage caused on the moon, in addition to the provisions of the Treaty on Principles Governing the Activities of States in the Exploration and Use of Outer Space, including the Moon and Other Celestial Bodies and the Convention on International Liability for Damage caused by Space Objects, may become necessary as a result of more extensive activities on the moon. Any such arrangements shall be elaborated in accordance with the procedure provided for in article 18 of this Agreement.

Article 15

1. Each State Party may assure itself that the activities of other States Parties in the exploration and use of the moon are compatible with the provisions of this Agreement. To this end, all space vehicles, equipment, facilities, stations and installations on the moon shall be open to other States Parties. Such States Parties shall give reasonable advance notice of a projected visit, in order that appropriate consultations may be held and that maximum precautions may be taken to assure safety and to avoid interference with normal operations in the facility to be visited. In pursuance of this article, any State Party may act on its own behalf or with the full or partial assistance of any other State Party or through appropriate international procedures within the framework of the United Nations and in accordance with the Charter.

2. A State Party which has reason to believe that another State Party is not fulfilling the obligations incumbent upon it pursuant to this Agreement or that another State Party is interfering with the rights which the former State has under this Agreement may request consultations with that State Party. A State Party receiving such a request shall enter into such consultations without delay. Any other State Party which requests to do so shall be entitled to take part in the consultations. Each State Party participating in such consultations shall seek a mutually acceptable resolution of any controversy and shall bear in mind the rights and interests of all States Parties. The Secretary-General of the United Nations shall be informed of the results of the consultations and shall transmit the information received to all States Parties concerned.

3. If the consultations do not lead to a mutually acceptable settlement which has due regard for the rights and interests of all States Parties, the parties concerned shall take all measures to settle the dispute by other peaceful means of their choice appropriate to the circumstances and the nature of the dispute. If difficulties arise in connexion with the opening of consultations or if consultations do not lead to a mutually acceptable settlement, any State Party may seek the assistance of the Secretary-General, without seeking the consent of any other State Party concerned, in order to resolve the controversy. A State Party which does not maintain diplomatic relations with another State Party concerned shall participate in such consultations,

at its choice, either itself or through another State Party or the Secretary-General as intermediary.

## Article 16

With the exception of articles 17 to 21, references in this Agreement to States shall be deemed to apply to any international intergovernmental organization which conducts space activities if the organization declares its acceptance of the rights and obligations provided for in this Agreement and if a majority of the States members of the organization are States Parties to this Agreement and to the Treaty on Principles Governing the Activities of States in the Exploration and Use of Outer Space, including the Moon and Other Celestial Bodies. States members of any such organization which are States Parties to this Agreement shall take all appropriate steps to ensure that the organization makes a declaration in accordance with the foregoing.

## Article 17

Any State Party to this Agreement may propose amendments to the Agreement. Amendments shall enter into force for each State Party to the Agreement accepting the amendments upon their acceptance by a majority of the States Parties to the Agreement and thereafter for each remaining State Party to the Agreement on the date of acceptance by it.

## Article 18

Ten years after the entry into force of this Agreement, the question of the review of the Agreement shall be included in the provisional agenda of the General Assembly of the United Nations in order to consider, in the light of past application of the Agreement, whether it requires revision. However, at any time after the Agreement has been in force for five years, the Secretary-General of the United Nations, as depository, shall, at the request of one third of the States Parties to the Agreement and with the concurrence of the majority of the States Parties, convene a conference of the States Parties to review this Agreement. A review conference shall also consider the question of the implementation of the provisions of article 11, paragraph 5, on the basis of the principle referred to in paragraph 1 of that article and taking into account in particular any relevant technological developments.

## Article 19

1. This Agreement shall be open for signature by all States at United Nations Headquarters in New York.
2. This Agreement shall be subject to ratification by signatory States. Any State which does not sign this Agreement before its entry into force in accordance with paragraph 3 of this article may accede to it at any time. Instruments of ratification or accession shall be deposited with the Secretary-General of the United Nations.
3. This Agreement shall enter into force on the thirtieth day following the date of deposit of the fifth instrument of ratification.
4. For each State depositing its instrument of ratification or accession after the entry into force of this Agreement, it shall enter into force on the thirtieth day following the date of deposit of any such instrument.

5. The Secretary-General shall promptly inform all signatory and acceding States of the date of each signature, the date of deposit of each instrument of ratification or accession to this Agreement, the date of its entry into force and other notices.

## Article 20

Any State Party to this Agreement may give notice of its withdrawal from the Agreement one year after its entry into force by written notification to the Secretary-General of the United Nations. Such withdrawal shall take effect one year from the date of receipt of this notification.

## Article 21

The original of this Agreement, of which the Arabic, Chinese, English, French, Russian and Spanish texts are equally authentic, shall be deposited with the Secretary-General of the United Nations, who shall send certified copies thereof to all signatory and acceding states.

IN WITNESS WHEREOF the undersigned, being duly authorized thereto by their respective Governments, have signed this Agreement, opened for signature at New York on

## Soviet proposal for a draft treaty on the prohibition of the stationing of weapons of any kind in outer space

The States Parties to this treaty,

Guided by the goals of strengthening peace and international security,

Proceeding on the basis of their obligations under the Charter of the United Nations to refrain from the threat or use of force in any manner inconsistent with the Purposes of the United Nations,

Desiring not to allow outer space to become an arena for the arms race and a source of aggravating relations between States,

Have agreed on the following:

Article 1

1. States Parties undertake not to place in orbit around the earth objects carrying weapons of any kind, install such weapons on celestial bodies, or station such weapons in outer space in any other manner, including on reusable manned space vehicles of an existing type or of other types which States Parties may develop in the future.

2. Each State Party to this treaty undertakes not to assist, encourage or induce any State, group of States or international organization to engage in activities contrary to the provisions of paragraph 1 of this article.

Article 2

States Parties shall use space objects in strict accordance with international law, including the Charter of the United Nations, in the interest of maintaining international peace and security and promoting international co-operation and mutual understanding.

Article 3

Each State Party undertakes not to destroy, damage, disturb the normal functioning or change the flight trajectory of space objects of other States Parties, if such objects were placed in orbit in strict accordance with article 1, paragraph 1, of this treaty.

Article 4

1. For the purpose of providing assurance of compliance with the provisions of this treaty, each State Party shall use the national technical means of verification at its disposal in a manner consistent with generally recognized principles of international law.

2. Each State Party undertakes not to interfere with the national technical

means of verification of other States Parties operating in accordance with paragraph 1 of this article.

3. In order to promote the objectives and provisions of this treaty, the States Parties shall, when necessary, consult each other, make inquiries and provide information in connexion with such inquiries.

Article 5

1. Any State Party to this treaty may propose amendments to this treaty. The text of any proposed amendment shall be submitted to the depositary, who shall promptly circulate it to all States Parties.

2. The amendment shall enter into force for each State Party to this Treaty which has accepted it, upon the deposit with the depositary of instruments of acceptance by the majority of States Parties. Thereafter, the amendment shall enter into force for each remaining State Party on the date of deposit of its instrument of acceptance.

Article 6

This treaty shall be of unlimited duration.

Article 7

Each State Party shall in exercising its national sovereignty have the right to withdraw from this treaty if it decides that extraordinary events related to the subject-matter of this treaty have jeopardized its supreme interests. It shall give notice to the Secretary-General of the United Nations of the decision adopted six months before withdrawing from the treaty. Such notice shall include a statement of the extraordinary events which the notifying State Party regards as having jeopardized its supreme interests.

Article 8

1. This treaty shall be open to all States for signature at United Nations Headquarters in New York. Any State which does not sign this treaty before its entry into force in accordance with paragraph 3 of this article may accede to it at any time.

2. This treaty shall be subject to ratification by signatory States. Instruments of ratification accession shall be deposited with the Secretary-General of the United Nations.

3. This treaty shall enter into force between the States which have deposited instruments of ratification upon the deposit with the Secretary-General of the United Nations of the fifth instrument of ratification.

4. For States whose instruments of ratification or accession are deposited after the entry into force of this treaty, it shall enter into force on the date of the deposit of their instruments of ratification or accession.

5. The Secretary-General of the United Nations shall promptly inform all

signatory and acceding States of the date of each signature, the date of deposit of each instrument of ratification or accession, the date of entry into force of this treaty as well as other notices.

Article 9

This treaty, of which the Arabic, Chinese, English, French, Russian and Spanish texts are equally authentic, shall be deposited with the Secretary-General of the United Nations, who shall send duly certified copies thereof to the Governments of the signatory and acceding States.

# Appendix 3

## *Glossary of terms*

| | |
|---|---|
| Accelerator | A machine for increasing the velocity of charged elementary particles and ions. |
| Amplitude modulation | The amplitude of the carrier wave is varied. |
| Apogee | That point in the orbit of an Earth satellite which is farthest away from the Earth. |
| Argument of perigee | The angle or arc, as seen from the focus of an elliptical orbit, from the ascending node to the closest approach of the orbiting body to the focus. |
| Aries, first point of | The point on the celestial sphere marking the Sun's position at the instant the Earth passes the vernal equinox. |
| Ascending node | When a satellite passes the Earth's equatorial plane from the south, the node is called an ascending node and when it passes from the north, a descending node. |
| Assured destruction | Unacceptable damage inflicted on an aggressor or aggressors at any time during the course of a nuclear exchange. |
| Atmospheric drag | Decelerating force due to air acting upon a satellite. |
| Bit | A unit of information; in particular a bit is a binary digit (0 or 1). |
| Bandwidth | (1) in an antenna, the range of frequencies within which its performance, in respect to some characteristics, conforms to a specified standard; (2) the number of cycles per second between the limits of a frequency band; (3) in information theory, the information-carrying capacity of a communication channel. |

| | |
|---|---|
| Booster | A rocket used to give extra power during lift-off or at another stage of a spacecraft's flight. |
| Byte | A unit of information equal to eight bits. |
| Capsule | A container carried on a rocket or spacecraft with the instruments or photographic films intended to be recovered after the flight. |
| Carrier rocket | A rocket vehicle used to carry an artificial Earth satellite into an Earth orbit after the initial boost. |
| Catalyst | A substance which by its mere presence alters the velocity of a reaction and may be recovered unaltered in nature or amount at the end of the reaction. |
| Centrifugal force | A force that appears to act outwards on a body that moves in a curved path. |
| Channel width | A band of frequencies radiated by a transmitter or which must be handled by a system to carry out a single conversion. |
| Computer compatible tape (CCT) | Digital data stored on magnetic tapes in such a way that it can be directly read by normal computers. |
| Contrast ratio | The ratio of the brightness of an object in a scene to that of the background. The ratio may depend on the spectral band in which it is measured. |
| Counterforce strategy | Nuclear war doctrine which emphasizes confining damage to enemy military targets and minimizing damage to civilian population and property. |
| Countervailing strategy | A strategy that requires not only the capability of assured destruction of the enemy but also plans for flexible, controlled use of strategic weapons against appropriate targets for any level of conflict. |
| Cryogenic propellant | Propellant at a very low temperature. |
| Data | The raw signals including imagery generated from an observation satellite. |
| Decay | Various forces acting on a satellite cause it to re-enter the Earth's atmosphere and eventually to burn up (or decay) in the atmosphere. |
| Deterrence | Measures taken to prevent opponents from initiating aggressive actions and to hinder escalation if such actions occur. |
| Doppler shift | The change in frequency with which energy reaches a receiver when the receiver and the energy source are in motion relative to each other. |
| Eccentricity | The ratio of the distance between the centre and focus of an ellipse to its semi-major axis. |
| Encryption | Transformation of data into an unusable form unless the rules of transformation are known (encryption key). |

| | |
|---|---|
| Ephemary | A table of predicted positions of bodies in the Solar System. |
| Escape stage | A small rocket engine attached to a satellite to provide it with additional thrust for separating it from the booster vehicle. |
| Flexible response strategy | A strategy which asserts that aggression should be appropriately matched, enabling smooth adjustment of response as situations change; it requires a wide range of options in response to potential aggression. |
| Frequency modulation | Changing the frequency of the carrier wave so that it varies in proportion to the amplitude of the signal. |
| Gain (of antenna) | A general term used to denote an increase in signal power in transmission from one point to another. An antenna gain is the ratio of the power transmitted along the beam axis to that of an isotropic radiator transmitting the same total power. |
| Gamma rays | Electromagnetic rays which have wavelengths in the range $10^{-10}$ to $10^{-13}$ metre. |
| Gravitation | The mutual attraction of bodies as a result of their mass; a mass has an associated gravitational field characterized by the fact that a force is exerted on a second mass placed in this field. |
| Ground track | The projected path traced out by a satellite over the surface of the Earth. |
| Guidance system | A system which directs the movements of a rocket or a spacecraft with particular reference to the selection of a flight path. |
| Housekeeping | Information representing the state of a spacecraft. |
| Image enhancement | The process of emphasizing certain aspects of imagery data not easily observable by the human eye. |
| Inclination (orbital) | The angle of inclination of the orbital plane of a satellite to the Earth's equatorial plane. |
| Instantaneous field of view (IFOV) | This is defined as the size of the spot on the ground which just fills the area of the sensor of a scanner at the instant of observation. |
| Ion | An atom or group of atoms with a net positive or negative charge; positive ions have a deficiency of electrons and negative ions have an excess. |
| Ionosphere | The atmospheric shell characterized by high ion density; its base is at about 75-km altitude and it extends to an indefinite height. |
| Kiloton | Measure of the explosive yield of a nuclear weapon equivalent to 1 000 metric tons of TNT. |

| | |
|---|---|
| Launcher | A rocket vehicle used to orbit a satellite. |
| Lifetime (of satellite) | The period of time during which instruments on board a satellite remain functioning, or the period from lift-off to decay of the satellite. |
| Megaton | Measure of the explosive yield of a nuclear weapon equivalent to one million metric tons of TNT (1 000 kilotons). |
| Microwave | Electromagnetic radiation with wavelengths in the range 0.01–0.3 metres, i.e., in the region between infra-red radiation and radio waves. |
| Modulation | The process of changing the waveform of one wave by combining it with another wave. |
| Multiple independently targetable re-entry vehicle (MIRV) | A missile payload comprising two or more warheads that can engage separate targets. |
| Multispectral scanner (MSS) | An instrument used to produce imagery data by scanning the Earth's surface below the satellite across the orbital path in two or more wavelength bands. |
| Neutron | An elementary particle with zero charge and a rest mass of $1.67492 \times 10^{-27}$ kg. |
| Node | The point at which the orbit of a satellite intersects some particular plane, such as the equatorial plane. |
| Nuclear yield | The explosive power of a nuclear bomb expressed in kilotons or megatons. |
| Orbit | The path of a satellite under the influence of the Earth's gravitational force; the path is a closed one so that the satellite returns periodically to the same point in the path. |
| Orbital elements | A set of six parameters defining the orbit of a satellite; these are the right ascension of the ascending node ($\Omega$), the argument of perigee ($\omega$), the orbital inclination ($i$), the semi-major axis of the orbit ($a$), the eccentricity of the orbit ($e$) and the time of perigee passage ($T$). |
| Parking orbit | An orbit of a spacecraft around a celestial body, used for assembly of components or to wait for conditions favourable for the departure from the orbit. |
| Payload | Load which a rocket or a spacecraft carries over and above what is necessary for the operation of the vehicle for its flight; for example, camera equipment, etc.. |
| Perigee | That point in the orbit of an Earth satellite which is closest to the Earth. |
| Period (orbital) | The time required for a satellite to go round the Earth once. |

| | |
|---|---|
| Phase modulation | Changing the phase of the carrier wave so that its difference from the unmodulated wave is proportional to the signal amplitude. |
| Photon | The elementary quantity or quantum of radiant energy. |
| Pixel or pel | Picture elements. |
| Plasma | A mixture of ions and electrons. |
| Polar orbit | Satellite orbit such that the orbital plane contains the Earth's axis of rotation. |
| Polymer | A compound whose molecules consist of repeated molecular units. |
| Population inversion | Under normal conditions, most atoms or molecules are in a low-energy state and little energy is emitted. If energy is added to a lasing medium, the atoms or molecules are excited so that more of them are at higher-energy levels than at lower-energy levels, a condition known as an inverted population or a *population inversion*; the process of producing large numbers of molecules in an excited state in this way is called pumping. |
| Precession | The fact that the Earth is flat at the poles causes the orbit of a satellite to turn round the Earth's axis while keeping its orbital inclination constant. This effect is known as precession. |
| Propellant | The fuel, including the oxidant, used in a rocket. |
| Radar cross section | The picture produced by recording radar waves reflected from a given target surface. |
| Radiometer | An instrument which detects natural electro-magnetic radiation (for example, in the infra-red or microwave regions). |
| Resolution | Ground resolution is the ground dimension equivalent to one line at the limit of resolution; photographic resolution is defined as the minimum observable spacing between black and white lines in a standard pattern. |
| Retrograde orbit | An orbit with an orbital inclination greater than 90° and less than 180°. |
| Return-Beam Vidicon (RBV) | A camera which essentially produces black-and-white or colour television images. |
| Right ascension | The angular distance (usually stated in time, 24 hours = 360 degrees) eastward along the celestial equator, measured from the first point of Aries to the meridian circle passing through any celestial object. |
| Rocket | A vehicle which is propelled by the recoil force produced when part of its mass is ejected at high velocity and which propels the satellite. |

| | |
|---|---|
| Satellite, artificial | A man-made spacecraft which revolves round a spatial body, such as the Earth. |
| Semi-major axis | One-half of the longest diameter of an ellipse. |
| Semi-minor axis | One-half of the shortest diameter of an ellipse. |
| Specific impulse | The ratio of the thrust produced to the rate at which the propellant is used up. |
| Spread spectrum | When transmitting many different signal waveforms in a wide band, power is spread thinly over the band so that narrow-band radios can operate within the wide-band without interference. |
| Sub-satellite | Often a satellite carries one or more smaller satellites, sub-satellites, which are orbited once the parent satellite is in its orbit. |
| Sun-synchronous orbit | The orbit of a satellite in which the orbital plane contains both the Earth and the Sun; in such an orbit a satellite crosses the Earth's equatorial plane at the same local time. |
| Synchronous orbit | The orbit of an artificial satellite which makes one revolution every 24 hours; if the orbit is in the same plane as the Earth's equator, the satellite remains stationary relative to the Earth. |
| Synthetic Aperture Radar (SAR) | This is a side-looking radar with a relatively short antenna which is made to behave like a very long one with a narrow beam; a long antenna can be synthesized by taking advantage of the motion of the platform (a satellite). |
| Telemetry | Transmission of data using microwaves (for example, housekeeping data, imagery data from satellite sensors, etc.), often using coded signals. |
| Terminal (ground) | A place or places on the Earth at which a transmitter or a receiver or both form a link between it and a communications satellite. |
| Tracking systems | Systems such as radars or optical telescopes which follow the movements of satellites in Earth orbit. |
| Trajectory | The dynamical path followed by an object under the influence of gravity and/or other forces. A satellite's orbit is a trajectory which is closed so that the satellite returns periodically to the same point. |
| Transponder | A combined receiver and transmitter the function of which is to transmit signals automatically when commanded. |
| Upper stage | A second or later stage in a multistage rocket. |
| Vernal equinox | That point of intersection of the ecliptic and the celestial equator occupied by the Sun as it changes from south to north declination. Also called the first point of Aries. |

| | |
|---|---|
| Vernier control rocket | Small rockets attached to a satellite to make small adjustments in the orbital flight of the satellite. |
| Vidicon tube | The vidicon tube is a part of the satellite television camera system whose main component is a layer of photoconductive material. The image of the object is formed on this layer and then converted into electrical signals. |
| Vulnerability | The susceptibility of any country, military force, or weapon system to any action by any means through which its effectiveness may be reduced. |

# Appendix 4

## *Abbreviations*

| | |
|---|---|
| A-1 | Vostok launch vehicle |
| A-2 | Vostok up-rated second stage |
| A/A-A | Atlas Agena-A |
| A/A-B | Atlas Agena-B |
| A/A-D | Atlas Agena-D |
| AFSATCOM | Air Force Satellite Communications System |
| ASAT | Anti-satellite |
| ATA | Advanced Test Accelerator |
| B-1 | Modified Sandal intermediate-range missile with an added upper stage |
| BMEWS | Ballistic Missile Early Warning System |
| BMD | Ballistic Missile Defence |
| C-1 | Skean intermediate-range missile plus upper stage |
| $C^3I$ | Command, Control, Communication and Intelligence |
| CCD | Charged Couple Device |
| CCT | Computer Compatible Tape |
| CEFSR | Committee for Evaluating the Feasibility of Space Rocketry |
| CEP | Circular Error Probable |
| CIA | Central Intelligence Agency |
| COMINT | Communications Intelligence |
| DARPA | Defense Advanced Research Projects Agency |
| DoD | Department of Defense |
| DSCS | Defense Satellite Communications System |
| EHF | Extremely High Frequency |
| EMP | Electromagnetic Pulse |
| ELDO | European Launcher Development Organization |
| ELINT | Electronic Intelligence |
| EORSAT | ELINT Ocean Reconnaissance Satellite |
| ESA | European Space Agency |

| | |
|---|---|
| ESRO | European Space Research Organization |
| ETR | Eastern Test Range |
| FOBS | Fractional Orbital Bombardment System |
| FLTSATCOM | Fleet Satellite Communications System |
| GOES | Geostationary Operational Environmental Satellite |
| GPS | Global Positioning System |
| HALO | High Altitude Large Optics |
| HI-CAMP | High Resolution Calibration Airborne Measurement Program |
| HOE | Homing Overlay Experiment |
| ICBM | Intercontinental Ballistic Missile |
| IFOV | Instantaneous Field of View |
| IGY | International Geophysical Year |
| IMEWS | Integrated Missile Early Warning System |
| IONDS | Integrated Operational Nuclear Detection System |
| IRBM | Intermediate-Range Ballistic Missile |
| IUS | Inertial Upper Stage |
| kt | Kiloton |
| LES | Lincoln Experimental Satellite |
| LoAD | Low Altitude Defense |
| LODE | Large Optics Demonstration Experiment |
| LOX | Liquid Oxygen |
| LTT | Long-Tank-Thor |
| LTTA | Long-Tank Thrust Augmented |
| LTTAT/A-D | Long-Tank-TAT Agena-D |
| MAD | Mutual Assured Destruction |
| MHIV | Miniature Homing Intercept Vehicle |
| MIRV | Multiple Independently Targetable Re-entry Vehicle |
| MOL | Manned Orbiting Laboratory |
| MPD | Magnetoplasmadynamic |
| MRBM | Medium-Range Ballistic Missile |
| MSS | Multispectral Scanner |
| Mt | Megaton |
| NASA | National Aeronautics and Space Agency |
| NNK | Non-Nuclear Kill |
| NOAA | National Oceanic and Atmospheric Administration |
| NORAD | North American Air Defense Command |
| PMTC | Pacific Missile Test Center |
| RBV | Return-Beam Vidicon |
| RORSAT | Radar Ocean Reconnaissance Satellite |
| SAINT | Satellite Inspection Technique |
| SALT | Strategic Arms Limitation Talks |
| SAMTEC | Space and Missile Test Center |
| SAR | Synthetic Aperture Radar |
| SCF | Satellite Control Facility |
| SDS | Satellite Data System |
| SERT | Space Electric Rocket Test |
| SLBM | Submarine-Launched Ballistic Missile |
| SOREP | Satellite Oceanographic Research Program |
| SPOT | Systéme Probatoire d'Observation de la Terre |
| SSTS | Space Shuttle Transportation System |

| | |
|---|---|
| T-3B/A-D | Titan-3B Agena D |
| T-3C | Titan-3C |
| T-3D | Titan-3D |
| TAD | Thrust-Augmented-Delta |
| TAT | Thrust-Augmented-Thor |
| Th/A-D | Thrust-Augmented-Thor Agena D |
| Th/A-A | Thor Agena-A |
| Th/A-B | Thor Agena-B |
| Th/A-D | Thor Agena-D |
| TIROS | Television and Infra-Red Observation Satellite |
| WSMR | White Sands Missile Range |
| WWMCCS | World-Wide Military Command and Control |

# Index